Ecological Studies

Analysis and Synthesis

Edited by

W. D. Billings, Durham (USA) F. Golley, Athens (USA)

O. L. Lange, Würzburg (FRG) J. S. Olson, Oak Ridge (USA)

H. Remmert, Marburg (FRG)

Volume 52

Air Pollution by Photochemical Oxidants

Formation, Transport, Control, and Effects on Plants

Edited by
Robert Guderian

Contributors

K. H. Becker, W. Fricke, R. Guderian, J. Löbel
R. Rabe, U. Schurath, D. T. Tingey

With 54 Figures

Springer-Verlag
Berlin Heidelberg New York Tokyo

Professor Dr. ROBERT GUDERIAN
Universität Essen
Institut für Angewandte Botanik
4300 Essen 1, FRG

ISBN 3-540-13966-4 Springer-Verlag Berlin Heidelberg New York Tokyo
ISBN 0-387-13966-4 Springer-Verlag New York Heidelberg Berlin Tokyo

Library of Congress Cataloging in Publication Data. Main entry under title: Air pollution by photochemical oxidants. (Ecological studies; v. 52) Bibliography: p. 1. Photochemical oxidants – Environmental aspects. 2. Plants, Effect of air pollution on. 3. Air-Pollution. I. Guderian, Robert. II. Series. TD885.A36 1985 628.5′32 84-23545

Typesetting, printing and binding: Brühlsche Universitätsdruckerei, Giessen
2131/3130-543210

Preface

Photochemical oxidants are secondary air pollutants formed under the influence of sunlight by complex photochemical reactions in air which contains nitrogen oxides and reactive hydrocarbons as precursors. The most adverse components formed by photochemical reactions in polluted air are ozone (O_3) and peroxyacetyl nitrate (PAN), among many other products such as aldehydes, ketones, organic and inorganic acids, nitrates, sulfates etc. An analysis and evaluation of the available knowledge has been used to characterize the relationships among emissions, ambient air concentrations, and effects, and to identify the important controlling influences on the formation and effects of photochemical oxidants.

The biological activity of photochemical oxidants was first clearly manifested during the early 1940's, when vegetation injury was observed in the Los Angeles Basin in the United States. Since that time, as a consequence of the increasing emissions of photochemical oxidant precursors, the photochemical oxidants have become the most important air pollutants in North America. In other parts of the world, for example South and Central America, Asia, and Australia, photochemical oxidants threaten vegetation, particularly the economic and ecological performance of plant life. According to my knowledge, the first observations of ozone and PAN injury to vegetation in Europe were made by Dr. Ellis F. Darley (Statewide Air Pollution Research Center, University of California, Riverside, California) during a study visit (1963/64) to the Federal Republic of Germany. In the subsequent years the exposure has increased; consequently in 1980 the Umweltbundesamt, Berlin, a central organization for directing the scientific deliberations of the federal government in the area of environmental protection, initiated the study *Luftqualitätskriterien für photochemische Oxidantien*. A group of authors from various scientific disciplines, who evaluated and interpreted research results from around the world concerning the formation, distribution, and effects of photochemical oxidants on organisms and materials, reached the conclusion that in Europe also ozone concentrations were sufficiently high to pose a risk to man and his environment.

Based on the ambient concentrations of photochemical oxidants, measured in various European countries, and the concentrations that cause effects, it appears that vegetation is particularly threatened by photochemical oxidants. Given current knowledge, photochemical oxidants should be considered as a primary cause of the new type of forest injury presently found in Central Europe.

The Air Quality Criteria volumes published by the Umweltbundesamt (Report 5/83, Erich Schmidt Verlag, Berlin 1983) were generally limited to environmental protection actions involving institutions within the German Federal Republic.

There has not been a comprehensive review of the European literature concerning photochemical oxidants. However, given their significance and importance as air pollutants in Europe, it is advisable that current scientific information should be made available to a larger circle of readers. To adequately described the scope of the photochemical oxidant problem and to provide a basis for further discussion of this important area of environmental protection, it is necessary to combine experimental results from Europe with those from other countries, particularly the United States.

The first chapter of this book, *Formation, Transport and Control of Photochemical Oxidants,* describes essential physicochemical properties of the compounds concerned and identifies emission sources of the photochemical oxidant precursors. Ambient air concentrations observed in different countries under various atmospheric conditions are documented, and the physicochemical reactions that lead to the formation of oxidants are explained. Laboratory and modeling efforts that provide an understanding of these reactions are shown, and atmospheric diffusion and transport models for examining source–receptor relationships are discussed. Finally, analytical techniques for monitoring the relevant components are described and suggestions for reducing ambient exposures from photochemical oxidants are given.

The section, *Effects of Photochemical Oxidants on Plants,* discusses the impact of the most important phytotoxic components, ozone and PAN, at the cellular, organal, organismal, plant community, and ecosystem levels, in relation to pollutant uptake, mode of action, and differential plant resistance, as well as the influence of the internal and external growth factors that modify the responses. Methods for detecting and evaluating air pollutant effects are reviewed. The dose–response relationships are used to derive air quality criteria which can provide an important perspective for developing air quality standards.

The original scope of this publication included a section on *Effects of Photochemical Oxidants on Humans and Animals.* However, this topic was reviewed in detail, in 1983 in *The Biomedical Effects of Ozone and Related Photochemical Oxidants,* Volume V, SD Lee, MG Mustafa, MA Mehlman (eds), Princeton Scientific Publishers, Princeton, New Jersey, USA, pp. 643.

The ultimate protection of humans, animals, plants, and materials from photochemical oxidant injury requires reduction of the ambient concentration of the specific air pollutant. The available emission estimates for the precursor compounds indicate that, at least for the foreseeable future, humans and the environment will continue to be impacted by photochemical oxidants. The material for this book is selected to provide a basis for preventive measures at the source of the precursor emission and at the site of impact. Another objective was to provide the increasing numbers of researchers, teachers, and students with information about photochemical oxidants and their effects.

Essen, January 1985   ROBERT GUDERIAN

Contents

Part 1 Formation, Transport, and Control of Photochemical Oxidants

By K. H. Becker, W. Fricke, J. Löbel, and U. Schurath
(With 35 Figures)

Part 2 Effects of Photochemical Oxidants on Plants
By R. Guderian, D. T. Tingey, and R. Rabe (With 19 Figures)

Part 1

Formation, Transport, and Control of Photochemical Oxidants

KARL H. BECKER

Physikalische Chemie/Fachbereich Chemie
Bergische Universität Wuppertal

WOLFGANG FRICKE

Essen

JÜRGEN LÖBEL

Verein Deutscher Ingenieure, Düsseldorf

ULRICH SCHURATH

Institut für Physikalische Chemie, Universität Bonn

Chapter 1.1 Introduction

This first part of the book summarizes the state of knowledge in various fields of chemistry, meteorology, and instrumental analysis concerning formation, transport, and control of photooxidants. Special emphasis is given to the situation in the Federal Republic of Germany and Western Europe. The oxidant problem in the USA and Japan, which has been covered by several reviews, is only briefly highlighted.

The photochemical oxidants and their precursors are individually characterized by their physicochemical properties in the first section. This is followed by a survey of precursor emissions including both, anthropogenic and natural sources. Results of long-term measurements of ozone, nitrogen oxides, and hydrocarbons from various parts of the Federal Republic of Germany are tabulated. Most of the available data on ozone are included, whereas the other data had to be restricted. This is followed by a brief summary of atmospheric measurements in Europe and overseas. One section is dedicated to airborne measurements and other more specialized investigations. The fundamentals of air chemistry and atmospheric transport – as far as needed to understand anthropogenic oxidant formation and abatement strategies – are discussed in close relation with smog-chamber experiments and kinetic models. The influence of meteorological parameters on oxidant formation is discussed separately. Several transport and dispersion models of photochemical oxidants and their precursors are briefly described.

The first section on monitoring of oxidants and precursors reviews the current analytical techniques. Secondly, suggestions are made for the planning of monitoring networks, selection of representative compounds, sites, and measuring periods.

In a final evaluation of the situation of oxidant pollution in Germany and Western Europe, the feasibility of abatement strategies by precursor reduction is assessed. No attempt has been made to compare national air quality standards, which do not exist for oxidants in any of the European countries.

Due to contributions from several research disciplines to the problem of atmospheric oxidants, the concentration units used in the literature are not consistent. In air chemistry molar mixing ratios are usually preferred, while state authorities often impose the use of mass concentration units in their monitoring networks. Therefore, the tabulated air quality data are given in mg m^{-3} or µg m^{-3}, while volume (molar) mixing ratios are widely used otherwise, in accordance with the original literature. For the more important compounds related with the oxidant problem, conversion factors between both units are compiled in the Appendix.

Chapter 1.2 Oxidants, Precursors, and Concomitant Species

1.2.1 Individual Compounds

The general terms "oxidants" and "photochemical air pollutants" comprise a large number of trace compounds which are formed as reaction products of certain primary pollutants, particularly NO, NO_2, and hydrocarbons, under the action of sunlight. Notable reaction products, which are also called secondary pollutants, are ozone, peroxyacetyl nitrate (PAN), higher oxides of nitrogen, aldehydes, and ketones, as well as several gaseous and/or particle-bound inorganic and organic acids.

The term oxidants is derived from the property of certain secondary pollutants to produce molecular iodine when bubbled through an aqueous solution of KI. The amount of iodine formed, which may be determined coulometrically or colorimetrically, is a measure of the amount of oxidant present in the air. The reaction of ozone with dissolved potassium iodide proceeds to stoichiometry under suitable conditions, but other oxidants like PAN and NO_2 are also detected to some extent. Negative interference is due to SO_2 and other reducing trace gases, which may be crucial in polluted air. The KI method is therefore said to yield "net" oxidant concentrations. Although this definition of "oxidants" and "oxidant concentrations" remains somewhat vague, it has been found to characterize the complex mixture of secondary pollutants in photochemical smog sufficiently well for many purposes.

Primary pollutants that are actively involved in the formation of photochemical oxidants via sunlight-induced reactions in the atmosphere are referred to as *precursors*. Notable precursors are NO, NO_2, and various hydrocarbons, which undergo reactions with OH radicals, oxygen atoms, and ozone, or are photodissociated in sunlight. Aldehydes, ketones, and other oxygenated organics are reaction products of hydrocarbons in the atmosphere, but may also be referred to as precursors when emitted as such, e. g., in the exhaust gas of motor vehicles. Methane and the halocarbons are too unreactive to contribute significantly to oxidant formation in polluted air, and are therefore not considered as precursors. Carbon monoxide, which reacts with OH radicals and is believed to be an important precursor of tropospheric ozone on a global scale, plays a minor role in polluted air. SO_2 is not considered as a precursor, since it does not enhance the rate of oxidant formation in the atmosphere. Its homogeneous removal rate by OH radicals is, however, enhanced in photochemical smog (D. F. Miller 1981), resulting in the formation of sulfuric acid, which ends up as particulate matter. Photochemical oxidants like hydrogen peroxide and ozone, and perhaps certain radicals

(Chameides and Davis 1982), rapidly oxidize SO_2 in the aqueous phase (cloud and rain droplets). This indicates that the formation of photooxidants and acid rain are interdependent processes.

1.2.2 Physicochemical Properties of Important Oxidants

1.2.2.1 Ozone

Pure ozone (chemical formula: O_3) is a pale blue gas at room temperature which liquefies below $-112\,°C$. The dark blue liquid is a dangerous explosive which can be stabilized by absorption on silical gel. At mixing ratios of a few 100 ppb, ozone has a characteristic irritating odor. Higher concentrations are extremely poisonous, causing lesions of the mucous membranes. It is about 50 times more soluble in water than oxygen (Ozone Chemistry and Technology 1959).

Molecular weight	48
Melting point	$-192.7\,°C$
Boiling point	$-111.9\,°C$
Gas density (0 °C, 1 bar)	$2.14\,g\,l^{-1}$
Solubility in water (0 °C, 1 bar)	49 ml/100 ml water

The temperature dependence of the solubility has been measured by Kosak-Channing and Helz (1983).

Optical Absorption in the UV (Hampson et al. 1973, Davenport 1982)

The absorption spectrum in the UV is reproduced in Chapter 1.4.1, Fig. 1.27. The absorption coefficient k, defined by Lambert-Beer's law,

$$lg(I_o/I) = k \cdot x \cdot p \cdot T_o/T, \quad T_o = 273\,K,$$

is $k = (130.6 \pm 1)bar^{-1}\,cm^{-1}$ at the wavelength 253.7 nm of mercury.

Thermochemical Properties (Hampson and Garvin 1978)

$$\Delta H_f^o(O_3\ gas) = 143\ kJ/mol$$

$$\Delta G_f^o(O_3\ gas) = 163\ kJ/mol$$

$$S^o(O_3\ gas) = 238.8\ J\ mol^{-1}\ K^{-1}.$$

Electrochemical Potential of Ozone (Handbook of Chemistry and Physics 1974)

$$O_3 + 2H^+ + 2e^- \rightarrow O_2 + 2H_2O \quad E_o = 2.07\ V$$
$$O_3 + H_2O + 2e^- \rightarrow O_2 + 2OH^- \quad E_o = 1.24\ V.$$

Ozone is one of the strongest oxidizing agents which reacts rapidly with various, particularly unsaturated, organic compounds, and destroys elastomers ("rubber cracking") by reaction with the double bonds.

Technical applications of ozone include the sterilization of drinking water, oxidative cleaning of odorous waste gases, and ozonation reactions in organic synthesis. A few vol. % of ozone can be generated continuously in an ozonizer, by means of silent discharges in dry air or oxygen. Light sources emitting UV radiation below 200 nm (e. g., low-pressure mercury lamps with quartz envelopes) can be used to make a few ppm ozone in air by the photolysis of molecular oxygen. The contribution of ozone technology to atmospheric ozone by direct emissions is entirely negligible. However, indoor pollution by ozone, e. g., in the vicinity of powerful light sources, has to be carefully controlled.

Large amounts of natural ozone, which are formed by the photolysis of molecular oxygen in the UV part of the solar spectrum (see Chap. 1.3.1.5), are transported downward into the troposphere and give rise to a natural background concentration in the order of 20–40 ppb at ground level (see Chap. 1.3.2.1).

1.2.2.2 Peroxyacyl Nitrates

The generalized chemical formula of these compounds is

$$RC\underset{OONO_2}{\overset{O}{\diagup}} \qquad (RC\overset{O}{\diagup} = acyl, aroyl) .$$

Peroxyformyl nitrate (R = H) is unstable and has not been detected in the atmosphere. The first more stable homolog of the series (R = CH$_3$) is peroxyacetyl nitrate = PAN, which is also the most abundant. Much lower concentrations of peroxypropionyl nitrate = PPN (R = C$_2$H$_5$) and peroxybenzoyl nitrate = PBzN (R = C$_6$H$_5$) have also been detected in the atmosphere.

Pure peroxyacetyl nitrate is a colorless liquid highly explosive at room temperature, nearly insoluble in water, but easily soluble in organic solvents.

Molecular weight	121
Boiling point	106 °C
Vapor pressure at 25 °C	38 mbar
(Bruckmann and Willner 1983)	

PAN hydrolyzes in alkaline medium, yielding a stoichiometric amount of nitrite (Nielsen et al. 1982). This can be used for quantitative analysis.

Gaseous PAN exists in equilibrium with its radical fragments,

$$CH_3C\underset{OONO_2}{\overset{O}{\diagup}} \underset{k_r}{\overset{k_f}{\rightleftharpoons}} CH_3C\underset{OO}{\overset{O}{\diagup}} \cdot + NO_2 \cdot$$

The forward rate constant k_f of its monomolecular decomposition, which is pressure-independent under ambient conditions (see Chap. 1.4.1), is strongly temperature-dependent:

$$k_f = 10^{15.8} \exp(-13,160/T) \, s^{-1}$$

(Schurath and Wipprecht 1980, Schurath et al. 1984). The rate constant k_f of PPN and its temperature dependence are very similar. The effective decay rate of PAN is much less than k_f, unless the peroxyacetyl radical is removed from the equilibrium by reaction with NO. Bruckmann and Mülder (1979) have measured a decay rate of 10%/month for 300 ppmV PAN in pressurized nitrogen at a storage temperature of 10°–15°C.

PAN has been prepared in the gas phase by the photolysis of ethyl nitrite in oxygen (Stephens 1969), by Cl_2 photolysis in the presence of NO_2 and acetaldehyde in air (Grosjean et al. 1984), or in the liquid phase by nitration of peracetic acid (Nielsen et al. 1982, Kravetz et al. 1980). The latter method is also suitable for the preparation of PPN, PBzN, etc. Essentially pure PAN has been obtained by selective extraction of the reaction product with a high boiling organic solvent (Holdren et Spicer 1984, Gaffney et al. 1984).

1.2.2.3 Hydrogen Peroxide

Pure hydrogen peroxide (chemical formula: H_2O_2) is a colorless liquid which explodes vigorously when contacted with organic matter or with traces of heavy metals. Its aqueous solutions are more stable, and are widely used for chemical and technical applications.

Molecular weight	34
Melting point	$-0.4°C$
Boiling point	$150.2°C$

Thermochemical Properties (Hampson and Garvin 1977)

$$\Delta H_f^o(H_2O_2 \text{ gas}) = -136.3 \text{ kJ/mol}$$
$$\Delta G_f^o(H_2O_2 \text{ gas}) = -105.5 \text{ kJ/mol}$$
$$S^o(H_2O_2 \text{ gas}) = 232 \text{ J mol}^{-1} \text{ K}^{-1}.$$

Gaseous hydrogen peroxide is a reaction product of HO_2 radicals in the atmosphere. It has also been detected in cloud and rain water; Henry's law constant was found to be 1.42×10^5 M atm^{-1} at 20 °C for atmospheric concentrations of H_2O_2 (Yoshizumi et al. 1984). SO_2 in cloud and rain droplets is rapidly oxidized by hydrogen peroxide.

1.2.2.4 Nitrogen Dioxide

Nitrogen dioxide (chemical formula: NO_2) is a brownish gas which has a characteristic odor. Pure NO_2 exists in equilibrium with its colorless dimer N_2O_4, but only the monomer is important in the atmosphere, where the dimer concentration is negligible.

Molecular weight	46
Melting point	$-11.3°C$
Boiling point	$21.15°C$.

The pure liquid or solid is colorless. When contaminated with N_2O_3, a greenish-blue color is obtained.

Thermochemical Properties (Hampson and Garvin 1978)

$$\Delta H_f^o(NO_2 \text{ gas}) = 33.1 \text{ kJ/mol}$$
$$\Delta G_f^o(NO_2 \text{ gas}) = 51.2 \text{ kJ/mol}$$
$$S^o(NO_2 \text{ gas}) = 240 \text{ J mol}^{-1} \text{ K}^{-1}.$$

The solubility of NO_2 in water is poor (Henry's law constant 7.5×10^{-3} M atm^{-1} at 295 K, Lee and Schwartz 1981). The dimer is much more soluble. NO_2 in aqueous solution is slowly converted to nitrous acid and nitric acid. The photochemical properties of NO_2 are discussed in Chapter 1.4.1, and the absorption spectrum is shown in Fig. 1.26.

1.2.3 Physicochemical Properties of Important Precursors and Concomitant Species of Photochemical Oxidants

The precursors of photochemical oxidants have already been defined in Section 1.2.1. The concomitant species discussed here are formed from the precursors or other primary pollutants via photochemically induced radical reactions, but do not exhibit oxidizing properties.

1.2.3.1 Nitric Oxide

Nitric oxide (chemical formula: NO) is the direct precursor of nitrogen dioxide. It is formed in the combustion of fossil fuels. High concentrations of NO are contained in car exhaust and in early plumes of power stations. Pure NO is a colorless gas which is weakly soluble in water (7.34 ml/100 ml water at room temperature).

Molecular weight	30
Melting point	$-163.6\,^{\circ}C$
Boiling point	$-151.8\,^{\circ}C$.

1.2.3.2 Hydrocarbons

Anthropogenic hydrocarbons in the atmosphere are due to evaporative losses and to incomplete combustion of fossil fuels. The *alkanes* (chemical formula: C_nH_{2n+2}; $n = 1, 2, 3, ...$) consist of straight or branched chains of single-bonded carbon atoms. Hydrocarbons containing at least one double bond between adjacent carbon atoms are called *alkenes* (chemical formula of the two simplest alkenes; ethene: C_2H_4, propene: C_3H_6). *Alkines* have at least one triple bond between adjacent carbon atoms. The simplest alkine is acetylene = ethine (chemical formula: C_2H_2). Benzene (chemical formula: C_6H_6) is the simplest *aromatic hydrocarbon*. It forms an equilateral planar six-membered ring. Other aromatic hydrocarbons derive from benzene by substitution of one or more hydrogen atoms by CH_3, C_2H_5, etc. The solubility of hydrocarbons in water is extremely low. Precursor hydrocarbons are colorless gases or liquids with widely differing chemical-kinetic properties.

Name and chemical Formula	Molecular weight	Melting point in °C	Boiling point in °C
Alkanes			
Ethane, C_2H_6	30	−183.3	− 88.6
Propane, C_3H_8	44.1	−189.7	− 42.1
n-Butane, C_4H_{10}	58.1	−138.4	− 0.5
Alkenes			
Ethene, C_2H_4	28	−169.2	−103.7
Propene, C_3H_6	42.1	−185.3	− 47.4
trans-2-Butene, C_4H_8	56.1	−105.4	+ 0.9
Aromatic hydrocarbons			
Benzene, C_6H_6	78.1	5.5	80.1
Toluene, $C_6H_5CH_3$	92.2	− 95	110.6
o-Xylene, $C_6H_4(CH_3)_2$	106.2	−25.2	144.4

1.2.3.3 Aldehydes and Ketones

Aldehydes (chemical formula: $RHC=O$, $R=CH_3$, C_2H_5, etc.) and ketones (chemical formula: $R_1R_2C=O$) should be considered as partially oxidized hydrocarbons. The aldehydes and ketones in the polluted atmosphere, which are mainly emitted by internal combustion engines, are due to incomplete combustion of fossil fuels. They are, however, also formed in the atmosphere by photochemically induced reactions of hydrocarbons.

Name and chemical formula	Molecular weight	Melting point in °C	Boiling point in °C
Formaldehyde, H_2CO	30	−118	−19.2
Acetaldehyde, CH_3CHO	44	−121	20.8
Acetone, CH_3COCH_3	58.1	− 95.4	56.2

The simple aldehydes and ketones are readily soluble in water. Pure monomer formaldehyde is a poisonous gas which is polymerized by traces of water (Walker 1975). Aldehydes and ketones absorb ultraviolet sunlight. The photochemistry of formaldehyde is discussed in Chapter 1.4.1, the absorption spectrum is shown in Fig. 1.28.

1.2.3.4 Aerosols

Aerosols are extremely small solid or liquid particles suspended in air. They are either emitted by natural or anthropogenic sources in particulate form, or are generated in situ by gas-to-particle conversion of low vapor pressure products of gas phase reactions. Small particles may grow by coagulation. Liquid phase reactions in cloud droplets, followed by evaporation, also contribute to particle growth. Aerosol particles smaller than about 1 μm in diameter behave more or less like gas molecules in the sense that their removal from the atmosphere by sedimentation is slow. Aerosol formation and growth is enhanced in photochemical smog by gas-to-particle conversion of SO_2 and other less well-characterized precursors. The impact of pre-existing aerosols on the formation rate of photooxidants is negligible according to smog chamber experiments discussed in Chapter 1.4.2.1. However, light scattering and absorption by aerosols may reduce the rate of photochemical processes near the ground.

The physicochemical properties of aerosols depend on particle diameter, surface area, structure, and chemical composition of substructures. These may be characterized by distribution functions, usually of the particle diameter as independent variable.

Chapter 1.3 Emissions
and Ambient Air Concentrations

1.3.1 Emissions

The impact of precursor emissions on oxidant formation in the atmosphere may in principle be assessed on the basis of emission inventories. In the assessment, a correlation must be found between emission rates and ambient air concentrations of precursors, which depend on transport, dispersion, and chemical conversion of the precursors in the atmosphere.

It should be extremely difficult, if not impossible, to assess the oxidant-forming potential of precursors on the basis of emission inventories, for the following reasons:

1. Emission inventories frequently refer to NO_x and "total hydrocarbons", or even less specifically to "total organic emissions". Detailed inventories of organic emissions are not generally available.
2. In some countries, emission inventories are incomplete, or not differentiated by emission height.
3. Most emission inventories are based on emission factors, rather than on measurements of actual source strengths.
4. The emission rates are long-term averages. Little information is available on diurnal and/or seasonal variations.

The contributions of natural versus anthropogenic precursors to oxidant formation are still debatable, at least on a global scale. Several authors (Ehhalt and Drummond 1982, Logan 1983) estimate that anthropogenic and natural NO_x emissions are roughly comparable on a global scale (see also Table 1.1). On the continents of the northern hemisphere, however, anthropogenic emissions are believed to exceed natural NO_x by about an order of magnitude (Logan 1983). Estimates of the global biogenic hydrocarbon emissions (excepting methane) range between 175 and 830×10^6 t a^{-1} (Rasmussen 1981, Zimmerman 1979), while the anthropogenic hydrocarbon emissions were estimated to amount to 67×10^6 t a^{-1} (Duce 1978). In the industrialized countries, where oxidant formation is of major concern, the natural components in ambient precursor concentrations are low and perhaps negligible.

For several greater urban areas in the Federal Republic of Germany, emissions by industry, trade, and transportation, and domestic emissions have been compiled in emission inventories. Total NO_x and organic emissions by area, summed over all source categories, are given in Table 1.2. The numbers in brackets pertain to emissions per inhabitant, in kg year^{-1}. The averages of these numbers, which

Table 1.1. Estimates of global NO_x emissions per year. (After Ehhalt and Drummond 1982)

Source	Emissions in 10^6 t Na^{-1}
Fossil fuel burning:	
Coal, lignite	5.5
Fuel oil	1.8
Natural gas	1.9
Automobiles	4.3
Soil release	5.5
Biomass burning	11.2
Lightning	5
Oxidation of NH_3	3.1
Miscellaneous	0.9
Total	39 ± 20 (estimated uncertainty)

are in the same order of magnitude for the areas considered, may be used to obtain estimates of NO_x and organic emissions in other areas for which no emission inventories have been published.

The emission ratios of organics/NO_2 in Table 2, last column, are surprisingly low. They do not correlate with observed ratios of nonmethane hydrocarbons/NO_x in ambient air, which are strongly dependent on effective source heights, meteorological parameters, and atmospheric lifetimes of the compounds. The emission ratios (between 0.2 and 1.9) in Table 1.2 are considerably lower than an estimate for the Greater London Area (Derwent and Hov 1979), which yielded emissions of 77,000 t a^{-1} NO_x (as NO_2), and 170,000 t a^{-1} nonmethane hydrocarbons in 1974/75, corresponding to an emission ratio of 2.2.

Table 1.2. Precursor emissions in t a^{-1} of NO_x and organics (excepting CH_4), for several greater urban areas of the Federal Republic of Germany. Numbers in brackets pertain to emissions per inhabitant in kg a^{-1}. Emission ratios of organics/NO_x are listed in the last column

Region, surface area, and number of inhabitants	NO_x (counted as NO_2) (Emissions per inh. in brackets)	Organics	Emission ratio, org./NO_x
Wiesbaden, 178 km^2, 0.24 Mio inhabitants	6,444 (27)	12,510 (52)	1.9
Cologne, 649 km^2, 1.41 Mio inhabitants	89,994 (64)	86,236 (61)	1.0
Duisburg–Mülheim–Oberhausen, 711 km^2, 1,26 Mio inhabitants	91,115 (72)	46,735 (37)	0.5
Dortmund, 712 km^2, 1.20 Mio inhabitants	95,610 (80)	19,192 (16)	0.2
Mainz, 222 km^2, 0.23 Mio inhabitants	12,004 (52)	4,233 (19)	0.4
Emissions per inhabitant (average, kg a^{-1})	(68)	(39)	0.57

References are given in Becker et al. (1983)

Table 1.3. Anthropogenic NO_x emissions in the Federal Republic of Germany, 1966–1978, in 1000 t a^{-1}. (Umweltbundesamt 1981)

Emission	Source category	1966	1970	1974	1978
NO_x	Power plants	650	820	920	940[a]
(as NO_2)	Industry	660	690	660	580
	Domestic heating, small trade	100	130	140	140
	Transporation	640	820	990	1,340
Total NO_x		2,050	2,450	2,700	3,000

[a] NO_x emissions (as NO_2) are estimated at 944,000 t a^{-1} for 1980 (VGB 1982)

Table 1.4. Anthropogenic emissions of organic compounds in the Federal Republik of Germany, 1966–1978 in 1,000 t a^{-1}. (Umweltbundesamt 1981)

Emission	Source category	1966	1970	1974	1978
Organic	Power plants	6	8	9	9
compounds	Industry	350	450	480	470
	Domestic heating, small trade	640	720	710	630
	Transportation	400	530	570	650
Total organics		1,400	1,700	1,800	1,750

Annual emissions of NO_x and organic compounds in the Federal Republic of Germany since 1966, distinct by source category, are listed in Tables 1.3 and 1.4. The data were published by the Umweltbundesamt in 1981. While organic emissions have been fairly constant over the years, NO_x emissions have increased by 50% since 1966.

1.3.1.1 Transportation

The composition of car exhaust is strongly dependent on the mode of driving (TÜV 1980). Table 1.5 lists the exhaust gas composition of an Otto engine (after Dulson 1978). Apart from the complete combustion products CO_2 and water vapor, significant amounts of incomplete combustion products (CO, H_2, hydrocar-

Table 1.5. Average exhaust gas composition of an Otto test engine. (After Dulson 1978; see also VDI-Richtlinie 2282)

Compound	% by Volume	Compound	% by Volume
CO_2	12.8	CO	2.3
H_2O	~10.5	N_2	76
O_2	1.0	H_2	0.4
NO_x	~ 0.5	Hydrocarbons	0.1

Table 1.6. Volatile organic emissions of an Otto engine (Dulson 1981)

Compound	% by Mass of total organic emissions	Compound	% by Mass of total organic emissions
Methane	7	2-Methylpentane	1.1
Ethine	10.9	3-Methylpentane	0.8
Ethene	15.7	n-Hexane	1.0
Ethane	1.6	Benzene	12.7
Propene	0.2	2-Methylhexane	0.7
Propane	1.1	3-Ethylpentane	0.6
Acetaldehyde	0.7	n-Heptane	0.4
n-Butane	1.8	Toluene	18.9
Butenes	0.7	1,1-Dimethylhexane	0.3
Acetonitrile	1.3	Ethylbenzene	2.1
Acetone	0.9	m-, p-Xylene	6.7
Isopentane	5.2	o-Xylene	1.8
n-Pentane	1.4	Trimethylbenzenes	4.0

bons) and of NO_x (essentially NO) from the high-temperature reactions of oxygen atoms with molecular nitrogen and other reactions are also emitted.

An analysis of hydrocarbons and other organic compounds contained in the exhaust gas of a four cylinder Otto engine is listed in Table 1.6 (Dulson pers. commun. 1983). The composition of the organic fraction of car exhaust in "on the road" condition, in which considerably more compounds have been detected, is quite variable, although a few compounds like ethene, toluene, ethine, xylenes, benzene, propene and i-pentane are usually found to dominate by mass (Nelson and Quigley 1984).

Tables 1.3 to 1.6 can be used to estimate annual emissions of individual compounds by transportation in the Federal Republic of Germany. A large proportion of the organic compounds contained in car exhaust is highly reactive under atmospheric conditions, and is known to enhance the rate of oxidant formation in smog chambers.

1.3.1.2 Industry

Approximately 90% of the NO_x from industrial processes is emitted as NO, mostly from stacks higher than 60 m. In the Federal Republic of Germany 20% is emitted from stacks higher than 120 m. Emission heights are further increased by plume rise due to thermal lift. These emissions can thus be transported and spread out over large areas, within the full depth of the mixing layer.

Emission heights of industrial organic emissions, by contrast, are usually less than 30 m, and the relative composition of organic emissions is highly variable. This is exemplified in Table 1.7, where several classes of organic emission are listed for five major source areas in the Federal Republic of Germany.

The first column includes mainly alkanes and alkenes; the second column includes mainly C_5–C_8 alkanes; the last column includes, in varying proportions, halogenated hydrocarbons, alcohols, esters, ketones, aldehydes, phenols, epoxides, amides, nitriles, ethers, amines, mercaptans, and carbon disulfide. Accord-

Table 1.7. Industrial organic emissions in five source areas of the Federal Republic of Germany, in t a^{-1}. (For references see Becker et al. 1983)

Source area	C$_1$ to C$_4$ hydrocarbons	Petrol hydrocarbons	Aromatics	Other volatile organics
Wiesbaden	700	1,289	655[b]	7,500
Cologne	29,780[a]	10,771	9,414[b]	29,960
Duisburg–Mülheim–Oberhausen	132,599[c]	7,289	1,927	7,846
Dortmund	259[c]	251	1,251[d]	2,607
Mainz	117[e]	819	149	1,445

[a] Mainly ethene
[b] Mainly toluene, xylene, benzene, in this order
[c] Mainly methane
[d] Mainly naphthalene, xylene, toluene, in this order

ingly, the oxidant-forming potential of industrial organic emissions cannot be assessed on the basis of total emission data, unless the detailed emission spectrum is also available.

1.3.1.3 Power Plants

The emissions of a heavy fuel fired power plant are presented in Table 1.8. From these data, an annual emission of only about 7 t a^{-1} total organics can be estimated. This must be compared with an annual emission of about 900 t a^{-1} NO. Obviously, organic emissions by power plants are relatively insignificant compared with other sources like car traffic, chemical industry, etc.

Typical concentration ranges of emissions from coal-fired power plants are given in Table 1.9. The emission of organic compounds amounts to less than a few ppm and is, therefore, entirely negligible. The flue gas concentrations depend strongly on the combustion conditions (burner type, temperature, air excess, etc.), and on fuel composition. Average emissions may be estimated by use of emission factors, which are defined either as pollutant mass flow per MW$_{thermal}$, or as mass ratio of pollutant emitted per fuel consumed (Gerold et al. 1980).

1.3.1.4 Domestic Heating and Small Trade

Emission heights in this source category are low, usually well below 30 m. Some 50 to 70% of the gaseous organic emissions are attributed to incomplete combustion of fossil fuels. Another large proportion, due to evaporative losses at filling stations, consists of light saturated hydrocarbons. Small amounts of tetrachloroethene and various solvents are emitted by dry cleaning and varnishing shops.

1.3.1.5 Natural Sources

Nearly all trace constituents of the atmosphere, or their precursors, possess both, anthropogenic and natural sources, with the notable exception of the fluoro-

Table 1.8. Flue gas analysis of an oil-fired power plant. (After
Jander 1979)

Compound	Volume mixing ratio, mass concentration
NO	350 to 300 ppm
NO_2	< 3 ppm
Propane	10 ppb
Butanes	8 ppb
Pentane to dodecane	5 to 10 $\mu g\ m^{-3}$
Ethene + ethine	16 ppb
Propene	11 ppb
Butenes	20 ppb
Pentenes	5 ppb
Benzene	9 ppb
Toluene	15 ppb
Xylenes	4 ppb
Trimethylbenzenes	6 ppb
Acetaldehyde	43 ppb
Methanol	490 ppb
Ethanol	146 ppb
Acetone	52 ppb
Acetic acid	149 ppb
Nitromethane	15 ppb
Acetonitrile	98 ppb
Total organics	2,095 $\mu g\ m^{-3}$

Burner temperature	1,600 °C
Flue gas emission rate	410,000 $m^3\ h^{-1}$
Oxygen contents	3% by volume
Oil consumption	96 t h^{-1} at full load (150 MW)
Sulfur contents of fuel	ca. 1.5%
Nitrogen contents of fuel	ca. 0.3%

carbons, which are entirely man-made. In view of existing natural sources, beneficial effects of oxidant precursor control measures on oxidant levels in the atmosphere, particularly of hydrocarbon emission control, have occasionally been questioned, or even denied altogether. It is therefore important to look upon source strengths of anthropogenic versus natural ozone precursors, and of ozone itself, in proper perspective. This shall be attempted in the following section. The reader is also referred to Section 1.3.2.1 which deals with the natural background of trace gases in the "clean" troposphere.

Table 1.9. Typical emission ranges of coal-fired power plants. (US
EPA 1977a; EPA/EPRI 1980)

Emission	Mixing ratio in ppm by volume
NO_x	50–1,000
CO	30– 300
SO_2	200–1,000, depending on sulfur content of the coal

Ozone

The solar spectrum below 190 nm is absorbed in the higher atmosphere above 40 km by the Schumann-Runge bands of oxygen, and below 175 nm by the Schumann-Runge continuum. The much weaker Herzberg continuum, which extends short of 240 nm, starts to contribute significantly to the photochemistry of molecular oxygen at lower altitudes, where an increasing fraction of the radiation above 220 nm is absorbed in the Hartley band of ozone (Turco 1975). The solar flux outside the atmosphere, integrated over all wavelengths shorter than 240 nm, amounts to approximately 2.3×10^{13} photons $cm^{-2} s^{-1}$ (Nicolet 1975, 1978). Since the photodissociation quantum yield of molecular oxygen is unity, except for a very low fluorescence yield in the Schumann-Runge band system (Turco 1975), approximately 5.2×10^{36} oxygen atoms per day (ca. 1.3×10^8 t d^{-1}) are produced in the entire atmosphere (cross-sectional area of the earth: $1.28 \times 10^{18} cm^2$). A large fraction of the atoms generate ozone by recombination with molecular oxygen. The observed ozone profile of the stratosphere represents a dynamic equilibrium between this enormous ozone source and a number of loss processes.

Although mass exchange between the stratosphere and the troposphere is extremely slow (transport time of gases through the tropopause region ca. 1–2 years (CH. Junge 1978), stratospheric ozone is continuously injected into the troposphere, where the mean vertical exchange time reduces to the order of one month (CH. Junge 1978). Loss of ozone in the natural troposphere occurs predominantly through reactions (1)–(4),

$$O_3 + h\nu\,(\lambda < 310\ nm) \longrightarrow O_2 + O(^1D), \tag{1}$$

$$O(^1D) + H_2O \longrightarrow 2\,OH,\ \text{in competition with} \tag{2}$$

$$O(^1D) + air \longrightarrow O(^3P) + air, $$

$$OH + O_3 \longrightarrow HO_2 + O_2, \tag{3}$$

$$HO_2 + O_3 \longrightarrow HO + O_2, \tag{4}$$

and by deposition at the earth's surface. Photodissociation of ozone at longer wavelengths, which yields $O_2 + O(^3P)$ in the electronic ground state (cf. Chap. 1.4.1), is quantitatively reversible at tropospheric pressures, and thus does not constitute a loss mechanism. Ozone destruction through reactions (3) and (4) occurs in competition with the reaction sequence (5)–(8), which results in ozone production:

$$OH + CO \xrightarrow{\ +O_2\ } HO_2 + CO_2, \tag{5}$$

$$HO_2 + NO \longrightarrow OH + NO_2, \tag{6}$$

$$NO_2 + h\nu\,(\lambda < 410\ nm) \longrightarrow NO + O, \tag{7}$$

$$O + O_2 \xrightarrow{\ +M\ } O_3 \tag{8}$$

Obviously, ozone in the troposphere is by no means an inert tracer, as was believed in earlier times (C. E. Junge 1962), but undergoes continuous destruction and re-formation. Recent model calculations (Fishman et al. 1979, Liu et al. 1980,

Logan et al. 1981, Hameed and Stewart 1983), on the basis of measured NO_x- and CO-profiles (Kley et al. 1981, Seiler and Fishman 1981), suggest that ozone destruction dominates in the equatorial troposphere, whereas net ozone production is believed to occur at mid-latitudes, particularly in the northern hemisphere. In support of this hypothesis, structures in the vertical profiles of O_3 and CO in the mid-latitudinal troposphere are often positively correlated (Fishman and Seiler 1983). It is possible that the ozone budget of the northern troposphere is already positively offset by anthropogenic CO, particularly from internal combustion engines, and by NO_x emissions of aircraft.

Ozone from the stratosphere can be mixed down into the troposphere quasi-continuously, or in discontinuous irregular intrusions of ozone-rich stratospheric air. Johnson and Viezee (1981) have recently investigated the mechanism of ozone intrusions into the troposphere in some detail, by means of ten aircraft ascents up to about 8 km altitude. They conclude from their measurements that intrusions of ozone-rich stratospheric air, so-called tropopause foldings (Danielsen 1968), are considerably more frequent than previously assumed, being probably associated with practically all low-pressure systems. The authors observed highest ozone concentrations between 240 and 400 ppb at an altitude of 6–8 km, decreasing to about 100–200 ppb at 2 km. The intrusions were between 100 and 300 km wide and extended several hundred km downwind. It remains a matter of controversy whether the exchange of air between the stratosphere and the troposphere occurs mainly by the tropopause folding mechanism, or whether slow steady exchange of air is equally important.

Hydrocarbons

On a global scale the biosphere is an important source of hydrocarbons in the atmosphere. Hydrocarbons are emitted mainly by the leaves of plants, with the exception of methane, which evolves as marsh gas in the anaerobic decomposition of biogenic material, e. g., in the digestive tract of ruminants. Methane is too unreactive to contribute significantly to the build-up of high oxidant mixing ratios in the atmosphere. Plant emissions are composed if isoprene, a large variety of terpenes, and other volatile compounds. The trapping, for GC-MS analysis, of these highly reactive species on a suitable pre-concentration column poses some problems (Knoeppel et al. 1980).

Airborne isoprene and terpenes do not behave very differently from reactive anthropogenic hydrocarbons when exposed to sunlight in the presence of nitrogen oxides: both ozone and other oxidants (PAN) are formed (Kamens et al. 1981, Gay and Arnts 1977). If natural sources contributed significantly to the hydrocarbon burden of the atmosphere, the consequences for oxidant formation, particularly in rural areas, would be serious: The efficiency of oxidant control strategies based on the reduction of hydrocarbon emissions without simultaneous NO_x reductions would be reduced.

Based on measured hydrocarbon emission rates of various leaf plants and conifers, Zimmerman et al. (1978) estimate a global biospheric emission rate of 8.3×10^{14} g C a^{-1}. More revealing than this global estimate are in situ measurements of biogenic hydrocarbons in the atmosphere of Florida (Lonneman et al. 1978), and in a forested area in the northwestern USA (Holdren et al. 1979). The measurements in Florida revealed that the concentration of natural hydrocarbons was always negligible, even in remote areas, compared with the concentrations of anthropogenic hydrocarbons. The measurements of Holdren et al. (1979) had enough sensitivity to detect isoprene and several terpenes inside a wood. Typical sum concentrations of these hydrocarbons amounted to less than 5 ppb C (ppb C see Chap. 1.3.2). Above the canopy, however, the concentrations fell below the detection limit of 0.01 ppb for individual terpenes. The authors conclude from their measurement that terpenes, considering their extremely low concentrations in the free atmosphere, do not contribute to photochemical ozone formation. This view is confirmed by more recent measurements in the United States (Roberts et al. 1983), and in Norway (Hov et al. 1983), although slightly higher sum concentrations (9–71 ppb C in June and August) were observed in the latter study. Yokouchi et al. (1983) report a pronounced seasonal variation of monoterpenes in the atmosphere of a pine forest.

Nitrogen Oxides

Extremely low concentrations of nitrogen oxides are found in the clean lower troposphere (see also Sect. 1.3.2.1), although a number of natural sources have been identified. Nitrogen oxides are by-products of biological processes (denitrification) in soils. It is known, however, that only minor amounts of NO are emitted into the atmosphere, the main product of denitrification being N_2O, which is chemically inert in the troposphere and does not contribute to the oxidant problem. Other natural sources are forest fires and lightning (Chameides et al. 1977, Hill et al. 1980). It is extremely difficult to quantify these sources. Lightning is most important in the tropical troposphere (Ehhalt and Drummond 1982). Ratsch and Tingley (1977) have reviewed the literature on NO/NO_2 emission rates of fertilized and unfertilized soils. They report typical fluxes of the order 0.01 (rarely reaching 0.1) $kg\ km^{-2}\ h^{-1}$, calculated as NO_2. Combining these data with a typical mixing height of 2 km and an estimated mean tropospheric residence time of NO_2 of 1 day yields a steady-state concentration of 0.1–1 ppb NO_x in the mixing layer, in fair agreement with available NO_x measurements over land surfaces in the clean remote troposphere. Since surface emission of NO_x by soil should occur rather evenly over vast areas, large local variations of the steady-state NO_x concentration due to biogenic emissions are unlikely in the clean remote atmosphere. Airborne measurements have revealed (Fricke 1980) that high oxidant levels are invariably linked with increased (≥ 10 ppb) NO_x concentrations in the mixing layer. It is therefore extremely unlikely that NO_x from natural sources makes a significant contribution to the oxidant problem in the densely populated European countries.

Most of the NO_x in the atmosphere, regardless of its origin, is converted to nitric acid which, when dissolved in droplets, contributes to the acidification of cloud and rain water. The same holds true for SO_2 and other sulfur-containing trace gases from natural sources, which can be converted to sulfuric acid either in the gas phase, or in droplets and on wet aerosols. However, the amount of free acid made from nonanthropogenic trace gases is very low in natural rainwater, and can be neutralized by ammonia and basic particulate matter. Distilled water in equilibrium with natural CO_2 in very clean air should exhibit pH 5.6. However, rainwater of pH 5 and less is typically found in the clean remote troposphere, due to acidic oxidation products of natural trace gases (e. g., sulfur dioxide and other sulfur compounds from volcanoes, fumarole fields, etc.), and/or incorporation of natural aerosol particles.

1.3.2 Immissions

The term immission denotes the ambient concentration of a trace gas in a receptor area, where it acts on man, animals, plants, and material. Immissions are much lower than typical source concentrations (e. g., in a plume from a chimney), due to dilution and chemical conversion of the pollutants during transport from the source to the receptor area. Since the concentrations of trace gases, particularly of short-living reactive compounds, can become extremely low when the distance between source and receptor area is large, extremely sensitive techniques are sometimes required for immission measurements.

Trace gas concentrations are often expressed in suitable units of molar mixing ratios (=volume mixing ratios): in ppb, or ppm (=1,000 ppb), or ppt (=0.001 ppb). One ppb denotes a molar mixing ratio of $1:10^9$. When molar mixing ratios must be converted to mass concentrations (e. g., $\mu g\ m^{-3}$), which are sometimes preferred in the technical literature, atmospheric pressure and temperature have to be taken into account:

$$1\ \mu g\ m^{-3} \cong \frac{8.32 \times 10^{-2}\,T}{M \times P}\ ppb$$

(T=temperature in Kelvin; M=molecular weight; p=pressure in bar).

Table 1.26 in the Appendix lists conversion factors for a number of trace gases which are relevant in photochemical smog.

Since the sensitivity of the flame ionization detector (FID), the most widely used hydrocarbon detector, is approximately proportional to the number of carbon atoms per molecule, the sum concentration of a complex carbon mixture which is measured unseparated, is often expressed in ppb C or ppm C, referring to the number of carbon atoms in the mixture. Molar (or volume) mixing ratios of individual hydrocarbons can be converted to these "methane equivalent" mixing ratios by multiplication with the number of carbon atoms per molecule. For hydrocarbon mixtures of unknown composition the conversion of ppb C (ppm C) to mass concentration units is not defined. In practice, the conversion equation given above can be preserved if the "molecular weight" M = 14 of a methylene group (—C—), the typical building block of a saturated hydrocarbon, is inserted (sometimes the molecular weight M = 16 of methane is used instead of the methylene group; however, this ambiguity is not really serious in view of the limited accuracy of most hydrocarbon immission measurements). The procedure fails when applied to organic heterocompounds (e.g., alcohols, ketones, amines, halocarbons, etc.). Nevertheless, "total organic emissions" (including undefined heterocompounds) are frequently expressed in mass units in the technical literature.

1.3.2.1 Natural Background Concentrations

The attempt is often made to distinguish between trace gas concentrations in anthropogenically polluted areas and so-called natural background concentrations, which exist in clean remote areas. The assignment of a hypothetical natural background concentration to a polluted area, which would be established if all anthropogenic sources were extinguished, is rather ambiguous. Unpolluted reference areas where background concentrations can still be measured (e. g., the SW coast of Ireland during westerly winds) differ from the comparison area in various ways. Furthermore, natural background concentrations are often so low that routine monitoring techniques are no longer adequate, and must be replaced by special procedures.

Ozone

Ozone is produced in the stratosphere, from where it mixes downward into the troposphere. Ozone concentration profiles which are obtained with balloon sonds three times a week at Meteorologisches Observatorium Hohenpeißenberg southwest of Munich, and are published every 6 months by the Deutscher Wetterdienst, have revealed that the ozone mixing ratio, which reaches 6 ppm around 25 km altitude, falls abruptly at the tropopause. Below the tropopause, typically at 12 km in Middle Europe, a nearly constant mixing ratio of 70 to 50 ppb is maintained throughout the free troposphere. Since the distinction between polluted and unpolluted areas fades out above the planetary boundary layer (ca. 2–3 km above the ground), 50–70 ppb pose an upper limit on the mixing ratio of natural ozone at ground level which is due to downward mixing of stratospheric air. Measurements of surface ozone in clean remote areas yield lower mixing ratios, because ozone is destroyed heterogeneously at the surface. Typical mean mixing ratios in clean air at ground level range between 20 and 40 ppb (Singh et al. 1978), with a relative maximum in spring, distinctly *before* the summer period when the photochemical activity is maximized. Practically all hourly averaged ozone mixing ratios in clean air fall short of 80 ppb.

At various places in Europe which are relatively unpolluted (Hohenpeißenberg, SW coast of Ireland), 80 ppb ozone are quite often exceeded during the summer months. Such events can be attributd to advection of "anthropogenic" ozone in polluted air masses (Attmannspacher et al. 1980). At the west coast of Ireland, where daily ozone maxima of 40–60 ppb are measured in air advected from the Atlantic, the occurrence of higher mixing ratios (at times in excess of 100 ppb ozone) was always found to coincide with maxima of the CCl_3F concentration. This halocarbon has been used as a tracer of anthropogenic pollution (Cox et al. 1975).

On very rare occasions, ozone peaks of short duration, sometimes in excess of 200 ppb, have been observed at remote mountain stations. Their occurrence is usually coupled with the passage of winterly cold fronts (Attmannspacher and Hartmannsgruber 1973, Attmannspacher 1976, 1977, Attmannspacher et al. 1980). Simultaneous measurements of the [7]Be activity, which serves as a tracer of stratospheric air (Reiter 1975), have revealed the stratospheric origin of an ozone event on Zugspitze (ca. 3,000 m altitude), where 200 ppb ozone were exceeded for 1 h at midnight (Singh et al. 1978). Apart from such measurements of

[7]Be and other radioactive Radon decay products originating in the higher atmosphere, ozone intrusions have also been inferred from vertical ozone profiles which were measured between Garmisch and the summit of the Zugspitze at regular intervals (Kanter et al. 1979, 1982). By contrast, not a single intrusion of stratospheric ozone has been identified in the Cologne-Bonn area, where up to five ozone analyzers were run continuously between 1974 and 1978, summer and winter. It seems that ozone intrusions from the stratosphere can be safely identified only on mountain tops and in clean remote areas (Viezee et al. 1983). This is in keeping with conclusions of R. T. Derwent et al. (1978), who identified and analyzed several intrusions of stratospheric air at rural stations, but not a single one in the Greater London Council area. In polluted areas the characteristic ozone peaks associated with stratospheric intrusions are flattened out too much by reactions with reducing pollutants, particularly with NO, before they can reach the surface. It is therefore extremely difficult, if not impossible, to identify ozone intrusions in polluted areas.

Nitrogen Oxides

Estimates of the global emission rate of anthropogenic nitrogen oxides in the 1970's centered around 17×10^9 kg N a^{-1}, and were usually ranked behind the global emission rate of natural NO_x (Ratsch and Tingley 1977). In 1982 Ehhalt and Drummond estimated a global emission rate of 39×10^9 kg N a^{-1}, more than half of which was attributed to anthropogenic sources. Since nearly all anthropogenic NO_x stems from the combustion of fossil fuels, most of which are consumed in the highly industrialized countries between 30° and 60° latitude in the northern hemisphere, atmospheric NO_x in this latitudinal band should be predominantly anthropogenic of origin. Nitrogen fixation by lightning was believed to make a substantial contribution to the natural NO_x budget of the atmosphere (Chameides et al. 1977, Hill et al. 1980), but other authors have re-evaluated the rate of nitrogen fixation by lightning to be only 1.8×10^9 kg N a^{-1} on the basis of recent laboratory experiments (Levine et al. 1981). Some NO is produced by the photolysis of nitrites in the surface waters of the oceans; however, this process was found to be entirely negligible as a source of NO_x in the atmosphere (Zafiriou and McFarland 1981).

Owing to the relatively short and variable lifetime of the nitrogen oxides under tropospheric conditions [for NO_2 a globally averaged residence time of 1.5 days is given in the recent literature (Baulch et al. 1982)], meaningful estimates of "typical" NO_x concentrations in polluted areas with strong point sources (chimneys, road traffic) cannot be given. All information on nitrogen oxide concentrations in such areas must be derived from in situ measurements.

The detection limit of commercially available NO detectors (chemiluminescence analyzers with hot catalytic NO_2–NO converters) is usually not much lower than 5 ppb. Airborne measurements with a commercial detector (Fricke 1980) have revealed that the NO_x concentration above the mixing layer is generally less than this limit, even over polluted areas.

Platt et al. (1979) used long path differential optical absorption to measure NO_2 concentrations in Dagebüll, north of Hamburg. The NO_2 concentration in air masses advected from N to NW ranged from 0.5 to 3 ppb. The authors conclude that only the lowest concentrations were unaffected by anthropogenic emissions, and were thus typical for clean air. NO_x concentrations between 0.2 and 0.6 ppb were measured at Fritz Peak Observatory in the Rocky Mountains (Kley et al. 1981), which should be typical of clean continental air. The NO_x concentration in clean marine air is even lower, according to recent measurements above the Pacific (McFarland et al. 1979) and the North Atlantic (Helas and Warneck 1981).

Hydrocarbons

Nearly all volatile hydrocarbons, with the exception of methane, ethane, and acetylene, react with OH radicals sufficiently rapidly to be classified as potential precursors of photooxidants in polluted air. The recognition that plants can evaporate terpenes, while low molecular weight hydrocarbons are probably emitted by soils (Van Cleemput et al. 1982), has caused considerable confusion with respect to the concentration of biogenic hydrocarbons in unpolluted air. Holdren et al. (1979) have been able to identify and measure several monoterpenes in a forest of the United States, with individual detection limits of 0.01 ppb using GC-MS coupling. One meter above the soil, the sum concentration of the terpenes amounted to 0.3 ppb (corresponding to 3 ppb C) inside the forest. The terpene concentration in the free atmosphere, above the canopy, was always lower than the detection limit. The authors conclude that natural terpenes are unimportant as precursors of photooxidants.

Rudolph et al. (1980) carried out numerous measurements of C_2–C_5 hydrocarbons in the Eifel mountains, which is a relatively unpolluted area in Germany. They measured individual concentrations between 0.1 ppb (C_2H_4 and C_3H_6) and 3 ppb (C_2H_6) in summer. The integrated concentration of all C_2–C_5 hydrocarbons amounted to 17 ppb C. In winter, slightly higher concentrations of the more reactive hydrocarbons C_2H_4 (1.5 ppb) and C_3H_6 (0.4 ppb) were detected. In clean marine air at the West Coast of Ireland and on the Atlantic, only the C_4 and C_5 concentrations were significantly lower than in the Eifel. The authors were led to the conclusion that seawater is a source of light hydrocarbons. Similar results have been obtained by Cofer (1982), who measured hydrocarbons in the marine boundary layer of the North and South Atlantic. An average sum concentration of light nonmethane hydrocarbons of 16 ppb C was obtained for the North Atlantic. Singh and Salas (1982) report a significant seasonal variability of the hydrocarbon concentrations over the Pacific.

In summary, concentrations of 10–20 ppb C gaseous nonmethane hydrocarbons seem typical of clean tropospheric air. Approximately one half consists of the relatively unreactive compounds ethane, propane, and acetylene, which can be transported over long distances and are thus partly anthropogenic, even in otherwise clean air.

Aldehydes

Formaldehyde, a product of the reaction chain initiated by OH radical attack on methane, is a regular trace constituent of the unpolluted troposphere which contains ca. 1,650 ppb methane. The distribution of formaldehyde in the troposphere has recently been reviewed by Lowe et al. (1981 a) and by Lowe and Schmidt (1983). The optical long path absorption measurements of Platt et al.

(1979) in air from the North Sea at Dagebüll near Hamburg (see above) have also yielded information on formaldehyde. Daytime concentrations ranged from 0.2 to 0.5 ppb, which was substantially lower than formaldehyde concentrations in Jülich near Aachen, where the anthropogenic component is likely to be high (Lowe et al. 1980). Clean marine air at the West Coast of Ireland contained approximately 0.3 ppb formaldehyde (Platt and Perner 1980), which agrees well with data obtained by Lowe et al. (1981 a) during an Atlantic crossing to Brazil. These authors absorbed formaldehyde in a solution of 2,4-dinitrophenyl-hydrazine. The resulting hydrazone was analyzed by liquid chromatography. A slightly different variant of the technique was developed by Neitzert and Seiler (1981), who measured between 0.4 and 3.8 ppb formaldehyde in the Hunsrück mountains at a so-called clean air station of the German Umweltbundesamt. This was significantly more formaldehyde than the 0.2–1 ppb measured in clean marine air at Cape Point in South Africa by the same authors. Photochemical smog reactions could not be held responsible for the enhancement of the formaldehyde concentration at the time of the measurements at Deuselbach (November). The good correlation of the formaldehyde concentration pattern with simultaneously measured SO_2 and CO concentrations was indicative of a common anthropogenic source of all three compounds.

Acetaldehyde is another intermediate in OH reactions with light hydrocarbons. Although acetaldehyde is of considerable importance as precursor of peroxyacetyl nitrate (PAN) in clean tropospheric air (see below), only a few preliminary field measurements in moderately polluted air in Germany have been reported. About one half of the measurements yielded acetaldehyde mixing ratios of less than 0.5 ppb (Schubert et al. 1984).

Peroxyacetyl Nitrate (PAN)

Singh and Hanst (1981), on the basis of a simple chemical steady state model, predicted that peroxyacetyl nitrate (PAN) is probably a natural constituent of the unpolluted troposphere. Precursors of PAN are naturally available, and the compound is considerably more stable in the colder regions of the higher troposphere than at ground level. The calculations predicted peak mixing ratios between 0.09 and 0.36 ppb PAN at an altitude of 9 km. Meanwhile, the first vertical profiles of the compound have been measured over the Pacific, off the Californian coast (Singh and Salas 1983), and in Germany (Meyrahn et al. 1984). Between 0.001 and 0.3 ppb PAN were measured above the Pacific up to 8 km altitude. Somewhat higher mixing ratios, up to 3 ppb in the lower 5 km of the continental troposphere in Germany, may have been partly due to anthropogenic precursors. The mean surface mixing ratio over the Pacific was approximately 0.03 ppb in August 1982. These measurements prove unambiguously that PAN is, in fact, a natural constituent of the unpolluted troposphere.

Other Compounds

Natural background concentrations of short-living trace compounds should show large regional variations, unless the sources are very evenly distributed over large areas. This is usually not the case for SO_2 and a few other reduced sulfur compounds (H_2S, $(CH_3)_2S$, CH_3SH), which react rapidly with OH radicals and/or

Table 1.10. Background concentrations of some trace gases in unpolluted air

Trace gas	Typical background concentrations in unpolluted air
Surface ozone	Mean concentration 20–40 ppb, daily maxima 40–60 ppb, 80 ppb rarely exceeded by intrusions of stratospheric air
Ozone above the mixing layer	50–70 ppb typical
$NO_x = NO + NO_2$	0.1–1 ppb
Volatile nonmethane hydrocarbons including terpenes	10–20 ppbC
Aldehydes	Formaldehyde 0.2–0.4 ppb, other aldehydes much lower
Peroxyacetyl nitrate (PAN)	Mean surface concentration 0.03 ppb in marine air in summer
H_2O_2	Distinctly less than 1 ppb (Kelly et al. 1979)
SO_2	0.03–0.2 ppb in marine air (Georgii 1982).Variable in clean continental air)

are converted to sulfate in cloud and fog droplets. SO_2 concentrations in the range 0.03 to 0.2 ppb have been measured over the Atlantic ocean (Georgii 1982). The background concentration of SO_2 over the polar ice caps is even lower. For reasons mentioned above, the natural component of the SO_2 mixing ratio must be highest on the continents, where the sources of natural sulfur compounds are largest in number and yield.

Typical background concentrations of some important trace gases in the unpolluted troposphere near ground level are summarized in Table 1.10.

1.3.2.2 Measurements in Densely Populated and Peripheral Areas

1.3.2.2.1 Federal Republic of Germany

Oxidants

As has already been mentioned in the previous sections, ozone measurements at meteorological and balneological stations have a long tradition in Europe, particularly in the Federal Republic of Germany. The meteorological measurements were performed mainly for the investigation of transport processes over larger regions (Regener 1938, Paetzold and Regener 1957, Ehmert 1949, 1952, C. E. Junge 1962) and for the determination of the loss rates of stratospheric ozone at the surface (Fabian and C. E. Junge 1970, Tiefenau and Fabian 1972). The ozone concentration near the ground which, according to these authors, is determined by transport from the stratosphere and decay at the surface, exhibits a latitudinal dependence showing a maximum at 60 ° in the northern hemisphere (Fabian et al. 1971).

Ozone formation in the troposphere which may also influence the ground level concentration had not yet been discussed at that time. In the course of these investigations, diurnal concentration profiles of surface ozone were measured at stations on Norderney, in Clausthal/Harz, in Lindau/Harz, and on the Zugspitze

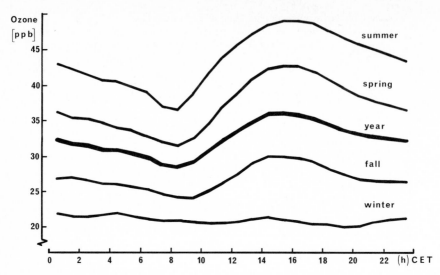

Fig. 1.1. Diurnal ozone concentrations at the station Hohenpeißenberg (station No. 41 in Fig. 1.3 and Table 1.11) averaged over the years from 1971 to 1977 for different seasons and as annual mean. (Attmannspacher 1977, Attmannspacher et al. 1980)

(Pruchniewicz 1970). Except for the Zugspitze, where higher daily maxima were found, the ozone concentrations were always less than 100 µg m^{-3}.

At the meteorological observatory on Hohenpeißenberg, measurements of ozone near the ground and vertical balloon soundings up to about 30 km in the stratosphere have been performed regularly since 1960 (Attmannspacher 1977, Attmannspacher et al. 1980).

Figure 1.1 shows diurnal profiles of ground level ozone on Hohenpeißenberg (approx. 1,000 m) by season and year averaged over the years 1971–1977. Monthly averages of the ozone concentrations from 1971–1982 at the same station are shown in Fig. 1.2. From this figure a slight increase of surface ozone can be seen over the last years. A similar trend of the ozone concentration near the ground has been observed at several stations in East Germany by Warmbt (1979, 1981).

The conclusion that this trend is due to anthropogenic emissions awaits further proof. At higher altitudes within the troposphere, the trend of increasing ozone concentrations during the last 10 years is even more pronounced (Attmannspacher 1983). This increase of tropospheric ozone seems to occur in parallel with a decrease of stratospheric ozone.

In the city of Bonn "aran" (= oxidant) measurements were carried out in 1950 by Emonds (1954). Emonds concluded from his measurements that the ozone deficiency in the city center relative to the periphery indicated a decrease of air quality.

Measurements in Berlin (Lahmann 1969) gave similar results: near roads with heavy traffic lower ozone concentrations were found than in typical residential areas. No evidence of anthropogenic ozone formation was obtained at that time.

In contrast, Jost (1970) found distinctly higher noon concentrations of ozone in the city of Frankfurt (July 27 and 28, 1967) than on a nearby mountain (Kleiner

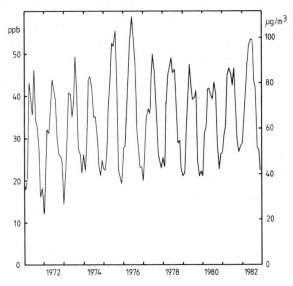

Fig. 1.2. Monthly means of the ozone concentration from 1971 to 1982 at the station Hohenpeißenberg (1971–1979: Attmannspacher 1981, 1981–1982: Attmannspacher pers. commun. 1983)

Feldberg, 825 m). Jost concluded that this excess of ozone within the city area was due to the formation of photochemical oxidants in the polluted surface air. In 1973, Becker and Schurath suggested on the basis of published measurements in the Netherlands (IG-TNO 1971, Guicherit et al. 1972) and England (Atkins et al. 1972, Derwent and Stewart 1973) that increased ozone levels due to photochemical reactions could be expected in the Federal Republic of Germany, taking into account the level of precursor pollution in urban areas.

In 1975, Becker and Schurath in Bonn, Deimel in Cologne, and Georgii in Frankfurt started a joint research project in the Cologne-Bonn area with supplementary measurements in Frankfurt, in order to establish the extent of photochemical oxidants formation in the polluted atmosphere (Becker and Schurath 1975, Georgii et al. 1977, Becker et al. 1979). Several measuring stations were set up in the Cologne-Bonn area for this project (see Fig. 1.3 and Table 1.11). Additional stations were operated on the Kleiner Feldberg (825 m) and in the city of Frankfurt. Supplementary airborne measurements were also carried out (Fricke 1980, Neuber et al. 1981, 1982). The results obtained in the Cologne-Bonn area were evaluated, taking meteorological considerations into account (Muschalik 1979, 1980), and more recently in a dispersion and transport model (Scherer and Stern 1982).

In addition, Doman and Jessel (1977) have published values of ozone concentrations at the isle of Sylt for the period between July 1971 and June 1974. Lahmann et al. (Lahmann et al. 1968, Lahmann 1969, Fett 1970) carried out ozone measurements in Berlin.

A recent compilation of monitoring stations which are operated by state authorities in the Federal Republic of Germany have been published (Bundesminister des Innern 1984).

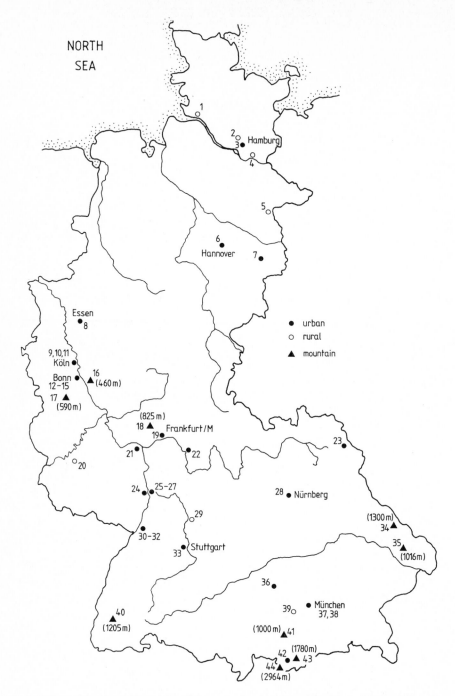

Fig. 1.3. Map of the Federal Republic of Germany showing the geographical position of the ozone monitoring stations from which the data are listed in Table 1.11. The numbering of the stations follows the north–south direction with indications of urban (●), rural (○) and mountain (▲) sites. For the mountain stations the height above sea-level is denoted in *parenthesis*

Table 1.11. Ozone concentrations in µg m^{-3} measured between 1975 and 1983 at different stations in the Federal Republic of Germany, see station No. on map in Fig. 1.3; *year:* annual mean; *summer:* summer mean (April–Sept.); *max:* maximum value with (*a*) 1/2 h; (*b*) peak; (*c*) 1 h; (*d*) 3 h averages

No.	Location	Foot-note	Time	1975	1976	1977	1978	1979	1980	1981	1982	1983
1	Westerbüttel	1	Year						032	040	041	046
			Summer						044	053	060	059
			Max (a)						148	199	196	196
2	Quickborn	2	Year					022	029	033	021	028
			Summer					031	042	046	031	039
			Max (b)					162	160	178	176	168
3	Hamburg	2	Year		017	017	017	015	024	027	028	025
			Summer		027	024	027	021	034	040	035	035
			Max (b)		152	100	138	130	136	168	160	168
4	Geesthacht	1	Year							032	038	040
			Summer							046	054	058
			Max (a)							169	197	224
5	Langenbrügge	3	Year					070	060	–		085
			Summer					086	078	–		099
			Max (c)					254	244	270		320
6	Hannover	4	Year					026	034	–		028
			Summer					038	049	047		033
			Max (a)					190	170	264		260
7	Braunschweig	4	Year					027	030	031		023
			Summer					047	046	047		026
			Max (a)					240	245	–		–
8	Essen	5	Year				–	–			–	
			Summer				036	042			042	
			Max (a)				226	270			300	
9	Köln-Eifelwall	6	Year	–	029	020	018	020	021	–	031	024
			Summer	042	044	023	026	030	032	034	048	038
			Max (a)	290	350	210	180	210	240	382	306	304
10	Köln-Rodenkirchen	5	Year				–	–	–	–		–
			Summer				046	046	044	039		038
			Max (a)				410	326	244	274		270
11	Köln-Godorf	6	Year	–	025	020	021	023	023	–	037	033
			Summer	036	038	028	030	032	036	050	052	051
			Max (a)	280	390	300	250	190	240	546	326	304
12	Bonn-Universität	6	Year	–	030	020	022					
			Summer	046	042	030	028					
			Max (a)	320	370	396	348					
13	Bonn-Friesdorf	5	Year				–	–	–	–		
			Summer				028	040	033	037		
			Max (a)				291	318	188	274		
14	Bonn-Röttgen	5	Year				–	–	–	–		
			Summer				036	039	047	051		
			Max (a)				392	306	234	302		

See footnotes page 32

Table 1.11 (continued)

No.	Location	Foot-note	Time	1975	1976	1977	1978	1979	1980	1981	1982	1983
15	Bonn-Venusberg	6	Year	–	036	036	038					
			Summer	060	060	052	052					
			Max (a)	320	380	410	410					
16	Ölberg	6	Year	–	053	047	060	060	046	–		
			Summer	062	078	056	068	066	060	064		
			Max (a)	260	340	260	340	350	250	318		
17	Michelsberg	6	Year		–	049	045					
			Summer		088	058	068					
			Max (a)		300	240	290					
18	Kl. Feldberg/Ts.	6	Year	–	070	050	058					
			Summer	078	094	068	080					
			Max (a)	220	310	240	320					
19	Frankfurt/M.-Westend	7	Year		039	020	022	023	032	028	033	
			Summer		068	032	034	035	048	043	054	
			Max (a)		409	211	231	278	315	347	366	
20	Deuselbach	3	Year						064	058	062	054
			Summer						087	072	078	067
			Max (c)						290	194	220	184
21	Mainz-Mombach	8	Year					056	032	032	035	027
			Summer					080	045	043	053	037
			Max (a)					410	319	217	268	187
22	Aschaffenburg	10	Year						–	028	030	029
			Summer						038	038	046	–
			Max (a)						258	274	240	210
23	Arzberg	10	Year							036	028	032
			Summer							–	037	050
			Max (a)							152	144	178
24	Ludwigshafen-Mundenheim	8	Year					034	028	027	034	037
			Summer					048	040	035	048	054
			Max (a)					289	218	197	268	278
25	Mannheim-Nord	11	Year	030	032	032	048	027	024	028	–	036
			Summer	042	066	047	063	043	036	044	059	062
			Max (d)	396	458	211	419	198	214	257	397	294
26	Mannheim-Mitte	11	Year	034	034	025	028	027	025		040	032
			Summer	057	053	043	034	046	038		057	–
			Max (d)	298	543	225	237	244	193		408	187
27	Mannheim-Süd	11	Year	034	046	015	027	043	033	031		
			Summer	052	075	024	032	071	054	043		
			Max (d)	326	432	143	352	395	272	228		
28	Nürnberg	10	Year						022	018	016	015
			Summer						030	024	026	023
			Max (a)						226	226	266	214
29	Heilbronn	11	Year						–	044	048	032
			Summer						062	062	054	051
			Max (d)						275	307	335	282

See footnotes page 32

Table 1.11 (continued)

No.	Location	Foot-note	Time	1975	1976	1977	1978	1979	1980	1981	1982	1983
30	Eggenstein	11	Year			038	049	035	032	031	042	046
			Summer			055	066	046	044	–	065	083
			Max (d)			296	383	230	390	–	407	428
31	Karlsruhe-Mitte	11	Year		028	027	034	021	–	028		019
			Summer		046	–	044	–	033	035		030
			Max (d)		358	409	299	232	287	245		222
32	Karlsruhe-West	11	Year		038	023	027	026	029	036	040	039
			Summer		061	026	035	039	042	050	059	–
			Max (d)		422	355	195	200	229	276	260	343
33	Stuttgart-Zuffenhausen	11	Year							038	031	021
			Summer							055	–	034
			Max (d)							313	363	254
34	Bodenmais	10	Year								053	053
			Summer								064	065
			Max (a)								133	145
35	Brotjacklriegel	3	Year						055	069	075	074
			Summer						063	088	085	096
			Max (c)						143	172	184	199
36	Augsburg	10	Year						018	012	016	–
			Summer						026	–	023	017
			Max (a)						130	–	156	126
37	München-Stachus	10	Year						018	012	008	–
			Summer						027	018	015	022
			Max (a)						170	174	140	152
38	München-Effnerplatz	10	Year						026	026	016	013
			Summer						040	036	026	017
			Max (a)						178	210	122	108
39	Oberpfaffenhofen	12	Year	027	032	032	028	025	034			
			Summer	041	050	043	040	039	047			
			Max (c)	196	200	170	166	170	190			
40	Schauinsland	3	Year						095		078	078
			Summer						113		094	094
			Max (c)						310		230	220
41	Hohenpeißenberg	13	Year	062	068	060	065	058	060	065	070	060
			Summer	083	086	074	080	074	073	078	086	073
			Max (c)	202	184	170	198	188	166	168	240	158
42	Garmisch-Partenkirchen	14	Year			052	053	046	043	047	054	050
			Summer			065	068	057	057	058	069	063
			Max (c)			184	178	143	163	172	198	175
43	Wank	14	Year			061	059	058	065	067	070	079
			Summer			070	068	067	072	076	082	093
			Max (c)			150	120	144	158	151	151	187
44	Zugspitze	14	Year			051	050	051	054	058	068	063
			Summer			060	054	060	061	065	077	073
			Max (c)			113	087	106	110	121	137	125

See footnotes page 32

Footnotes to Table 1.11: (also used in Tables 1.17 and 1.18)

1 Gewerbeaufsichtsamt Itzehoe (pers. commun.)
2 Deutscher Wetterdienst, Meteorologisches Observatorium Hamburg, 1976–1983; Winkler (1980)
3 Monatsberichte aus dem Metznetz des Umweltbundesamtes/Monthly Reports; Grosch (pers. commun. 1984)
4 Niedersächsisches Landesverwaltungsamt, Monats- und Jahresberichte: Umweltschutz in Nieder-sachsen – Reinhaltung der Luft/Monthly and Annual Reports; Müller (pers.. commun. 1984)
5 Monatsberichte der Landesanstalt für Immissionsschutz des Landes Nordrhein-Westfalen über die Luftqualität an Rhein und Ruhr/Monthly Reports; Ixfeld (1977), Bruckmann and Eynck (1979, 1980a), Bruckmann and Langensiepen (1981), Bruckmann et al. (1982)
6 Georgii et al. (1977), Becker et al. (1979), Deimel (1982a, b, pers. commun. 1984)
7 Monatsberichte aus dem Meßnetz des Umweltbundesamtes/ Monthly Reports; Rudolf (pers. commun. 1983)
8 Zentrales Immissionsmeßnetz – ZIMEN – des Landesgewerbeaufsichtsamtes für Rheinland-Pfalz/ Monthly Reports
9 Lufthygienische Monatsberichte der Hessischen Landesanstalt für Umwelt/Monthly Reports
10 Bayerische Landesanstalt für Umweltschutz, Lufthygienische Jahresberichte, R. Oldenbourg Verlag, München und Wien/Annual Reports; Michetschläger et al. (1978), Strauß (1980)
11 Monatsberichte der Landesanstalt für Umweltschutz Baden-Württemberg/Monthly Reports; Obländer and Siegel (1977)
12 Paffrath et al. (1983)
13 Deutscher Wetterdienst – Sonderbeobachtungen des Meteorologischen Observatoriums Hohen-peißenberg/Semi-annual Reports; Attmannspacher (pers. commun. 1984)
14 Kanter et al. (1982), Reiter and Kanter (1982a, b)

All available data on ozone concentrations near ground in the Federal Republic of Germany have been listed in Table 1.11. The data were taken from stations which have been operated continuously over longer time periods using compara-ble analytical techniques. The data in units µg m^{-3} are given as annual mean (year), mean values over the summer period from April to September (summer), and as yearly maxima for integration times of (a): ½ h, (b): peak, (c): 1 h, or (d): 3 h. The stations are numbered in order from the North Sea to the Alps. Their positions are shown on a map of the Federal Republic of Germany in Fig. 1.3.

The numbering of the stations according to their positions in Fig. 1.3 has been used in Tables 1.11 and 1.13–1.19. On this map urban (●), rural (○), and moun-tain (▲) stations have been distinguished. For mountain stations above 1,000 m altitude the mass concentrations were converted to a standard pressure of 1 atm. Some data of certain stations had to be corrected for calibration errors, leading to some differences between the present data and previous publications (Becker et al. 1983).

Figure 1.4 illustrates diurnal ozone cycles measured from March to April and June to September, and averaged over the years 1977–1980, at the valley-based station Garmisch (740 m) and two mountain stations, on Wank (1,780 m) and Zugspitze (2,964 m) (Reiter and Kanter 1982a). It can be seen from the figure that at noon on sunny days, in spring as well as in summer, the station in the valley exhibits higher ozone concentrations than either of the two mountain stations. For the summer months, this is in agreement with the previously mentioned ob-servation of Jost (1970) and with results from the Cologne-Bonn area stations, Bonn-Universität, Bonn-Venusberg, and the hill-based station Ölberg, see Fig. 1.5 (Georgii et al. 1977). The results of Reiter and Kanter (1982a) show that

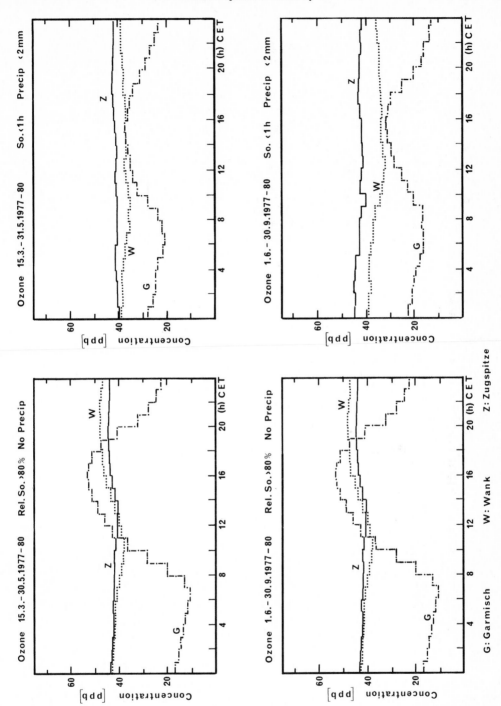

Fig. 1.4. Diurnal profiles of ozone concentrations averaged from 1977 to 1980, at the valley-based station Garmisch (740 m), the mountain-based station Wank (1,780 m), and the mountain-based station Zugspitze (2,964 m), on sunny days (*below*) and less sunny days (*above*). (Reiter and Kanter 1982a)

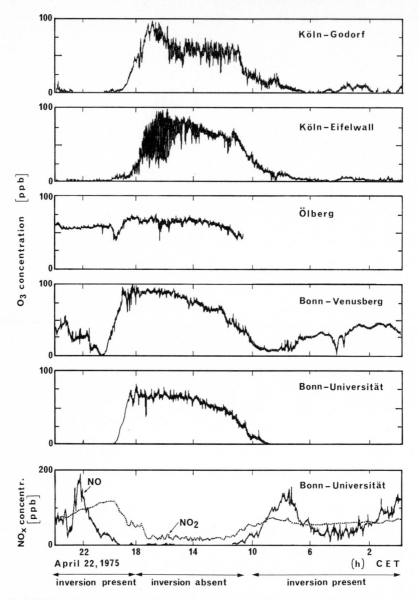

Fig. 1.5. Diurnal ozone concentration on April 4, 1975 at the stations Köln-Godorf, Köln-Eifelwall, Ölberg and Bonn-Universität. (Georgii et al. 1977)

in Garmisch the oxidant concentration at lower altitudes remains higher even for an average over 4 years when data of sunny days only are considered (Fig. 1.4). From the measurements at the stations Garmisch, Wank, and Zugspitze for the period 1977–1981 (Reiter and Kanter 1982b, Kanter et al. 1982) a relative frequency distribution (1-h averages) can be constructed as shown in Fig. 1.6. From this figure it can be seen that ozone at a location like Garmisch, which is strongly

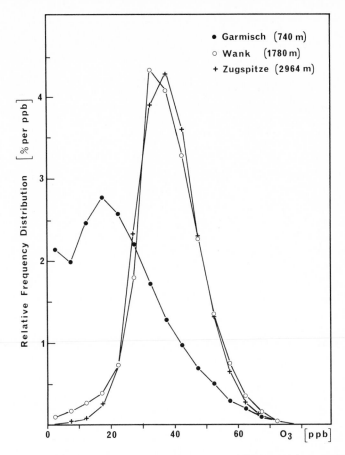

Fig. 1.6. Frequency distribution of the ozone concentrations (1 h average) between 1977 and 1981 at the stations Garmisch (●), Wank (○), and Zugspitze (+). (Kanter et al. 1982, Reiter and Kanter 1982 b)

influenced by anthropogenic precursors, is characterized by a frequency distribution which differs remarkably from the distributions of the more remote mountain stations on Wank or Zugspitze. Despite certain limitations, the ozone values of the two mountain stations can be considered as being fairly representative of the free atmosphere. These stations are much less frequently cut off from the free troposphere by inversion layers during the night, and because of this the destruction of ozone by nitrogen monoxide is much weaker during the night than in the polluted valley. On the other hand, on sunny days the polluted air in the valley produces "extra" (anthropogenic) ozone, raising its level at noon above that in the free atmosphere. The mountain stations show a fairly symmetric frequency distribution which peaks at about 40 ppb, with a halfwidth of about 20 ppb.

Table 1.12 presents the few data which up to now are available on the PAN concentration in air near ground (Schurath 1979 b, Bruckmann and Mülder 1979, Löbel et al. 1980).

Table 1.12. Maximum PAN concentrations

Location and time period	Max. PAN-values (1/2 h)	Ref.
Bonn, 1976	25 µg m^{-3} (5.0 ppb)	[1]
Essen, 1978	18 µg m^{-3} (3.6 ppb)	[2]
Köln, 1979–1980	30.5 µg m^{-3} (6.1 ppb)	

[1] Schurath (1979b); Löbel et al. (1980)   [2] Bruckmann and Mülder (1979)

In the course of a special investigation carried out by Mihelcic (pers. commun. 1981) between April 16 and 17, 1980 in Deuselbach/Hunsrück a maximum sum concentration (1-h average) of 1.9 ppb peroxyradicals, RO_2, has been observed.

It is not possible to assume comparable error limits and representativity of the air samples for all the results presented in the following tables. Measurements near heavy car traffic will generally give lower ozone values than measurements aside in residential or peripheral areas. Ozone destruction at the ground, on the surface of buildings, and in the sampling lines of the analyzers, also results in lower ozone values. The data of Table 1.11, however, should give a general view of the ozone load in ground level air layers at different locations in the Federal Republic of Germany over a period of several years. So far, the highest ozone levels have been observed in the area of Mannheim; even averages over 3, 12, and 24 h show relatively high values. The diurnal cycle with the highest ½ h mean of ozone ever registrated in Germany is shown in Fig. 1.7. In Mannheim-Mitte, between 14.00 and 16.00 h of June 23, 1976, a broad ozone shoulder culminating in a ½ h mean concentration of 664 µg m^{-3}, was superimposed on a plateau of about 300 µg m^{-3}, which persisted for the entire day. The measurements at Mannheim, Karlsruhe, and with certain restrictions, those at Frankfurt indicate that, compared to other regions of Germany and Europe, the Upper Rhine Valley may be regarded as particularly endangered by photochemical pollution.

The monthly ozone maxima from the stations at Mannheim, plotted as 3 h mean values for the years 1975–1979 in Fig. 1.8, illustrate, however, that high ozone levels occur only in summer. Winkler (1980), on the other hand, interprets his ozone data (station No. 3 in Table 1.11) to demonstrate that increased ozone concentrations due to anthropogenic precursors do not occur in Hamburg even in summer. This is in contrast to observations from densely populated coastal areas in the Netherlands (TNO 1978), and to measurements at the neighboring stations Westerbüttel, Quickborn, and Geesthacht. Compared to other parts of Germany, the ozone levels in Hamburg and peripheral locations, however, are relatively low. Elevated ozone concentrations have also been found in the Cologne-Bonn area in summer, especially in 1976, comparable with the results from the Upper Rhine Valley. The diurnal concentration profiles of O_3, NO, and NO_2 for three consecutive days in the summer of 1981 from the stations Köln-Eifelwall (center of the city), Köln-Godorf (in the vicinity of petrochemical plants), and Ölberg (hill-based station south of Bonn near the Rhine valley) are shown in Fig. 1.9. In Godorf, the ozone concentration exceeded 520 µg m^{-3} (260 ppb) for several hours on August 15. The gradual increase becomes already noticeable at noon on August 14; on August 16, the high ozone levels have disappeared.

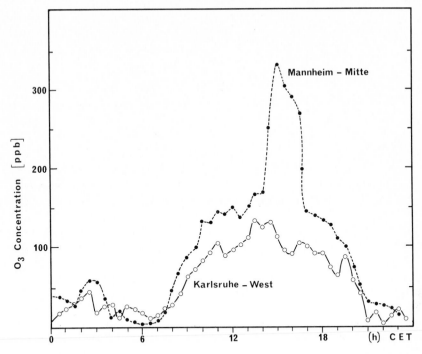

Fig. 1.7. Diurnal ozone concentration on June 23, 1976 at the stations Mannheim-Mitte and Karlsruhe-West. (Obländer and Siegel 1977)

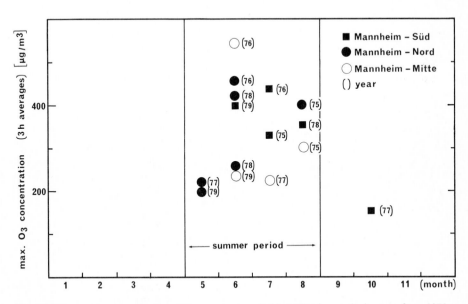

Fig. 1.8. Max. ozone concentrations (3 h averages, see Table 1.11) of a particular year from 1975 to 1979 at the stations Mannheim-Nord, Mannheim-Mitte, and Mannheim-Süd plotted versus the months in which the values were measured; ( ) indicates the year of the ozone maxima

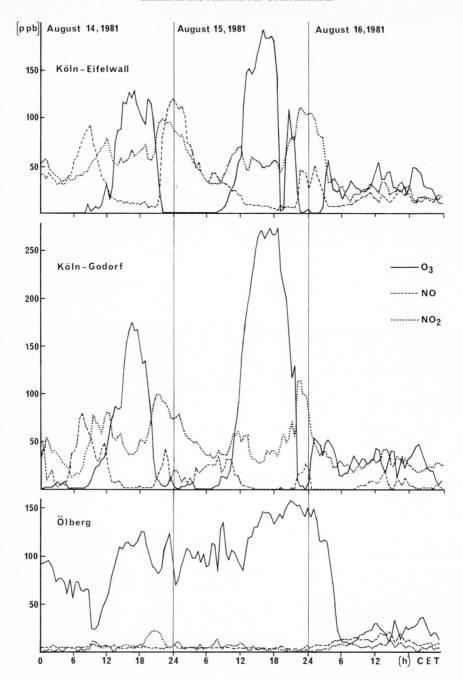

Fig. 1.9. Diurnal concentration profiles of O_3, NO, and NO_2 from August 14 to 16, 1981 at the stations Köln-Eifelwall, Köln-Godorf, and Ölberg. (Deimel 1982 a)

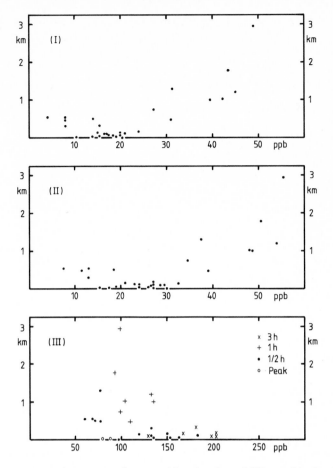

Fig. 1.10. Ozone concentrations measured at ground-based stations of different altitudes above sea level in 1982; data from Table 1.11 (**I**) annual mean; (**II**) summer mean (April–Sept.); (**III**) max. values

Table 1.11 also shows that relatively high annual (year) and especially summer (summer) mean values of ozone occur in peripheral and remote areas, while the very high peak values remain restricted to densely populated urban areas. This is again made clear in Fig. 1.10. Here it is also apparent that the annual as well as the summer mean values exhibit a dependence on altitude, probably with a maximum in the upper range of medium altitude mountains, at about 1,000 m.

Table 1.13 lists transgression frequencies of some threshold concentrations of ozone in the Cologne-Bonn area and at the Kleiner Feldberg/Taunus in the years 1975–1981. The frequency of exceeding higher threshold values varies greatly from year to year, depending on the large-scale weather situation (Großwetterlage).

Because of the high ozone levels in Mannheim in 1976, Table 1.14 gives the monthly maxima from the station Mannheim-Mitte as 3 h, 12 h and 24 h mean values, the monthly mean values, and the 50 and 95 percentiles. Table 1.15 gives

Table 1.13. Frequency of exceeding different thresholds of the ozone concentration values ($^1/_2$ h averages) in different years

No.	Station	Year	Number of values			
			> 75 ppb	> 80 ppb	> 100 ppb	> 150 ppb
9	Köln-Eifelwall	1975	158	97	32	–
		1976	289	205	63	6
		1977	(4)	(4)	(1)	(–)
		1978	18	9	–	–
		1979	50	35	2	–
		1980	68	50	5	–
		1981	177	147	48	9
11	Köln-Godorf	1975	57	40	9	–
		1976	141	100	40	4
		1977	16	15	8	–
		1978	19	11	3	.
		1979	27	18	–	–
		1980	91	73	14	–
		1981	369	311	172	37
16	Ölberg	1975	134	97	14	–
		1976	352	244	74	4
		1977	104	59	17	–
		1978	368	242	34	2
		1979	(300)	(239)	(118)	(3)
		1980	233	142	42	–
		1981	(369)	(310)	(108)	(6)
17	Michelsberg/	1976	735	517	168	–
	Bad Münstereifel	1977	37	19	8	–
		1978	286	222	89	–
18	Kleiner Feldberg/	1975	147	91	6	–
	Taunus	1976	872	620	146	2
		1977	51	36	4	–
		1978	388	285	106	2

Numbers in parenthesis: frequent failures of the measuring devices, therefore lower limits

Table 1.14. Maximum monthly ozone concentrations with mean values and 50 and 95 percentiles in 1976 at the measuring station, Mannheim-Mitte (No. 26); units: $\mu g\ m^{-3}$

Month 1976	Max. O_3-values			Mean	50%	95%
	(3 h)	(12 h)	(24 h)			
Jan.	–	–	–	–	–	–
Feb.	95	67	50	8	0	44
March	207	144	96	36	24	113
April	200	151	106	42	23	139
May	171	128	108	48	43	136
June	543	327	235	103	80	272
July	360	332	248	55	24	197
August	233	178	130	47	23	171
Sept.	167	95	71	25	6	109
Oct.	194	68	44	11	0	54
Nov.	60	49	37	6	0	28
Dec.	52	39	35	5	0	28
Year	543	332	248	34	10	148

Table 1.15. Maximum monthly NO_x concentrations[a] with mean values and 50 and 95 percentiles in 1976 at the measuring station, Mannheim-Mitte (No. 26), units: mg m^{-3}

Month 1976	Max. NO_x-values			Mean	50%	95%
	(3 h)	(12 h)	(24 h)			
Jan.	0.350	0.260	0.240	0.050	0.040	0.160
Feb.	0.410	0.350	0.300	0.110	0.080	0.250
March	–	–	–	–	–	–
April	0.260	0.210	0.180	0.080	0.070	0.190
May	0.230	0.160	0.130	0.070	0.060	0.140
June	0.130	0.100	0.090	0.050	0.040	0.090
July	0.250	0.080	0.060	0.030	0.030	0.050
August	0.170	0.090	0.090	0.060	0.060	0.130
Sept.	0.070	0.050	0.050	0.020	0.020	0.050
Oct.	0.290	0.200	0.200	0.050	0.030	0.180
Nov.	0.370	0.310	0.250	0.100	0.080	0.260
Dec.	0.610	0.490	0.460	0.180	0.140	0.044
Year	0.610	0.490	0.460	0.070	0.050	0.220

[a] Sum of $NO + NO_2$ measured as NO_2

Table 1.16. Maximum monthly concentrations of the total hydrocarbons (C_nH_m) with mean values and 50 and 95 percentiles in 1976 at the measuring station, Mannheim-Mitte (No. 26), units: mg m^{-3}

Month 1976	Max. C_nH_m-values			Mean	50%	95%
	(3 h)	(12 h)	(24 h)			
Jan.	2.0	1.9	1.9	1.4	1.4	1.7
Feb.	2.1	1.8	1.8	1.5	1.5	1.8
March	2.0	1.8	1.7	1.3	1.3	1.7
April	5.5	5.5	5.5	1.5	0.9	5.5
May	1.5	1.3	1.2	0.9	1.0	1.2
June	1.7	1.6	1.5	1.1	1.1	1.4
July	1.9	1.9	1.8	1.0	1.0	1.3
August	1.6	1.4	1.3	1.1	1.1	1.3
Sept.	–	–	–	–	–	–
Oct.	1.7	1.6	1.6	1.1	1.1	1.5
Nov.	1.5	1.4	1.3	1.1	1.0	1.3
Dec.	1.6	1.5	1.4	0.9	1.1	1.4
Year	5.5	5.5	5.5	1.2	1.1	1.6

the respective values for the nitric oxides, and Table 1.16 lists those for the total hydrocarbons. No obvious correlation can be seen between the high ozone levels and the respective precursor concentrations from these tables, except that the precursor levels, in general, were quite high during that month. The plot of different ozone data from Table 1.14 in Fig. 1.11 demonstrates that even the monthly mean value is not appropriate to characterize the ozone load near the ground. The 12-h maxima and the monthly 95 percentiles are in notable agreement throughout the year.

Figure 1.12 shows the mean diurnal ozone cycles from the measuring stations in the Cologne-Bonn area and the Kleiner Feldberg in the summers 1976 and

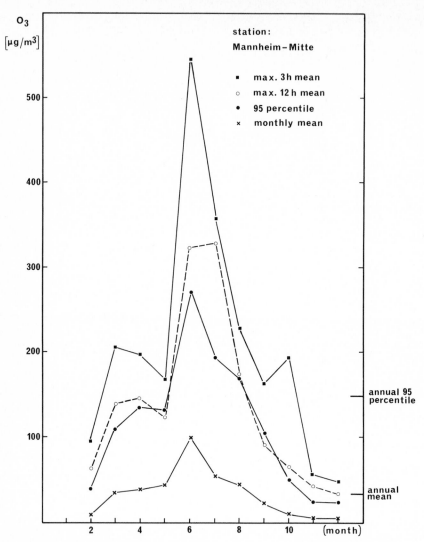

Fig. 1.11. Representation of different time-averages of ozone concentrations measured in a particular month of 1976 at the station Mannheim-Mitte; (■) max. 3 h mean, (○) max. 12 h mean, (●) 95 percentile, (×) monthly mean; the annual 95 percentile and the annual mean are also indicated in the figure

1977. The 3-h maxima for several years in Mannheim are given in Fig. 1.13. Both figures demonstrate the strong influence of meteorological factors on ozone formation, in agreement with Chapter 1.4.3, where the dependence of high ozone levels on the large-scale weather situation in Middle Europe is discussed.

It cannot always be decided to what extent ozone concentrations near the ground are influenced by intrusions of stratospheric air. Attmannspacher (1977) attributes short-term concentration peaks at Hohenpeißenberg to such intrusions. According to Reiter and Kanter (1982a), the influence of the stratosphere

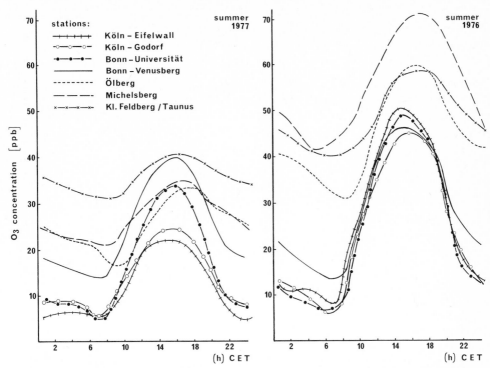

Fig. 1.12. Diurnal variation of the ozone concentration at the stations Köln-Eifelwall, Köln-Godorf, Bonn-Universität, Bonn-Venusberg, Ölberg, Michelsberg, and Kl.-Feldberg (see Table 1.11) averaged over the summer period (April–Sept.) of 1976 and 1977. (Deimel 1979)

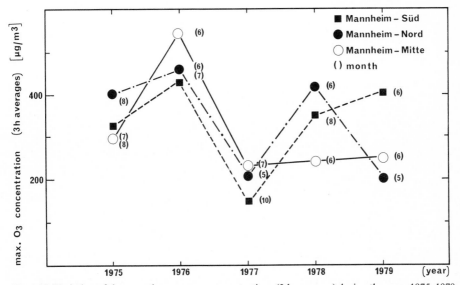

Fig. 1.13. Variation of the annual max. ozone concentrations (3 h averages) during the years 1975–1979 at the stations Mannheim-Nord (●), Mannheim-Mitte (○), and Mannheim-Süd (■); ( ) indicates the months in which the max. values were measured

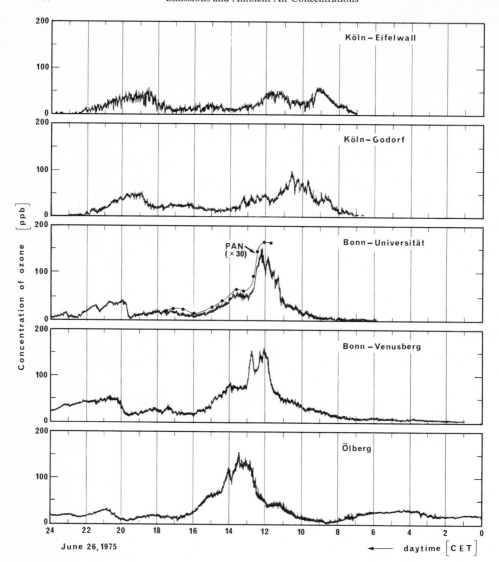

Fig. 1.14. Diurnal ozone concentrations on June 26, 1975 at the stations Köln-Eifelwall, Köln-Godorf, Bonn-Universität, Bonn-Venusberg, and Ölberg during the movement of air masses from Cologne to Bonn; the concentration of PAN is scaled up by a factor of 30 (Schurath 1979b). Wind direction between 8.00 and 14.00 CET: N/NW; wind speed: 0,5–1.0 m s^{-1}

on ozone in the mixing layer is less than had previously been assumed. Observations at various altitudes, Fig. 1.4 and 1.5, provide at least qualitative proof of photochemical ozone formation near the ground. Advective transport processes, however, may have also played a role. Peak ozone values (½-h average) of over 300 μg m^{-3} at valley-based stations cannot be attributed to ozone transport from the stratosphere (Sects. 1.3.1.5 and 1.3.2.1).

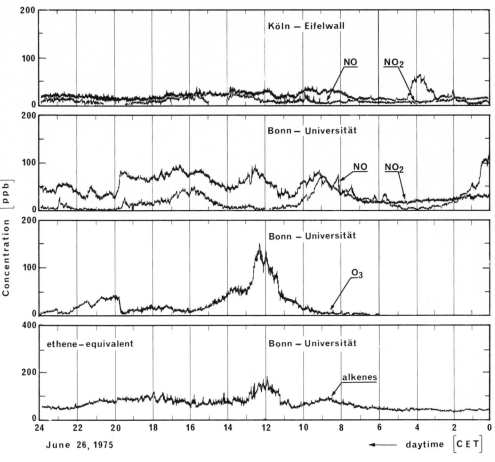

Fig. 1.15. Diurnal concentration profiles of O_3, NO, NO_2, and alkenes (ethene equivalents) on June 26, 1975 at the station Bonn-Universität. (Schurath 1979 b)

At the station Bonn University, for diurnal ozone cycles similar to those registrated in Mannheim, Fig. 1.7, the simultaneous occurrence of ozone and olefin peaks was repeatedly observed (Schurath 1979 b); Fig. 1.14 and 1.15 illustrate these findings. An analysis of the time-concentration profiles recorded at all the stations in the Rhine valley between Cologne and Ölberg on the same day indicates that high ozone concentrations near the ground were correlated with high levels of reactive hydrocarbons in polluted air parcels which moved horizontally through the valley: The occurrence of ozone peaks at the various stations shown in Fig. 1.14 corresponded to the velocity by which air masses were moved from Cologne to Bonn on the respective day. Under constant wind conditions, the ozone values peaked in the area of Bonn after a delay of about 2 h with respect to Cologne. These measurements and observations in Mannheim seem to indicate that very high ozone peaks at ground level, which are superimposed on a fairly high and more persistent ozone level, can build up from locally emitted precursors

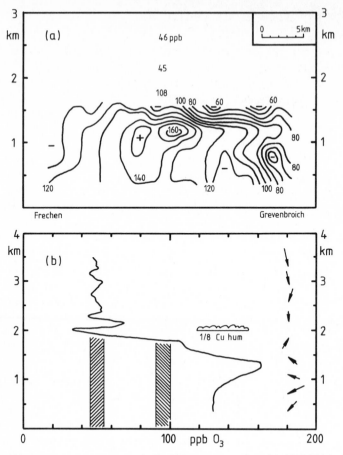

Fig.1.16a, b. Airborne ozone measurements in the Cologne area on June 26, 1976 (Fricke 1980): **a** O_3 concentration cross-sections down-wind of Cologne between Frechen and Grevenbroich, 14.00–14.30 CET. **b** Vertical profile in the zone of max. O_3 concentration, ▨ concentration above the mixing layer ▨ advective contribution within the mixing layer; wind vectors were taken from balloon soundings north of Cologne

under special weather conditions which also favor ozone formation on a much larger scale.

The processes leading to the formation and transport of regionally or locally produced oxidants can be analyzed more clearly by airborne measurements. The ozone profiles described in Section 3.2.3 (Fig. 1.21) demonstrate that, while flying through an olefin plume at an altitude of 450 m above ground level, simultaneous increase of the ozone concentration is observed (Fricke 1980). The horizontally limited ozone plume is still discernible at an altitude of 900 m above ground level. However, when flying through a power plant plume (Fig. 1.21) a horizontally limited decrease of the ozone concentration due to reaction with nitrogen oxide was observed. Airborne measurements also reveal ozone peaks superimposed on large-scale high background levels. Depending on the weather conditions, enriched ozone layers within the mixing layer were observed by Fricke on his flights

(Fricke 1980, Fricke and Rudolf 1977), which exhibited a large-scale and, in part, regional structure (Figs. 1.16 and 1.20). The observations in the Cologne-Bonn area have been confirmed by more recent flights in the same region (Neuber et al. 1982) and in the Rijnmond area in the Netherlands (Van Duuren et al. 1982).

It may therefore be assumed that under certain weather conditions photochemical oxidants are formed mainly in layers of polluted air, which may persist for several days at an altitude of several hundred meters in stable high pressure systems. Ozone is thus formed and preserved at these altitudes. The heating of the air near the ground at midday results in a vertical air exchange, bringing ozone down from this reservoir. The dependence of temperature on the intensity and duration of solar irradiation, and the temperature dependence of the oxidant forming reactions enhance this, resulting in an ozone peak near the ground at noon (Bruckmann and Eynck 1980a, Bruckmann and Langensiepen 1981). The ozone-enriched air layers probably extend over large parts of Western Europe during stable large-scale weather conditions (Guicherit and Van Dop 1977, Fricke and Rudolf 1977), obscuring small-scale variations due to local or regional precursor sources. These may, however, further increase the oxidant concentration near the ground under suitable conditions, at least for short periods of time. High peak concentrations in excess of 400 μg m^{-3} ozone are quite rare events at ground level, but their dependence on precursor concentrations is more obvious. From airborne measurements on June 24 in 1976 (Fig. 1.16), Fricke (1980) estimated that the ozone concentration of roughly 350 μg m^{-3} at about 1,000 m above ground in the Cologne area consisted of approximately one third natural ozone from above the mixing layer, one third of a large-scale advective anthropogenic contribution, and one third ozone resulting from the urban precursor plume of Cologne.

Photooxidants formed in the urban plume could be followed by aircraft (Fricke 1980) downwind of Cologne for as far as 100 km, mainly in a westerly direction, under high pressure weather conditions. Preliminary model calculations by Scherer and Stern (1982), based on field measurements in the Cologne-Bonn area (Becker et al. 1979, Fricke 1980, Neuber et al. 1982), suggest that the Ruhr region should also be included in these considerations as an important source of precursors.

The formation of PAN occurs in parallel to that of ozone, according to the few measurements which have been carried out up to now, Table 1.12 and Fig. 1.14. It may, therefore, be concluded that among a sufficient concentration of total hydrocarbons always enough precursor hydrocarbons are available for the production of PAN. Figure 1.17 illustrates diurnal profiles at the station Bonn-Universität for ozone, PAN, ethene, and propene (Löbel et al. 1980). Similar results have been obtained by Bruckmann and Mülder (1979), which additionally reveal a slightly delayed decrease of PAN relative to ozone after the daily maximum.

Precursors

The following tables show to what extent various areas of the Federal Republic of Germany are polluted by oxidant precursors: nitrogen oxides and hydrocarbons. The data were taken, as far as possible, from stations which are already listed in Table 1.11, with the same numbering and references. All concentrations

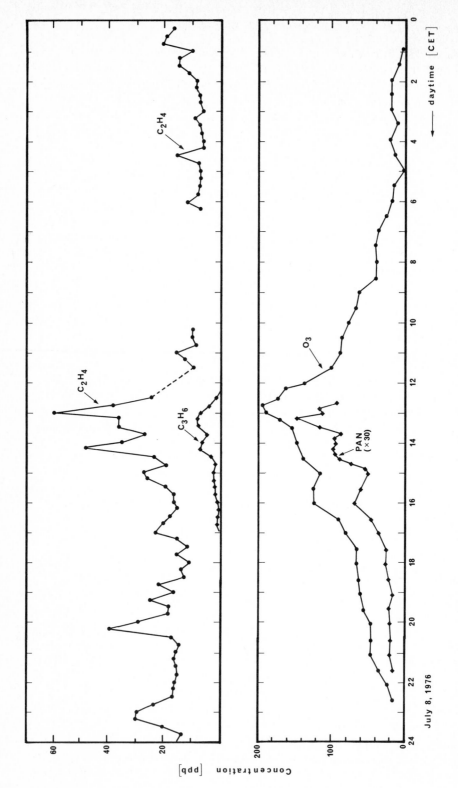

Fig. 1.17. Diurnal concentration profiles of ozone, PAN, ethene and propene on June 8, 1976 at the station Bonn-Universität. (Schurath 1979 b, Löbel et al. 1980)

Table 1.17. Hydrocarbon concentrations in mg m^{-3} measured between 1975 and 1980 at different stations in Germany; *year*: annual mean, *max*: maximum value with (a) $^1/_2$-h and (d) 3-h averages

No.	Location	Foot-note[a]		Time	1975	1976	1977	1978	1979	1980
9	Köln-Eifelwall	6	Ethene	Year				0.016		
				Max (a)				0.122		
			Propene	Year				0.006		
				Max (a)				0.087		
11	Köln-Godorf	6	Ethene	Year					0.148	0.084
				Max (a)					1.800	1.200
			Propene	Year					0.049	0.045
				Max (a)					0.800	1.100
			C_nH_m	Year					–	1.592
				Max (a)					6.300	8.400
21	Mainz-Budenheim area	8	C_nH_m less CH_4	Year					0.100	0.090
				Max (a)					1.680	4.270
22	Aschaffenburg	10	C_nH_m	Max (a)	Since Febr. 1977					5.840
23	Arzberg	10	C_nH_m less CH_4	Max (a)	Since August 1980					1.660
24	Ludwigshafen-Frankenthal area	8	C_nH_m less CH_4	Year					0.140	0.100
				Max (a)					3.560	4.950
25	Mannheim-Nord	11	C_nH_m	Year	2.300	1.400				
				Max (d)	12.200	14.300				
26	Mannheim-Mitte	11	C_nH_m	Year	1.100	1.200				
				Max (d)	9.200	5.500				
27	Mannheim-Süd	11	C_nH_m	Year	1.300	1.300				
				Max (d)	5.800	4.300				
28	Nürnberg	10	C_nH_m less CH_4	Max (a)	Since July 1977					4.620
36	Augsburg	10	C_nH_m less CH_4	Max (a)	Since August 1978					6.480
37	München-Stachus	10	C_nH_m less CH_4	Max (a)	Since Febr. 1978					5.490
38	München-Effnerplatz	10	C_nH_m less CH_4	Max (a)	Since Jan. 1978					5.430

[a] See footnote of Table 1.11

in the tables are expressed in mg m^{-3}. The few cases where total nitrogen oxides were measured as NO_2 are marked. Total hydrocarbon concentration, C_nH_m, or total hydrocarbons less methane (nonmethane hydrocarbons), C_nH_m less CH_4, are usually measured with a flame ionization detector (FID) after certain sampling and separation procedures. The FID signal is assumed to be proportional to the number of C atoms in the hydrocarbon mixture. When a hydrocarbon mixture is measured in units of ppm C or ppb C with a properly calibrated detector, the result can be converted to mass concentration units referring to CH_2 or CH_2 (Sect. 1.3.2). The hydrocarbon data from the Bavarian monitoring network, orig-

Table 1.18. Nitrogen oxide concentrations in mg m^{-3} measured between 1975 and 1980 different stations in Germany; *year*: annual mean, *max*: maximum value with (a) ½-h, (c) 1-h, and (d) 3-h averages

No.	Location	Foot-note[b]		Time	1975	1976	1977	1978	1979	1980
9	Köln-Eifelwall	6	NO	Year		0.043	0.040	0.033	0.034	0.038
				Max (a)		0.748	0.570	0.478	0.425	0.715
			NO$_2$	Year		0.042	0.042	0.050	0.055	0.053
				Max (a)		0.369	0.218	0.254	0.428	0.397
11	Köln-Godorf	6	NO	Year		0.031	0.030	0.029	0.026	0.029
				Max (a)		0.434	0.398	0.280	0.315	–
			NO$_2$	Year		0.050	0.036	0.029	0.032	0.055
				Max (a)		0.294	0.432	0.283	0.407	–
12	Bonn-Universität	6	NO	Year		0.024	0.018	0.021		
			NO$_2$	Year		0.040	0.038	0.057		
16	Ölberg	6	NO	Year					–	0.005
				Max (a)					0.085	0.104
			NO$_2$	Year					–	0.023
				Max (a)					0.101	0.151
19	Frankfurt/M-Mitte (different from Frankfurt/M-Westend	9	NO	Year			0.029	–	–	–
				Max (a)			0.308	0.403	0.568	0.620
			NO$_2$	Year			0.038	–	–	–
				Max (a)			0.305	0.914	0.524	0.229
21	Mainz-Budenheim area	8	NO	Year					0.033	0.040
				Max (a)					1.256	1.104
			NO$_2$	Year					0.036	0.040
				Max (a)					0.344	0.296
24	Ludwigshafen-	8	NO	Year					0.036	0.050
				Max (a)					1.225	0.946
			NO$_2$	Year					0.036	0.050
				Max (d)					0.296	0.422
25	Mannheim-Nord	11	NO	Year	0.100[a]	0.070[a]		0.030	0.070	0.040
				Max (d)	0.910[a]	0.510[a]		0.320	–	0.420
			NO$_2$	Year					0.030	0.050
				Max (d)					–	0.840
26	Mannheim-Mitte	11	NO	Year	0.120[a]	0.070[a]		0.040	0.030	0.040
				Max (d)	0.880[a]	0.610[a]		0.290	–	0.550
			NO$_2$	Year					0.010	
27	Mannheim-Süd	11	NO	Year	0.100[a]	0.080[a]		0.050	0.040	0.050
				Max (d)	1.290[a]	1.290[a]		0.600	–	0.790
			NO$_2$	Year					0.010	0.040
				Max (d)					–	0.250
28	Nürnberg	10	NO	Max (a)	Since August 1977					0.725
			NO$_2$	Max (a)	Since August 1977					0.420
30	Eggenstein	11	NO	Year				0.030	0.050	0.040
				Max (d)				4.000	–	0.750
			NO$_2$	Year					0.030	0.050
				Max (d)					–	0.350

Table 1.18 (continued)

No.	Location	Foot-note[b]		Time	1975	1976	1977	1978	1979	1980
31	Karlsruhe-Mitte	11	NO	Year		0.160[a]		0.060		0.100
				Max (d)		1.360[a]		0.790		0.910
			NO_2	Year						0.60
				Max (d)						0.480
36	Augsburg	10	NO	Max (a)	Since Jan. 1978					1.525
			NO_2	Max (a)	Since Jan. 1978					0.802
37	München-Stachus	10	NO	Max (a)	Since Jan. 1978					0.950
			NO_2	Max (a)	Since Jan. 1978					0.344
38	München-Effnerplatz	10	NO	Max (a)	Since August 1977					1.750
			NO_2	Max (a)	Since August 1977					0.764
39	Hohenpeißenberg	13	NO	Max (c)					0.048	0.036
			NO_2	Max (c)					0.084	0.117

[a] Sum of $NO + NO_2$ measured as NO_2 [b] See footnotes of Table 1.11

inally given in ppm C, have been converted to mg m^{-3} referring to CH_4 in Table 1.17. The data from other networks were already reported in mass concentration units. In the case of single component measurements the units mg m^{-3} (μg m^{-3}) or ppm (ppb) may be interconverted using the conversion factors of Table 1.26 in the Appendix.

When comparing the concentrations of the individual components with the concentrations of mixtures given in ppm C or ppb C, the volume mixing ratio of the individual component should be multiplied by the number of C atoms of the respective component (Sect. 1.3.2).

From Table 1.18 it is apparent that NO as well as NO_2 concentrations may reach levels of 1 mg m^{-3} in many populated and industrialized areas for short periods of time. The annual mean value of each of the oxides can exceed 0.050 mg m^{-3} in such areas. The concentration ratio (NO)/(NO_2) is very dependent on the meteorological and chemical conditions of the atmosphere near ground. For peak values this ratio is mostly greater than 1, whereas the ratio becomes smaller for concentration values averaged over longer time periods.

Maximum hydrocarbon concentrations (less methane) of over 5 mg m^{-3} have been observed (Table 1.17). The high peak concentrations in the industrial areas of Mainz and Ludwigshafen may be influenced by releases of natural gas. Similar high values have, however, also been found in industrialized regions of Bavaria and in the Cologne area, as the data of Table 1.17 demonstrate. With respect to the formation of photooxidants, the concentration ratio (nonmethane hydrocarbons)/(nitrogen oxides) is of particular importance (see Chap. 1.4.2.3).

At the two Cologne stations, No. 9 and No. 11, the important reactive hydrocarbons ethene and propene have been measured individually. Of particular interest is the station Köln-Godorf in the vicinity of petrochemical plants. Table 1.19 gives more information on the hydrocarbon composition measured at that station in 1977 and 1978 (Dulson 1979, 1981).

Table 1.19. Mean concentration values averaged over the measuring period in $\mu g\ m^{-3}$ of different components of reactive hydrocarbons during calm winds ($\leq 1.5\ m\ s^{-1}$) at the station Köln-Godorf (No. 11) in a distance of about 0.5 km from a petrochemical plant. (Dulson 1979, 1981)

Time period	Total C	Ethane	Ethene	Ethine	n-Butane	Propene	$\varDelta$
Oct.–Dec. 1977	1,718	10.7	63.0	18.2	22.5	22.3	1,583
Jan.–March 1978	1,891	23.2	44.9	15.3	17.2	19.1	1,783
April–June 1978	–	19.2	67.8	12.1	26.2	74.3	–
July–Sept. 1978	2,599	22.7	168.5	19.1	35.0	75.1	2,279

$\varDelta$: total C less separated components

Table 1.20. Airborne measurements at Köln-Godorf, 400–600 m above ground level on August 1, 1977 (Neuber et al. 1981); concentration in $\mu g\ m^{-3}$

Wind direction	Ethine	Ethene	Propene	Propane
Upwind	–	84.2	0.9	5.5
Downwind	21.6	2561	0.5	3.7

In addition to the data from Table 1.19, Table 1.20 presents some results from airborne measurements above Köln-Godorf in 1977 (Neuber et al. 1981). The strong source of reactive hydrocarbons in this area and its impact on the formation of photooxidants is also demonstrated in Fig. 1.21 of Section 1.3.2.2.3 as mentioned previously.

Additional airborne measurements above the Ruhr area (Neuber et al. 1982) and measurements from the ground in industrialized sites of the Rhine-Ruhr region (Ixfeld 1977, Frohne and Schneider 1977), as well as in the city of Hamburg (Nassar and Goldbach 1977), give further information on the pollution by hydrocarbons.

It can be seen that wherever high oxidant loads are observed in densely populated areas in summer, relatively high precursor concentrations are present as well.

A direct relationship between the ozone maxima and the precursor concentrations can be observed at ground stations only in exceptional cases since, during the formation of photooxidants under the influence of variable meteorological conditions, the areas with high ozone levels are removed from the precursor source areas by transport processes (Ahrens 1983). In situations of short reaction times and slow air movement, relations may be found between high precursor levels and the formation of oxidants within a spatially limited region. Ambient air concentrations of NO show an inverse behavior in this regard since NO reacts fast with O_3, Figs. 1.5, 1.9, 1.15, and 1.21. This has also been observed by recent airborne measurements (Neuber et al. 1982). In the presence of high levels of reactive hydrocarbons, as illustrated in the diurnal profiles in Figs. 1.14, 1.15, and 1.17, it can be seen that the maximum ozone concentration is attained within a few hours, depending on the particular meteorological conditions of the day, as already mentioned. In all cases of high oxidant, remarkable amounts of reactive

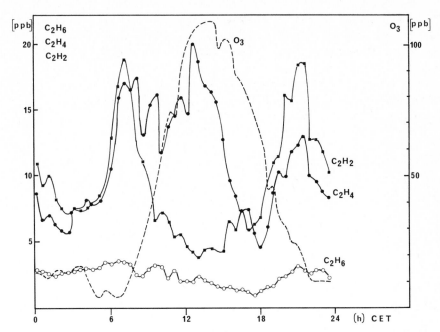

Fig. 1.18. Diurnal concentration profiles as monthly mean of ozone, ethane, ethene, and ethine in August 1976 at the station Karlsruhe. (Obländer and Siegel 1977)

alkenes were observed during the day which, in simulation experiments, also quickly led to the formation of oxidants. The monthly means of the diurnal concentration profiles of ozone and different hydrocarbon components at a station in Karlsruhe in August 1976 are shown in Fig. 1.18 (Obländer and Siegel 1977). It can be learned from this figure that even for the monthly mean, the maximum of the ozone concentration occurs with about a 2 h delay in relation to the peak value of the reactive hydrocarbon ethene. The concurrence of the ozone maximum with the minimum of the less reactive hydrocarbon ethine at noon indicates a breaking-up of low inversion layers, resulting in a stronger vertical mixing.

Airborne measurements in the Cologne-Bonn area, Fig. 1.21, carried out while flying through an alkene-enriched plume which had stationarily built-up above Godorf across a spatially limited section, yielded the highest ozone values (Fricke 1980). Further measurements from the air by Neuber et al. (1982) over the Ruhr region confirmed the relationship between high levels of reactive hydrocarbons and high ozone values in locally and regionally limited areas. A statistical analysis of the data from ground-based stations supports the assumption that locally emitted hydrocarbons at the southern border of Cologne contribute significantly to the ozone load in the Cologne area (Bruckmann and Langensiepen 1981).

More precise trajectory analyses of the air masses in the Cologne-Bonn region and in the Upper Rhine Valley are necessary in order to improve the understanding of the influence of local, regional, and remote sources of precursors on the formation of oxidants in these areas. Particularly, the transport processes in the upper part of the mixing layer should be investigated more carefully, since this layer strongly influences the diurnal course of the oxidant, as well as the precursor concentrations near the ground, by vertical air exchange at midday. For a given precursor load, the extent of the oxidant formation is very dependent on the variable meteorological conditions. Further measurements in a carefully chosen area over a sufficient time period, and mathematical transport models also including chemical processes, are necessary for the development of a more quantitative relationship between oxidant and precursor concentrations.

While considering the photochemical oxidants problem in the polluted atmosphere, it should be remembered that other chemical processes, such as the formation of acids from the oxides of nitrogen and sulfur, are also accelerated by the oxidants and their concomitant species.

1.3.2.2.2 European Countries and Overseas

Measurements of surface ozone have a long tradition in European countries, but were mostly undertaken in order to delimit the natural ozone contents of the air. In the last century, the first systematic ozone determinations were reported from Krakow (Karlinsky 1874), Emden (Prestel 1874), and Egypt (Zittel 1874). These investigators used a rather primitive version of the KI method in their field studies, which dates back to Schoenbein (1844), the discoverer of ozone; he found that KI-starch paper turns blue when exposed to ozone. The blue tint was taken as a relative measure of the ozone dosage. From these measurements, which were reported in relative units, an annual trend of surface ozone can already be derived, showing a maximum in March. Zittel (1874) observed furthermore that the ozone concentration was usually higher in the desert than in the Nile valley, which he attributed to the different surface properties.

At the end of the last century, Levy, of the Meteorological Observatory Montsouris in Paris, adopted an improved wet chemical technique (a solution of K_3AsO_3/KI) for his first quantitative determinations of surface ozone concentrations. He found mixing ratios between 7 ppb in February and 3.5 ppb in December. Although the absolute values may be in error, considering the well-known difficulties inherent in wet chemical determinations of very low ozone concentrations, it is worth noting that the highest ozone mixing ratios at the time were not observed during the summer, but generally in spring, in parallel with the annual trend in stratospheric ozone. The formation of anthropogenic ozone, unsuspected at the time, can thus be safely excluded on account of these historical data.

The vertical distribution of ozone (concentration increasing with altitude), its dependence on the weather situation, and the fact that the spectrum of the sun is abruptly cut off below 295 nm, led Hartley (1881) to conclude that the upper atmosphere is a source of ozone. This idea obtained further support from numerous other observations until it was definitely confirmed by Fabry and Buisson (1921), who carried out the first quantitatively correct determination of the ozone column. It is for this reason that surface ozone has traditionally been looked upon as a reliable tracer of clean stratospheric air. This view is exemplified by the "aran" (= ozone) measurements in the Bonn metropolitan area, and their interpretation, by Emonds in 1954 (cf. Sect. 1.3.2.2.1).

The fact that "anthropogenic" ozone far in excess of the natural background concentration in surface air can evolve from photochemically induced reactions of NO_x and hydrocarbons has been well established for the Los Angeles basin at the latest since the end of the 1940's. Somewhat later, ozone measurements were initiated in the Netherlands (Wisse and Velds 1970) and in Berlin (Lahmann et al. 1968, Lahmann 1969) at approximately the same time, in order to assess the formation of "anthropogenic" ozone in these areas. After 1 year of ozone monitoring in Berlin, by the side of two roads which differed considerably in traffic density, the authors arrived at the conclusion that the requirements for photochemical ozone formation, prevailing in the Los Angeles basin, were nonexistent in Berlin at the time of the measurements. Their findings have often been invoked as evidence that photochemical ozone from anthropogenic precursors is unlikely to be observable in Middle Europe, owing to the comparatively low levels of insolation. It is now well established that the insolation in Middle Europe is quite often strong enough for the rate of ozone formation by photochemical reactions in polluted air to more than outweigh its rate of destruction by reactions with reducing emissions, particularly with NO. Daily doses of photochemically active sunlight in Northern Europe become comparable, in summer, to daily doses in Southern California, although the noontime intensities are substantially lower (Schjoldager et al. 1978).

After a few exploratory measurements in Vlaardingen in 1968 (Wisse and Velds 1970), continuous ozone measurements were taken up in Delft from 1969 onwards. Already in the first year ozone mixing ratios in excess of 200 ppb were observed and interpreted in terms of photochemical reactions involving anthropogenic precursors (Guicherit et al. 1972, Guicherit 1973, 1975). A larger number of monitoring stations for oxidant precursors (NO_x, GC-separated light hydrocarbons) and ozone (at least three analytical techniques, cf. Guicherit et al. 1972) are maintained in the Netherlands. All ozone analyzers are uniformly calibrated against a standardized KI method. A good account of the results obtained in the Netherlands between 1969 and 1977 is given in an article of Bos et al. (1978) in the TNO-Report *Photochemical Smog Formation in the Netherlands*. According to this report, the hourly average ozone concentration reached 270 ppb at Vlaardingen in 1976. This particular year is renowned for the hitherto most spectacular oxidant episodes in several European countries.

The establishment of new measuring stations for ozone, precursors, and other oxidants in Europe, as well as the continuation and expansion of existing monitoring networks, was fostered by the Environmental Directorate of the OECD, who initiated an investigation into the problem of photochemical oxidants and their precursors in the European countries in 1973.

Even before this initiative, high ozone mixing ratios had been measured in England, at times exceeding 100 ppb, 1971 at Harwell, and 1972 in London (Atkins et al. 1972, Derwent and Stewart 1973). Chemiluminescence analyzers calibrated against the KI method have been used from the beginning. Particularly high ozone levels were reached in 1976, like everywhere else in Europe, at three sites in London, at Stevenage, and at Harwell (highest hourly average 258 ppb) (Apling et al. 1977).

Ozone measurements in France were carried out at Nice in 1973, and at Martigues (Rhone delta, under strong influence of oil refineries). The maximum hourly average at Nice amounted to 170 ppb ozone. Bos et al. (1978), quoting from OECD sources, report maximum hourly averages of more than 200 ppb ozone at Martigues, and 180 ppb at Port du Bouc, in 1975. Ozone was also measured in Paris, and since January 1976 at Vert-Le-Petit, 35 km south of Paris. The data collected in 1976, which include ozone mixing ratios up to 215 ppb, have been analyzed by Bénarie et al. (1979). The authors conclude that these high ozone mixing ratios cannot be accounted for by photochemical conversion of local precursors (i. e., NO_x and hydrocarbons from the greater Paris area), but must have been advected from an unidentified distant ozone source.

From the 1983 European International Campaign of Fos-Berre, a highly industrialized coastal area near Marseille, 12 days of continuous ozone data are available from eight instruments. They were positioned inside a half-circle of ca. 50 km radius around the main precursor source area. Important meteorological parameters, such as wind field and mixing height data, were also measured. Very recently a preliminary data analysis has been presented (Perros and Toupance 1984).

In most other European countries, measurements of ozone and oxidant precursors were started or expanded in response to the OECD initiative. The OECD Report of 1975 lists four monitoring stations in Sweden (Stockholm, Gothenburg, and Rorvik on the Swedish West Coast), where ozone has been measured with chemiluminescence analyzers (calibration against the US-KI-method) since 1973. At Gothenburg an hourly maximum of 100 ppb ozone occurred already in 1973. The measurements at Gothenburg and Rorvik continue; during the summer 1979 at Rorvik, hourly averages of 100 ppb ozone were reached and considerably exceeded on a number of days (see also Grennfelt 1977, Nielsen et al. 1981, Hov 1984).

The first measurements of surface ozone in Italy were made in 1974 with a galvanometric instrument in a suburban part of Rome. The maximum hourly mean in 1974 amounted to 140 ppb ozone. In 1977 ozone measurements were taken up in Northern Italy at the Nuclear Research Center of Ispra (Stangl et al. 1980). Ozone and precursors have been measured since 1978 in the heavily industrialized Adriatic coastal area near Ravenna (Giovanelli et al. 1982). The measurements revealed pronounced photochemical ozone formation in the entire area. The concentration-time profiles of ozone depended quite distinctly on the prevalent wind direction. Elevated ozone concentrations inland during the night were rationalized by a land-sea-breeze effect, well known from the Los Angeles basin. The interpretation obtained further support from ozone measurements aboard a drilling platform 20 km offshore.

Norway started to measure ozone in 1975 at several stations in the Greater Oslo Fjord area. The instruments used were chemiluminescence analyzers (Rhodamine B and ethene), and a Dasibi (optical absorption at 253.7 nm). All instruments were calibrated against the KI method. The hitherto highest hourly mean in Norway, of 199 ppb ozone, was reported from the northermost station of the area in 1979. Hourly averages at the other five stations reached between 70 and 100 ppb in the same year (Schjoldager and Stige 1980). The results of these measurements are documented in several publications (Schjoldager et al. 1978, 1983, Schjoldager 1979).

The first continuous ozone measurements in Yugoslavia were made in summer 1975 on a tall building in the city center of Zagreb. The results have been analyzed and published (Božičević et al. 1976a, b, 1978, Cvitaš et al. 1979). The diurnal profile of the ozone concentration, averaged over the entire measuring period (1 May–30 September 1975) shows a plateau of approximately 60 ppb between 11 and 16 h (standard deviation ± 20 ppb). The maximum hourly mean amounted to 140 ppb ozone. Later on (October 1976–September 1977), ozone was measured on the island of Krk in the Northern Adriatic Sea, and in the city of Split on the mid Adriatic coast (April–October 1979) (Cvitaš and Klasinc 1979, Novak and Sabljić 1981, Deželjin et al. 1981, Butković et al. 1983). During winter on the island of Krk, a near-constant ozone mixing ratio of about 20 ppb was found, day and night. In early July, an ozone episode was observed on the island. The episode (ozone maximum 200 ppb, persistently more than 140 ppb for 32 h) ended as abruptly as it had started. The concentration-time profile was not indicative of local ozone formation, with maxima occurring irregularly at all times, e. g., also very early in the morning. Since very few precursors are emitted in the environs of the measuring station, while petrochemical plants are situated in the coastal area and on other parts of the island, transport was held responsible for the ozone episode (Butković et al. 1983).

In Greece, ozone was measured in summer 1982 with a chemiluminescence analyzer (KI calibration, comparison with a Dasibi) (Güsten 1983 pers. commun.), in a joint Greek-German measuring program. The instrument was assembled in four places in and near Athens for continuous monitoring periods of between 4 and 8 days. In the city center, 250 ppb ozone were exceeded for 5 h in four days. North of the city, more than 100 ppb ozone were measured several times, while the mixing ratio on the island of Aiyina remained constantly below 60 ppb for 4 days.

In Austria, ozone has been measured continuously since 1976 at up to four sites in Greater Vienna. The chemiluminescence analyzers were calibrated by the KI method. During the summer of 1979, 200 ppb ozone have been exceeded for a total 14 h at Ilmitz, 65 km south of Vienna.

The Spanish Department of Environmental Health has measured ozone on the University Campus of Madrid since February 1982 (De La Serna 1983 pers. commun.). Evaluated data up to June 1982 are available. The diurnal ozone profiles (monthly 90 percentiles of hourly means) exhibit a maximum after 1600 h, sometimes as late as 1800 h.

The OECD and others have attempted to compare, in tabulated form, maximum hourly averages and/or transgression frequencies of certain threshold concentrations in the European countries from which data are available. This procedure requires a careful selection of the measuring sites to be compared, since ozone concentrations are strongly influenced by nearby precursor sources (e. g., ozone destruction by NO, cf. Chap. 1.5.2.2) and orographic peculiarities (e. g., enhancement or impediment of vertical exchange). Airborne measurements of the vertical ozone distribution (van Dop et al. 1977, Fricke and Rudolf 1977, Fricke 1980) have revealed that surface ozone is usually lower, in terms of mixing ratios, than ozone in the mixing layer aloft (e. g., between 500 and 1,500 m aboveground), where the anthropogenic contribution can be segregated from the natural ozone background much more clearly than at ground level (cf. Sect. 1.3.2.2.3). Several studies have fostered the hypothesis that, under certain weather conditions, anthropogenic ozone is formed on a European scale, and can be transported over long distances (Cox et al. 1975, Guicherit and van Dop 1977, Apling et al. 1977, Schjoldager et al. 1983).

Among the non-European countries, the US still hold a leading position in the field of oxidant research, which has focused recently on questions of oxidant and precursor transport, and on the issue of the natural ozone background on which the anthropogenic component is superimposed (e. g., Spicer et al. 1979). In Canada, a large number of ozone-measuring stations have been installed since the early 1970's (cf. the OECD Report of 1976 on Photochemical Oxidants).

According to a brief report on the extent of oxidant pollution in Mexico City (Bravo et al. 1979), ozone maxima in the order of 470 ppb have been observed.

A considerable oxidant problem exists in Japan, not only in the Tokyo Bay Area, where oxidant measurements were started in 1967. The results have been published in Japanese by the Tokyo Metropolitan Research Institute for Environmental Protection. In 1973 the Air Quality Bureau of the Environmental Agency of Japan issued a monograph in English, entitled *References on Photochemical Air Pollution in Japan*. Mizogushi et al. (1977) report a study on the effect of atmospheric oxidants on school children in Tokyo. During the 61 days of the investigation, 200 ppb ozone had been exceeded on several occasions. Airborne measurements in the Tokyo Bay Area have revealed (like analogous studies in the USA and in Germany, cf. Fricke 1980) that ozone is formed aloft in the mixing layer during transport of polluted air masses (Wakamatsu et al. 1983, Uno et al. 1984). This process cannot be studied as clearly by ground-based measurements.

The detection of high ozone concentrations in the Sydney area in Australia (Post 1979, 1981, Post and Bilger 1978) illustrates that oxidants, and particularly ozone, are inevitably formed wherever the precursors NO_x and NMHC are available in sufficient concentrations, provided that certain meteorological conditions prevail (see also Hyde and Hawke 1977).

Other Photooxidants

Measurements of peroxyacetyl nitrate (PAN) in Europe were first reported from South England (Penkett et al. 1975). Up to 8.9 ppb PAN had been detected at Harwell under atmospheric conditions favoring the formation of photochemical oxidants. Diurnal profiles of the trace compound had been measured in the Netherlands as early as 1972, but were first published in 1976 (Nieboer and van Ham 1976). These authors observed an approximately linear correlation between the concentrations of ozone and PAN for a given wind direction. The dependence of the concentration ratio on the wind direction led them to conclude that PAN is formed from very specific precursors. It was suggested that PAN, in the absence of any seizable natural sources, is a more reliable indicator of anthropogenic oxidant pollution than ozone, which has a strong natural source in the stratosphere. However, owing to the complexity of the analytical procedure, and because of the lack of a simple but reliable calibration technique (Bruckmann and Mülder 1979),

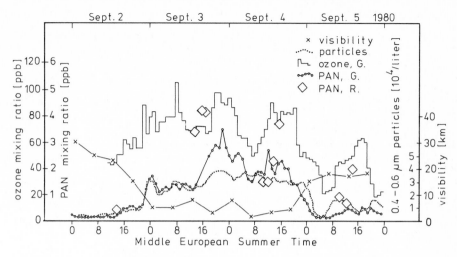

Fig. 1.19. Profiles of PAN concentration, ozone concentration, particle number density, and visibility at Göteborg (*G.*), and at Risø (*R.*) near Copenhagen, 2–5 September 1980. (Grennfelt et al. 1982)

PAN measurements have become considerably less popular than ozone measurements.

During the summer months of 1980, PAN was measured simultaneously near Kopenhagen in Denmark, and in Göteborg in Sweden. The measurements covered a period of 44 working days (Grennfelt et al. 1982). The maximum PAN mixing ratio per day varied between 0.1 and 4 ppb. Figure 1.19 shows concentration profiles of PAN and ozone, and a few other parameters, for an oxidant episode of four days' duration. It was concluded from trajectory analyses and model calculations (Hov 1984) that the high PAN mixing ratios did not arise from locally emitted precursors, but were due to long-range transport of PAN and its precursors from England and parts of the Continent across the North Sea.

1.3.2.2.3 Mobile Measurements and Special Investigations

Ozone reacts extremely rapidly with NO (reaction times typically shorter than 1 min under atmospheric conditions), which is emitted both by strong steady point sources (e. g., power stations) directly into the mixing layer, and by numerous unsteady mobile sources (motor vehicles) at ground level. Furthermore, ozone is rapidly destroyed at the earth's surface by heterogeneous processes. This is the reason why ground-based stationary measurements do not convey a true picture of the large-scale horizontal and vertical ozone distribution in the atmosphere. Stationary measurements of the precursors NO, NO_2, and NMHC at ground level, on the other hand, yield little information on their concentrations a few hundred meters aloft where anthropogenic ozone is predominantly formed. Aircraft have therefore been used quite early, particularly in the United States (Edinger 1973, Gloria et al. 1974, Calvert 1976a, b), to probe the horizontal and vertical distribution of ozone and its precursors in the atmosphere.

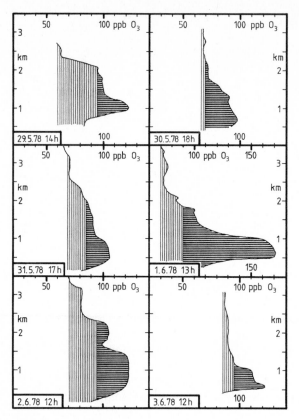

Fig. 1.20. Vertical profiles of the ozone concentration in the Cologne-Bonn area, 29 May–3 June, 1978; ▨ excess ozone in the mixing layer; ▆ locally formed excess ozone in the Cologne-Bonn area

   In Germany, airborne measurements of ozone and precursors have been carried out by the Department of Meteorology and Geophysics at the University of Frankfurt. The investigation concentrated on the Bonn-Cologne-Ruhr area and its luff- and lee-side wind fields. A summary report of the flights between 1975 and 1978 was compiled by Fricke (1980). Preliminary reports of the more recent flights, which include airborne samplings of the C_2 to C_8 hydrocarbons in different air masses, are also available (Neuber 1982, 1983, Neuber et al. 1981, 1982).

   Notable results of the airborne measurements in the Cologne-Bonn area may be summarized as follows (Fricke 1980):

1. During summer periods of high pressure with low wind speeds, the vertical ozone distribution exhibits a layered structure. The ozone concentration within the mixing layer is distinctly higher than in the free troposphere aloft. This is illustrated by six vertical profiles in Fig. 1.20.
2. The anthropogenic ozone excess within the mixing layer (hatched areas in Fig. 1.20) may be further subdivided into a large scale anthropogenic component (vertically hatched) which is also present upwind of particular precursor areas, and regional or local maxima (horizontally hatched). These maxima are

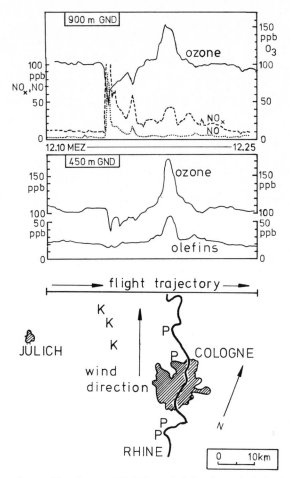

Fig. 1.21. Concentration profiles of ozone, NO, NO$_x$, and olefins downwind of Cologne, 450 and 900 m above ground. The NO$_x$ maxima and the corresponding ozone minima are attributed to the plumes of large lignite-burning power stations (K power station, P petrochemical plant)

clearly superimposed, for tens of kilometers downwind of strong precursor sources, on top of the large scale ozone field in the mixing layer.

3. Ozone mixing ratios at the ground, where the losses occur, are always lower than in the mixing layer aloft, where anthropogenic ozone is predominantly formed, stored, and transported (Fig. 1.21). The ozone concentration at ground level depends mainly on the downward mixing velocity of ozone-rich air from the mixing layer above. This explains the strong influence of meteorological parameters on surface ozone (Muschalik 1980).

These results agree well with airborne measurements which have been carried out in the Netherlands since 1972 (van Duuren et al. 1982).

Fricke (1980) and Neuber (1982, 1983) have also measured nitrogen oxides (NO and NO$_x$) during most of their flights. For example, Fig. 1.22 shows concentration profiles of the nitrogen oxides, 400 and 800 m above ground, which were

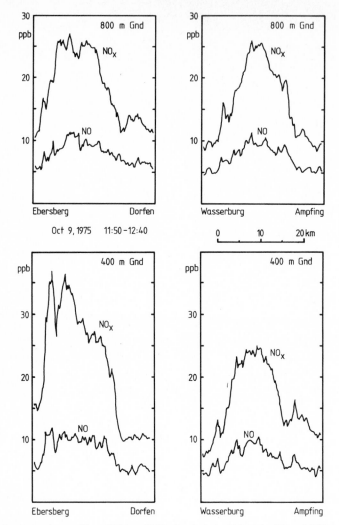

Fig. 1.22. Concentration profiles of NO and NO_x downwind of Munich, 400 and 800 m above ground. The flight trajectories are shown in Fig. 1.23. (Fricke, unpublished results)

obtained downwind of Munich. The flight pattern is outlined in Fig. 1.23. The width of the plume corresponds to the diameter of Munich. The highest concentrations were measured on the lower traverse closest to Munich. The following data and assumptions are input to a rough estimation of the NO_x mass flow in the Munich plume:

Length of one plume traverse	25 km
Vertical extension of the plume ($\cong$ depth of the mixing layer)	1 km
Mean NO_x concentration inside the plume	20 ppb
Mean NO_x concentration outside the plume	10 ppb
Mean wind velocity	5 m s^{-1}
Angle between wind direction and traverses of the plume	45°.

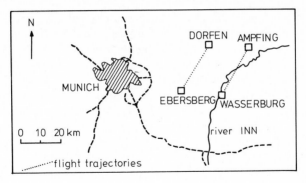

Fig. 1.23. Flight trajectories downwind of Munich, 9 October 1975. Wind direction 270 ° (westerly)

This yields an approximate flux of 1.5 kg s^{-1} NO$_x$ due to emissions in the city of Munich, calculated as NO$_2$. According to an emission inventory [1], the NO$_x$ emission rate of car traffic in Munich amounts to 0.2 kg s^{-1} (annual average). When other NO$_x$ sources (domestic heating, industrial burners, etc.) are taken into account, an order-of-magnitude agreement between the emission inventory and the estimate based on the plume profiles may be achieved. It should, however, be kept in mind that the emission rate derived from the emission inventory represents an annual average, which may well be lower than the mid-day emission rate at the time of the flights.

The following general conclusions can be drawn from airborne measurements of the horizontal and vertical concentration profiles of nitrogen oxides in Germany.

Except for very high NO$_x$ concentrations in the plumes of power plants and other prominent point sources, NO$_x$ concentrations in the mixing layer are usually much lower than typical surface concentrations. However, average concentrations of surface NO$_x$ at a few remote nonurban sites agree well with mixing layer data (Deimel 1979, and other unpublished measurements on a mountain near Bonn). The top of the mixing layer is characterized by a distinct decrease of the NO$_x$ concentration below the detection limit of the analyzers.

Up to now, no commercial instruments are available for continuous airborne measurements of reactive hydrocarbons. On some of his campaigns Fricke has also flown the prototype of a continuous chemiluminescence analyzer which is selective for unsaturated hydrocarbons (Schurath et al. 1976). Simultaneous measurements of ozone and olefins in the plumes of cities and industrial plants, at times of strong oxidant formation in the mixing layer, have revealed that high ozone concentrations ("regional" or "local" ozone, cf. Fig. 1.20) frequently coincided with high olefin concentrations. A few typical profiles downwind of Cologne are shown in Fig. 1.21.

Concentrations of individual hydrocarbons in the mixing layer can be determined in air samples collected aboard aircraft for later analysis by gas chroma-

1 Räumliche Erfassung der Emissionen ausgewählter luftverunreinigender Stoffe aus Industrie, Haushalt und Verkehr in der Bundesrepublik Deutschland 1960–1980, Battelle-Institut e. V., Frankfurt/ Main, August 1976

tography in the laboratory (Gloria et al. 1974, Calvert et al. 1976a). This method
was adopted by Neuber in flights across the Bonn-Cologne-Ruhr area. Sampling
time was 3 to 5 min, corresponding to flight paths of 9 to 15 km. His results
(Neuber 1982, 1983) sketch a rather inhomogeneous picture of the sum concentrations of C_2 to C_8 hydrocarbons in the atmosphere 300 m above ground, and
of the relative amounts of aromatics, alkanes, and alkenes present. While the upwind concentrations southwest of the Cologne-Bonn area amounted to 20 or less
ppb C NMHC (cf. Sect. 1.3.2.1, which reports hydrocarbon measurements in the
Eifel mountains southwest of the area), average NMHC concentrations of
290 ppb C were found downwind of Cologne and Leverkusen, and up to
2,690 ppb C downwind of the Ruhr. Aromatic hydrocarbons and alkanes were
dominant in the Ruhr plume, whereas the hydrocarbons northeast of Cologne
and Leverkusen contained a large fraction of olefins.

Special Investigations

In addition to the precursors NO, NO_2, and hydrocarbons, and to the most
important oxidants ozone and PAN, many other often more reactive species are
known to be involved in oxidant formation which cannot, however, be easily detected with conventional analytical techniques. Furthermore, the existence of
short-lived intermediates has been predicted by the reaction mechanisms developed for modeling photooxidant formation in the atmosphere. Recently, considerable progress has been made in measurements of such "exotic" species, leading
to important improvements and corrections of the chemical models.

Detection of OH Radicals in the Atmosphere

The most important reaction chain carrier in the atmosphere is the OH radical
which reacts more or less rapidly with practically all trace gases including CO and
methane (notable exceptions are N_2O and the Freons). The average OH concentration in the atmosphere governs the lifetime and spatial distribution of all those
trace gases which react readily with OH radicals, other homogeneous and/or heterogeneous loss mechanisms being less efficient. The large majority of the hydrocarbons and the CO molecule fall into this category.

Average OH concentrations as function of latitude, time of the year, and altitude have been modeled, e.g., by Wofsy (1976) and by Crutzen and Fishman
(1977). These calculations have been supplemented by measurements of the latitudinal distribution of ^{14}CO in the troposphere. ^{14}CO is produced by cosmic rays
in the higher atmosphere, transported downward into the troposphere, and removed nearly exclusively by reaction with OH radicals (Volz 1979, Volz et al.
1981). The observed ^{14}CO distribution could be fitted with a two-dimensional
model calculation, scaling the calculated OH concentration with a constant factor. The annually and globally averaged OH concentration thus obtained
amounts to 6.5×10^5 radicals per ccm for the troposphere. Chameides and Tan
(1981) have analyzed the uncertainty of calculated global OH concentrations, and
of latitudinal and vertical OH distributions.

The first direct measurements of OH radicals in the atmosphere were reported
by Wang and Davis (1974). The authors found concentrations of the order of 10^8
OH cm^{-3} which cannot be reconciled with current atmospheric models. It was

shown later, however, that these extremely high OH concentrations were an artifact of their laser-induced fluorescence technique, which produces its own OH radicals. Perner et al. (1976) reported OH measurements in Jülich by a long-path UV absorption technique. Of 33 measurements between April and November, only four exceeded the estimated detection limit of 4×10^6 OH cm^{-3}. This technique was further improved in the following years for measurements in Jülich and Deuselbach (Hübler et al. 1982). An entirely different chemical trapping technique for OH (addition of ^{14}CO) was adopted by Campbell et al. (1979) at 47 ° northern latitude. They found noon-time concentrations around 2×10^6 OH cm^{-3} in July and August.

The results of their own and other published OH measurements in the atmosphere have been critically reviewed by Perner and Hübler (1982).

Measurements of NO_3 in the Atmosphere

The residence time of NO_2 in the atmosphere, as a limiting factor in the formation of anthropogenic ozone in polluted air masses, is of considerable importance. However, the rates and mechanisms of its removal from the atmosphere are not fully understood. NO_2 is lost by reaction with OH (formation of nitric acid vapor), by dry deposition, and to a minor extent by conversion to organic nitrates. Furthermore, NO_2 is known to react slowly with ozone. The reaction product NO_3 is rapidly photolyzed in daylight (photolysis frequency approximately $0.1 s^{-1}$), or converted back to NO_2 by reaction with NO. At night NO_2 can only be lost by dry deposition or formation of NO_3, provided that a loss mechanism exists for NO_3 and/or for N_2O_5, which rapidly establish an equilibrium with NO_2.

NO_3 was first detected in the atmosphere by differential optical absorption in the red, at night in Los Angeles (Platt et al. 1980a). Concentrations in excess of 0.3 ppb were observed. Later, NO_3 was also found at night in Jülich and in Deuselbach in the Hunsrück mountain, although at lower concentrations than in Los Angeles (Platt 1981). The measurements revealed that the reaction mechanisms of NO_2, NO_3, and N_2O_5 have not yet been fully elucidated. The formation of NO_3 by the reaction of NO_2 with ozone appeared to be an efficient night-time removal mechanism for odd nitrogen. The more recently observed lifetime dependence of NO_3 on relative humidity (Platt et al. 1984) suggests the involvement of N_2O_5 removal on wet aerosols and/or surfaces in the loss mechanism. The homogeneous reaction of N_2O_5 with water vapor is far too slow to be important under atmospheric conditions (Tuazon et al. 1983). Furthermore, NO_3 can be removed from the atmosphere at night by its rapid reactions with naturally emitted hydrocarbons (Atkinson et al. 1984).

Measurements of HONO in the Atmosphere

In many chemical kinetics models of the atmosphere (e. g., Falls et al. 1979), HONO (nitrous acid) formation from NO, NO_2, and water vapor plays an important role as a source mechanism for OH radicals:

$$HONO + h\nu \rightarrow HO + NO.$$

Without this OH source ozone formation would proceed far too slowly in these models.

HONO is photolyzed approximately eight times more slowly than NO_2 in the atmosphere (Schurath et al. 1981). Laboratory studies have revealed that the formation of HONO from nitrogen oxides and water vapor occurs heterogeneously, but the mechanism is not fully understood (Sakamaki et al. 1983). Although HONO is known to be formed in smog chambers, its presence in the free atmosphere remained a matter of controversy until, very recently, Kessler et al. detected the molecule by differential optical absorption in the atmosphere of Jülich, Cologne, Riverside, and Los Angeles (Platt et al. 1980b, Kessler et al. 1981, 1982, Harris et al. 1982, and references cited therein). Up to 2.2 ppb HONO were measured during the night in Jülich. In Los Angeles, the night-time concentration reached about 8 ppb.

In Jülich, daytime concentrations between 0.5 and 2 ppb HONO were detected in January. HONO was found to accumulate in polluted air during the night, providing an extremely powerful photochemical OH source immediately after sunrise. About 2 h later, most of the nitrous acid has disappeared, and other OH radical sources become more efficient. The mechanism of HONO formation in the polluted atmosphere could not be derived from these measurements.

Measurements of H_2O_2 in the Atmosphere

Model calculations predict that hydrogen peroxide (H_2O_2) is formed as a product of HO_2 radical reactions in the atmosphere, e.g., in

$$HO_2 + HO_2 \rightarrow H_2O_2 + O_2.$$

Hydrogen peroxide has strong oxidizing properties and must thus be counted among the photooxidants. Measurements of H_2O_2 in the atmosphere provide indirect insight into the concentration of HO_2 radicals which cannot be measured directly in the troposphere at present. HO_2 radicals convert NO to NO_2, and are thus important precursors of anthropogenic ozone in polluted air.

An early colorimetric technique for H_2O_2 trace analysis made use of the well-known liquid-phase reaction with titanium sulfate. In 1978 a much more sensitive wet chemical chemiluminescence technique with luminol was introduced by Kok et al. (1978a). Field measurements with this novel technique in California, in the months of July and August, yielded hydrogen peroxide concentrations in the atmosphere in the range 10 to 30 ppb. At the same time, peak ozone concentrations between 150 and 200 ppb were observed (Kok et al. 1978b). Much lower H_2O_2 concentrations, between 0.3 and 3 ppb, were measured in February at a rural location (Kelly et al. 1979).

More recent laboratory tests (Zika and Saltzman 1982) have revealed that the wet chemiluminescence method of Kok yielded H_2O_2 concentrations which were either too high or too low, depending on the amount of ozone present in the aspired air. It was speculated that H_2O_2 could be formed from ozone in the cloud droplet or rain droplet phase (Heikes et al. 1982). This hypothesis seems to be supported by H_2O_2 analyses in rain water (Zika et al. 1982). The mechanism opens up new perspectives in the atmospheric chemistry of SO_2, since H_2O_2 is capable of oxidizing dissolved SO_2 below pH 5.5. More recently, however, Ten Brink et al. (1984) have speculated that surface reactions are involved in the formation of artifact H_2O_2 from ozone.

Measurements of Nitric Acid in the Atmosphere

Formation of gaseous nitric acid (HNO_3), by recombination of NO_2 with an OH radical, is an important loss mechanism for atmospheric NO_x:

$$NO_2 + OH \rightarrow HNO_3.$$

In contrast to H_2SO_4, HNO_3 has a substantial vapor pressure in humid air, and can thus exist in gaseous form in the atmosphere, until it is removed by rainout, dry deposition, or incorporation in aerosol particles. In smog chambers, reasonable nitrogen balances can only be achieved when the formation of nitric acid is taken into account, which poses analytical problems (Spicer 1977). HNO_3 is at least partially reduced to NO in thermal NO_x converters, and can interfere with NO_2 measurements, particularly in smog chambers (Henrich et al. 1982).

Concentrations of gaseous HNO_3 in the free troposphere have been measured by more conventional techniques (Spicer 1977), and by long-path infrared spectrometry (Tuazon et al. 1981, Hanst et al. 1982). Although a concentration maximum of 49 ppb HNO_3 has been measured in the Los Angeles Basin (Tuazon et al. 1981), concentrations of 10 ppb and less seem to be more common in photochemical smog (Hanst et al. 1982, Spicer 1977). Considerably lower HNO_3 levels were measured in relatively unpolluted rural air: in February, the mixing ratio of HNO_3 was always less than 20% of the NO_x mixing ratio (Kelly et al. 1979).

Chapter 1.4 Air Chemistry and Dispersion

Photochemical oxidants are called *secondary* pollutants. They are formed in polluted air only under certain meteorological conditions, during hours and days of transport, in a complex sequence of photochemically induced reactions from a multicomponent mixture of anthropogenic precursors. Ozone, the most easily measured photochemical pollutant, is also formed in the stratosphere and mixed down into the troposphere, where it gives rise to a natural background concentration. "Natural" ozone molecules are chemically and physically identical with "anthropogenic" ozone molecules which prevail in photochemical smog. These considerations cause certain difficulties in the development of suitable strategies for the reduction of oxidant pollution:

1. The measures to be taken depend upon the scale on which they should be most effective (local ozone reduction versus large-scale reductions), and upon the type of effect required (elimination of very high peak concentrations of short duration; reduction of long-term averages; reduction of a specific photochemical oxidant, e. g., peroxyacetyl nitrate).
2. Justification of measures imposed upon the emitters of specific precursors is extremely difficult, and presupposes long-term field measurements of oxidants, precursors, and meteorological parameters.
3. Beneficial effects which are expected from measures taken against specific precursors cannot, at present, be quantitatively predicted. It is equally difficult, after such measures have been taken, to obtain irrefutable proof of the effect, owing to the year-to-year variability of meteorological parameters (Schurath 1979 b).

In order to alleviate an understanding of the complex mechanism of oxidant formation in the atmosphere, the problem is simplified here by considering the chemical transformations in the gas phase separately, uncoupled from meteorological factors like transport, mixing and dilution of the precursors, intermediates and products, which will be introduced in Sections 1.4.3 and 1.4.4. A model of the chemical transformations is built up from elementary photochemical and chemical steps for which quantitative data exist or can be obtained in the laboratory. The linking together of these elementary steps, which constitute the chemical model, produces a set of coupled differential equations which can be solved with available numerical methods, yielding the concentration-time profiles of all molecules, radicals, and atoms involved in the mechanism.

In the atmosphere, chemical reactions are unseparably coupled with transport, mixing, and dilution. Comparison of a chemical model with the real world is thus only feasible if the model is coupled into an atmospheric dynamics model of equally high sophistication. Today, however, the coupling of a detailed chemical model with a detailed atmospheric dynamics model is feasible only at the price of simplifations of either model.

Another possibility of testing a chemical reaction mechanism against measurements consists of eliminating the effects of transport, mixing, and dilution, which is possible in principle in a smog chamber. However, the inevitable presence of the chamber walls, particularly in small smog chambers, causes serious problems of contamination and heterogeneous reactions. It is therefore essential that models in atmospheric chemistry be built on a theoretically and experimentally sound basis, as furnished by gas phase reaction rate theory and countless laboratory studies of the rates and mechanisms of elementary reactions.

1.4.1 Physicochemical Basis of Atmospheric Chemistry

Chemical reactions of gaseous compounds in the atmosphere can be classified as

a) bimolecular reactions,
b) ter- and monomolecular reactions,
c) photochemical reactions,
d) heterogeneous reactions.

Bimolecular reactions are the most important by number. In the elementary step of a bimolecular reaction, a molecule, radical or atom A collides with another molecule, radical, or atom B, yielding products C and D if the collision was successful:

$$A + B \rightarrow C + D. \tag{1}$$

Gas kinetic collision theory is therefore essential for an understanding of the rates and mechanisms of chemical transformations in the atmosphere.

The rate r of a bimolecular reaction is defined as the derivative with respect to time of a reactant (or product) concentration, due to this particular reaction. It follows from collision theory that r is proportional to the concentrations of both reactants:

$$r = -\frac{d(A)}{dt} = -\frac{d(B)}{dt} = \frac{d(C)}{dt} = \frac{d(D)}{dt} = k(A)(B).$$

The proportionality factor k is called the *rate constant* of the bimolecular reaction. Its dimension is (concentration^{-1} time^{-1}). A bimolecular reaction is fully characterized by its particular rate constant which depends on temperature only. The temperature dependence can be cast into a two-parameter function, usually to a high degree of accuracy, which is better known as the Arrhenius equation:

$$k_{(T)} = Z \exp(-E/RT) = Z \exp(-E'/T).$$

The parameters are called the pre-exponential factor Z [dimension of (concentration^{-1} time^{-1})], and the activation energy E [dimension of (energy mol^{-1})], commensurate with the energy units of the gas constant R. Z and E are temperature-independent for all practical purposes. In the more recent literature, the activation energy E is often replaced by the more convenient parameter $E' = E/R$, of di-

Table 1.21. Conversion factors for rate constants of bimolecular reactions. A rate constant given in units A may be converted to units B by multiplying it with the factor listed in line A, column B. Temperatures T are in Kelvin, pressures in bar. (1 bar $\cong 0.987$ atm $\cong 750$ Torr)

A \ B	$cm^3\ mol^{-1}\ s^{-1}$	$dm^3\ mol^{-1}\ s^{-1}$	$cm^3\ s^{-1\,a}$	$ppm^{-1}\ min^{-1}$
$cm^3\ mol^{-1}\ s^{-1}$	1	10^{-3}	1.66×10^{-24}	7.21×10^{-7} p/T
$dm^3\ mol^{-1}\ s^{-1}$	10^3	1	1.66×10^{-21}	7.21×10^{-4} p/T
$cm^3\ s^{-1\,a}$	6.02×10^{23}	6.02×10^{20}	1	4.34×10^{17} p/T
$ppm^{-1}\ min^{-1}$	1.39×10^6 T/p	1.39×10^3 T/p	2.30×10^{-18} T/p	1
$(mm\ Hg)^{-1}\ s^{-1}$	6.24×10^4 T	62.4 T	1.04×10^{-19} T	4.50×10^{-2} p

[a] These units are sometimes termed "molecular units", which may be emphasized by writing (cm^3 molecule^{-1} s^{-1}) instead of, more correctly (cm^3 s^{-1})

mension (Kelvin^{-1}). In terms of gas kinetics, an upper limit Z_{max} is set to the pre-exponential factor by the collision rate of the reactants. This upper limit amounts to about 3×10^{-10} cm^3 molecule^{-1} s^{-1}, or 2×10^{14} cm^3 mol^{-1} s^{-1}, for typical molecular diameters and masses of atmospheric reactants. Pre-exponential factors of important bimolecular atmospheric reactions are usually between one and three orders of magnitude smaller than Z_{max}. Only a few extremely fast atom-radical and radical-radical reactions approach the gas kinetic limit.

Rate constants for a large number of aeronomically important reactions have been measured in the laboratory at room temperature, or over an extended range of temperatures. Critically evaluated rate constants and available Arrhenius parameters have been published in tabulated form. These tables are regularly updated (Hampson et al. 1973, Hampson and Garvin 1978, Hampson 1980, Baulch et al. 1980, 1982, Jet Propulsion Laboratory 1981).

Several units of reaction rate constants are in use in the literature. Some useful conversion factors for bimolecular reactions are summarized in Table 1.21. Rates of bimolecular reactions, $A + B \rightarrow C$, are sometimes expressed in terms of mixing ratios. An appropriate rate constant k' may be defined as follows:

$$\frac{d(A)}{dt} = (A)(B)k = 10^{-6}\gamma_B(A)(M)k = 10^{-6}\gamma_B(A)k'.$$

γ_B denotes the mixing ratio of species B in ppm, $(M) = 7.236 \times 10^{21} \times$ p/T is the number density of air molecules at pressure p (in bar) and temperature T (in Kelvin). To convert the so-defined rate constant k' (in units of ppm^{-1} time^{-1}) into a conventional rate constant in units of concentration^{-1} time^{-1}, the temperature and pressure dependence of (M) must be taken into account.

One of the rare examples of an elementary termolecular reaction in which three colliders undergo chemical change is the oxidation of nitric oxide by molecular oxygen:

$$NO + NO + O_2 \rightarrow NO_2 + NO_2. \tag{2}$$

The reaction rate r is defined as

$$r = -1/2\frac{d(NO)}{dt} = -\frac{d(O_2)}{dt} = 1/2\frac{d(NO_2)}{dt}$$
$$= k(NO)^2(O_2).$$

The square dependence of the reaction rate on the nitric oxide concentration explains why this reaction is important only at high NO concentrations, e. g., in a concentrated plume, becoming negligibly slow in the free atmosphere. Rate constants k of termolecular reactions have dimensions of (concentration^{-2} time^{-1}). They are only weakly temperature-dependent. When parameterized in Arrhenius form (which is feasible only over limited temperature ranges), small *negative* activation energies are usually obtained.

More important in atmospheric chemistry are termolecular recombination reactions of atoms and small radicals, e. g.,

$$O + O_2 + M \rightarrow O_3 + M, \qquad (2a)$$

where M denotes an inert third collider (usually N_2 or O_2 in the atmosphere), which does not undergo chemical change in the collision. It stabilizes the energy-rich collision pair $(O \cdot O_2)^*$, which would otherwise re-dissociate, by removing some of the energy of the newly formed chemical bond in the collision. Since the concentration of air as a third body is usually constant, one defines a pseudo-bimolecular rate constant k_r for termolecular recombination reactions,

$$k_r = k(M),$$

where k is the "true" termolecular rate constant in this simple treatment. The pressure dependence of k_r is in fact more complex than this equation would suggest (Troe 1979). For practical applications in atmospheric chemistry, a parameterization of the pressure dependence is available (preface in Baulch et al. 1980, 1982).

The reverse of molecule formation by recombination of atoms or radicals, Reaction (2a), is the decomposition of a thermally labile molecule, e. g., of peroxyacetyl nitrate (PAN), into its radical fragments:

$$CH_3C\overset{\displaystyle O}{\underset{\displaystyle OONO_2}{\diagup\diagdown}} \longrightarrow CH_3C\overset{\displaystyle O}{\underset{\displaystyle OO}{\diagup\diagdown}} + NO_2. \qquad (3)$$

This is termed a monomolecular reaction, since it follows simple first order kinetics at constant pressure,

$$r = -\frac{d(PAN)}{dt} = k_d(PAN),$$

which integrates to an exponential decay law,

$$(PAN)_t = (PAN)_{t=0} \exp(-t \cdot k_d).$$

The rate constant k_d has dimension of (time^{-1}). Its strong temperature dependece is very well reproduced by the Arrhenius equation, the activation energy E being of the order of the dissociation energy of the bond being broken. When a molecule decomposes irreversibly at the rate given above, its half-life $t_{1/2}$ is independent of initial concentration:

$$t_{1/2} = (\ln 2)/k_d.$$

Many decompositions of thermally labile molecules are reversible under suitable conditions, establishing a dynamic equilibrium:

$$PAN \rightarrow CH_3CO_3 + NO_2 \quad k_d, \qquad (3)$$

$$CH_3CO_3 + NO_2 \rightarrow PAN \quad k_r, \qquad (4)$$

$$\mathbb{K} = \frac{(PAN)}{(CH_3CO_3)(NO_2)} = k_r/k_d.$$

$\mathbb{K}$ is the pressure-independent thermodynamic equilibrium constant. Since k_r is a function of pressure, as outlined above, k_d must also be pressure-dependent. The theoretical treatment and parametrization of both pressure dependencies are in fact fully analogous (Troe 1979, preface in Baulch et al. 1980, 1982).

Molecules which are not transparent throughout the visible and ultraviolet spectrum of sunlight may be photodissociated in the atmosphere, provided that the energy of the absorbed photon exceeds the dissociation energy of a molecular bond. A typical photochemical reaction may be regarded as a sequence of elementary steps:

1. Excitation of an absorber molecule A by a photon of energy $h\nu$:

$$A + h\nu \rightarrow A^*. \qquad (5)$$

2. Monomolecular dissociation of the excited molecule A^* into fragments F_1 and F_2 (= photodissociation):

$$A^* \rightarrow F_1 + F_2. \qquad (6)$$

3. When the rate of photodissociation of the excited molecule A^* is not exceedingly fast, stabilization by collisions is also possible:

$$A^* + M \rightarrow A + M. \qquad (7)$$

The energy dependence (= wavelength dependence) of Reactions (6) and (7) must of course be taken into account.

Anthropogenic oxidant precursors and numerous other pollutants are stable molecules which react neither with each other nor with other stable constituents of the atmosphere. However, photodissociation converts stable molecules into extremely reactive photofragments (atoms and/or radicals with unpaired electrons) which initiate chains of radical-molecule reactions in the polluted atmosphere, e.g.:

$$H_2CO + h\nu \quad \rightarrow \quad H + HCO$$

$$H + O_2 + M \quad \rightarrow \quad HO_2 + M$$

$$HCO + O_2 \quad \rightarrow \quad HO_2 + CO$$

$$HO_2 + NO \quad \rightarrow \quad OH + NO_2$$

$$OH + \begin{Bmatrix} \text{hydrocarbons, CO, numerous} \\ \text{other pollutants} \end{Bmatrix} \rightarrow \text{radical} + \text{molecule}.$$

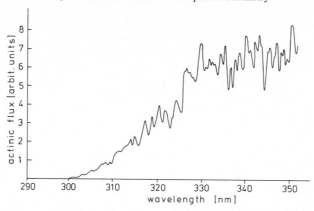

Fig. 1.24. Daylight spectrum below 350 nm. Solar elevation 52 ° (zenith angle z = 38 °). Ozone column at the time of the measurement ca. 320 Dobson. Spectral resolution 0.4 nm. (Bahe et al. 1979a)

Note that NO is converted to NO_2 in the sequence. The chain character of the scheme arises because each radical which reacts with a stable compound creates another radical. A particularly important chain carrier is the extremely reactive OH radical which involves numerous otherwise stable pollutants (e. g., CO) in the chain. The length of the radical chains is limited by termination reactions which annihilate radicals, e. g.:

$$OH + NO_2 + M \rightarrow HNO_3 + M \ (= \text{stable products}),$$
$$HO + HO_2 \quad \rightarrow H_2O + O_2 \ (= \text{stable products}).$$

These radical losses must be compensated by photodissociations of stable molecules. Clearly, daylight is essential for the initiation and maintenance of chemical reactions in the atmosphere.

Only those molecules can be photodissociated in the atmosphere which are not totally transparent to sunlight in the ultraviolet, where the photon energy becomes commensurate with the dissociation energy of chemical bonds. Figure 1.24 shows that the light intensity at ground level (and throughout the troposphere) decreases rapidly below 350 nm, owing to the filtering effect of the ozone layer. Although there is no sharp cutoff at 300 nm, as can be seen in a logarithmic plot of the cutoff region (Fig. 1.25) the photon flux is negligible below 295 nm. The cutoff region depends on the solar zenith angle, and varies with season as function of the ozone column.

The most important light-absorbing molecules in the troposphere are NO_2 (Fig. 1.26), ozone (Fig. 1.27), formaldehyde (Fig. 1.28), higher aldehydes, and ketones (for absorption spectra see Calvert and Pitts 1966), and HONO (Stockwell and Calvert 1978). The attenuation of a parallel monochromatic light beam of intensity I_0 by an absorbing species of partial pressure p or concentration c, after a path length of x cm, is usually given by Lambert-Beer's law:

$$I = I_0 \exp(-x \cdot p \cdot k)$$
$$= I_0 \exp(-x \cdot c \cdot \sigma)$$
$$= I_0 10^{-x \cdot c \cdot \varepsilon}.$$

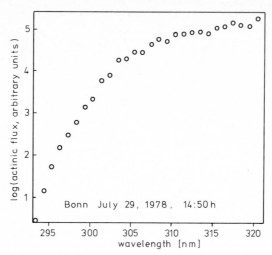

Fig. 1.25. Intensity distribution of sunlight (direct component) below 320 nm. Ozone column ca. 325 Dobson at the time of the measurement. Spectral resolution 0.4 nm. (Bahe et al. 1979a)

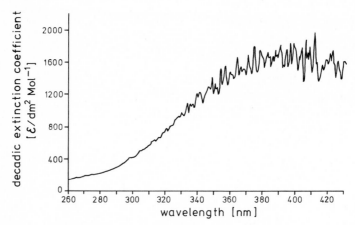

Fig. 1.26. Absorption spectrum of NO_2 below 430 nm

k is the absorption coefficient, σ the absorption cross section (usually in cm^2/molecule), and ε the extinction coefficient of the absorber, at the specified wavelength. Some useful conversion factors for applications of Lambert-Beer's law are listed in Table 1.22.

The photolysis of NO_2 is chosen to define the rate of a photochemical reaction:

$$NO_2 + h\nu \rightarrow NO + O. \tag{8}$$

This important process, which is followed by the rapid recombination [Reaction (2a)] of O with O_2, is the immediate cause of "anthropogenic" ozone formation in the troposphere. The reaction rate is defined in accordance with monomolec-

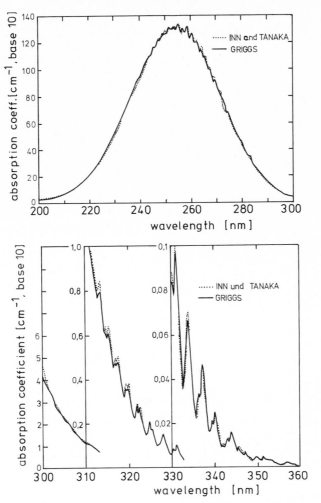

Fig. 1.27. Absorption spectrum of ozone below 360 nm. Absorption coefficients in cm^{-1} (atm at $273\ K)^{-1}$, base 10; cf. Table 1.22. (Hampson et al. 1973)

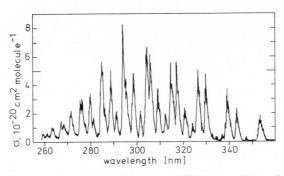

Fig. 1.28. Absorption spectrum of formaldehyde at 296 K between 260 and 360 nm. (Bass et al. 1980a, b)

Table 1.22. Conversion factors for optical absorption coefficients, absorption cross-sections, extinction coefficients, etc. To convert from units A to units B, multiply by the factor listed in line A, column B (Baulch et al. 1980)

A \ B	Cross-section: cm^2 molecule^{-1}, base e	(atm at 273 K)$^{-1}$ cm^{-1}, base e	Liter mol^{-1}cm^{-1}, base 10	cm^2 mol^{-1}, base 10
(atm at 273 K)$^{-1}$ cm^{-1}, base e	3.72×10^{-20}	1	9.73	9.73×10^3
(atm at 273 K)$^{-1}$ cm^{-1}, base 10	8.57×10^{-20}	2.303	22.414	2.24×10^4
(Torr at 273 K)$^{-1}$ cm^{-1}, base e	2.83×10^{-17}	760	7.39×10^3	7.39×10^6
cm^2 mol^{-1}, base 10	3.82×10^{-24}	1.03×10^{-4}	10^{-3}	1
cm^2 molecule^{-1}, base e	1	2.69×10^{19}	2.62×10^{20}	2.62×10^{23}
Liter mol^{-1} cm^{-1}, base 10	3.82×10^{-21}	0.103	1	1,000

ular reactions:

$$r = -\frac{d(NO_2)}{dt} = j(NO_2).$$

As distinct from rate constants of true monomolecular reactions, the proportionality factor j [dimension (time^{-1})] is called the photolysis frequency of the photochemical reaction. It is a function of light intensity I_λ, cross-section σ_λ of the absorber, and quantum yield ϕ_λ:

$$j = \int_\lambda I_\lambda \phi_\lambda \sigma_\lambda d\lambda. \qquad (9)$$

The quantum yield ϕ of a photochemical Reaction (6) [or (8) in the above example] is defined as the fraction of molecules excited in Reaction (5) by absorption of a photon which undergo photodissociation. The quantum yield ϕ of a photochemical reaction can be reduced below unity by competitive reactions like (7), and may be wavelength-, pressure-, and temperature-dependent. An upper wavelength limit of ϕ is set by the dissociation energy of the bond to the broken, which corresponds to about 400 nm in reaction (8).

Leighton (1961) has already used Eq. (9) to calculate photolysis frequencies of important atmospheric absorbers. He derived simple formulae for actinic flux calculations (actinic flux = light intensity I_λ in units of (photons cm^{-2} nm^{-1} s^{-1}), integrated over all directions) as function of solar zenith angle, which can nowadays be replaced by more sophisticated model calculations. Recently, a number of field measurements of important photolysis frequencies have become available (Bahe et al. 1979b, 1980, Schurath et al. 1981, Marx et al. 1983). The results are summarized in Table 1.23 and Fig. 1.29.

Heterogeneous reactions (adsorption on solid surfaces including aerosols, uptake by the biosphere, dissolution in cloud and rain droplets) can significantly shorten the atmospheric lifetimes of certain trace gases. For example, ozone in remote unpolluted areas is mainly destroyed heterogeneously at the earth's surface. The ground deposition of a trace gas may be parameterized by its deposition velocity v_D, which derives from the concentration gradient close to the surface (Chamberlain 1960, Whelpdale 1982):

$$v_D = F/c_{(z)} \text{ (cm s}^{-1}).$$

Table 1.23. Photolysis frequencies of important photodissociation reactions in the atmosphere as function of solar zenith angle z, based on measurement under cloud-free conditions. (Schurath et al. 1981, Marx et al. 1983)

Photochemical process	Linear approximation $\ln{(j, s^{-1})}$	$j\ (s^{-1})$ at noon, summer solstice	Reference data (s^{-1})	
$NO_2 + h\nu \rightarrow NO + O$	$-0.64 \sec{(z)} - 4.14$	7.8×10^{-3}	7.5×10^{-3}	a
$CH_3ONO + h\nu \rightarrow CH_3O + NO$	$(-0.64 \sec{(z)} - 5.87)^a$	1.4×10^{-3}	6.3×10^{-3}	b
$HONO + h\nu \rightarrow OH + NO$	$(-0.64 \sec{(z)} - 6.2)^{a,\,b}$	$9.9 \times 10^{-4\,b}$	2.5×10^{-3}	c
$H_2CO + h\nu \rightarrow$ all products	$-1.05 \sec{(z)} - 8.56$	5.9×10^{-5}	5.7×10^{-5}	d
$H_2CO + h\nu \rightarrow H + HCO$	$-1.34 \sec{(z)} - 9.64$	1.5×10^{-5}	2.2×10^{-5}	d
$CH_3CHO + h\nu \rightarrow CH_3 + HCO$	$-1.21 \sec{(z)} - 11.16$	3.4×10^{-6}	3.9×10^{-5}	e
$C_2H_5CHO + h\nu \rightarrow$ products	$(-1.20 \sec{(z)} - 12)$	1.6×10^{-6}	4.3×10^{-5}	e
$O_3 + h\nu \rightarrow O_2 + O(^1D)$				
(325 Dobson O_3)	$-1.55 \sec{(z)} - 9.0$	2.2×10^{-5}	3.1×10^{-5}	f
(390 Dobson O_3)	$-1.59 \sec{(z)} - 9.59$	1.15×10^{-5}	2.1×10^{-5}	f

a Derwent and Hov (1979), b Taylor et al. (1980), c Cox and Derwent (1977), d Calvert (1981), e Cox et al. (1981), f Dickerson (1980), Dickerson et al. (1982)

[a] Assuming that the z-dependence is the same as for NO_2
[b] Estimate based on relative absorption reates of HONO and CH_3ONO given by Cox and Derwent (1977), and assuming the same z-dependence as for NO_2

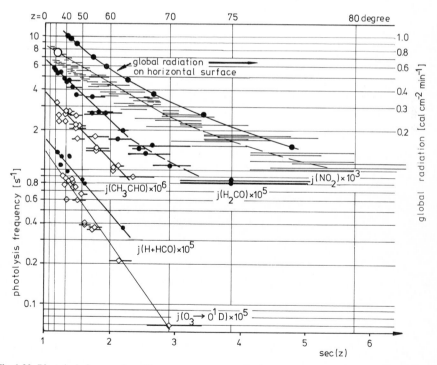

Fig. 1.29. Photolysis frequencies of important trace gases in the atmosphere as function of solar zenith angle. (Schurath et al. 1981, Marx et al. 1983)

[F = flux perpendicular to the surface; $c_{(z)}$ = concentration at a reference height z].

Deposition velocities are highly variable, depending on the type and momentary condition of the surface onto which deposition occurs. Deposition velocities of ozone have been published by several authors (Galbally 1971, Garland and Penkett 1976, Garland and R. G. Derwent 1979). Peroxyacetyl nitrate (PAN) exhibits deposition velocities of 1.6 cm s^{-1} on soil and 0.5 cm s^{-1} on grass (Garland and Penkett 1976). Deposition of PAN onto water surfaces is extremely inefficient, and a deposition velocity of 0.006 cm s^{-1} has been measured (Schurath et al. 1984).

The following deposition velocities were used in recent model calculations of R.G. Derwent and Hov (1979, 1980):

$$\text{Ozone} \quad v_D = 0.6 \, \text{cm s}^{-1}$$

$$\text{SO}_2 \quad v_D = 0.8 \, \text{cm s}^{-1}$$

$$\text{NO}_2 \quad v_D = 0.1 \, \text{cm s}^{-1}$$

$$\text{PAN} \quad v_D = 0.2 \, \text{cm s}^{-1}.$$

The authors assume that ozone and sulfur dioxide do not differ significantly with respect to heterogeneous removal. Very little is known about deposition velocities of other photochemical pollutants (H_2O_2, HNO_3, N_2O_5), and estimates based on the physical properties of the compounds have been used in model calculations.

1.4.2 Simulation

1.4.2.1 Experimental Investigations

In the 1940's a new class of secondary air pollutants with oxidizing properties and specific environmental effects (plant injury, eye irritation, visibility reduction by aerosols) was first observed in the Los Angeles Basin. The pollutants were formed only in intense sunlight, and were therefore called photochemical oxidants. Their identities and formation mechanisms were at first not well understood. In the early 1950's, smog-chamber experiments provided the only tool for unraveling the dependence of oxidant formation on precursor concentrations (NO_x, hydrocarbons) and on light intensity. A clear distinction between "oxidants", i.e., trace gases which release molecular iodine from a KI solution, and ozone could be made only after the invention of the chemiluminescence detector (Nederbragt et al. 1965).

A smog chamber encloses a fixed volume of air with known pollutant contents, which can be irradiated with artificial light sources or with genuine sunlight. The concentrations of the precursors (NO, hydrocarbons), intermediates and products (NO_2, aldehydes, ozone, PAN, etc.) are measured during irradiation, continuously if possible, or discontinuously at short intervals. Smog chambers differ widely with respect to volume, wall material, intensity and spectral distribu-

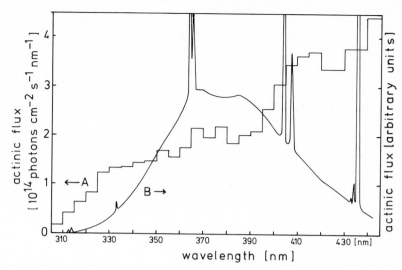

Fig. 1.30. *Curve A*, spectral distribution of daylight, solar zenith angle z=40 °. (Peterson 1976). *Curve B*, relative spectral distribution of light inside a smog chamber. The chamber was illuminated with fluorescent lamps (Osram L 65–80 W/70) through its Duranglass walls. (Schurath 1979a)

tion of the light sources, temperature control, and analytical equipment (e. g., destructive methods, or in situ spectroscopic techniques like multipath FTIR).

Very voluminous smog chambers cannot be evacuated, and must be conditioned by flushing with purified or ambient air, depending on the type of experiment intended. Such smog chambers are usually irradiated from one or several sides through Pyrex plates or transparent plastic film. Another smog chamber variant is entirely fabricated from thin transparent Teflon or Tedlar film, and can be inflated with a blower like a balloon. Such smog chambers can be built on a very large scale (e.g., over 300 m³ volume, Jeffries et al. 1976b). When exposed to sunlight in the open air, the intensity and spectrum of the light inside the chamber differ very little from those outside. Smaller glass chambers or (Teflon-coated) stainless steel reactors are often evacuable, rendering possible irradiation experiments in rapid succession. Cylindrical steel and glass chambers have been irradiated with the filtered parallel beam of a high pressure Xe arc system (Akimoto 1979b, Bruckmann and Eynck 1980b, Bruckmann et al. 1980). Borosilicate glass chambers can be irradiated through the glass wall with UV fluorescent tubes which emit strongly in the 300–400 nm region. However, UV radiation below 350 nm is increasingly attenuated by the borosilicate glass, which cuts off at about 310 nm (Becker et al. 1978). The spectral distribution inside such a smog chamber is compared with a UV spectrum of natural sunlight at ground level in Fig. 1.30 (Schurath 1979a, see also Fig. 1.24 for a spectrally resolved UV spectrum of sunlight below 350 nm).

According to Table 1.24, OH and HO_2 radicals are predominantly made in the atmosphere by the photolysis of ozone and of aldehydes, at wavelengths shorter than 310 nm and 330 nm, respectively. The intensity ratio of radiation below 330 nm, and of radiation in the NO_2 photolysis region (330–400 nm), is

Table 1.24. Important photochemical radial source in the atmosphere, their wavelength limits, and typical photolysis frequencies for summer conditions in Middle Europe

Photochemical process and important secondary reaction	Wavelength limit (nm)	Photolysis frequency (s^{-1})
$NO_2 + h\nu \rightarrow NO + O$ $O + O_2 + M \rightarrow O_3 + M$	ca. 400	0.0078
$HONO + h\nu \rightarrow OH + NO$	ca. 330	ca. 0.001
$H_2CO + h\nu \rightarrow H + HCO$ $H + O_2 + M \rightarrow HO_2 + M$ $HCO + O_2 \rightarrow HO_2 + CO$	ca. 330	1.5×10^{-5}
$CH_3CHO + h\nu \rightarrow CH_3 + HCO$ $CH_3 + O_2 + M \rightarrow CH_3O_2 + M$ $HCO + O_2 \rightarrow HO_2 + CO$	ca. 320	3.4×10^{-6}
$O_3 + h\nu \rightarrow O_2 + O(^1D)$ $O(^1D) + H_2O \rightarrow OH + OH$ $O(^1D) + M \rightarrow O + M$	ca. 310	ca. 2×10^{-5}

usually much lower in smog chambers than in the free atmosphere (see also Akimoto et al. 1979). It is therefore difficult to compare experiments in different smog chambers with each other and with the real world, even if the light intensity in the smog chambers is given quantitatively in terms of the NO_2 photolysis frequency.

A serious problem, particularly in small smog chambers, is created by wall reactions and contamination effects by outgassing wall material (van Ham 1978, Lonneman et al. 1981). The possibility of radical creation by photochemical surface reactions has been suggested (Carter et al. 1981, 1982), in order to explain the observed rates of chain initiation in smog chambers. Other authors have found that HONO is formed heterogeneously from NO_2 on smog chamber walls. HONO is an efficient OH radical source when photolyzed in the NO_2 photolysis range (Sakamaki et al. 1983).

In spite of all these drawbacks, smog chambers are still the only systems to date in which oxidant formation can be investigated in a homogeneous air sample as function of precursor composition and concentration. Such experiments yield comparison data which are needed in validations of chemical models (Whitten et al. 1980).

Smog-chamber irradiations of NO (or NO_x) and hydrocarbons, or simply car exhaust, in air at constant light intensity, proceed in three typical phases:

I *Induction phase:*
  Rapid $NO \rightarrow NO_2$ conversion, beginning consumption of the reactive hydrocarbons, buildup of aldehydes;
II *Most active phase:*
  Accumulation of oxidants (ozone, PAN), rapid consumption of the reactive hydrocarbons and of NO_2;

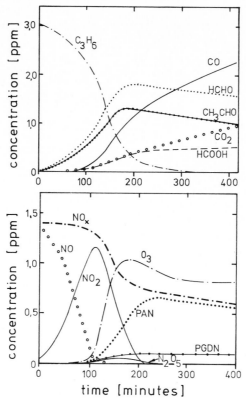

Fig. 1.31. Concentration profiles of reactants and products in a smog chamber experiment at constant light intensity. (Akimoto et al. 1980). *PGDN* polypropyleneglycol 1,2-dinitrate

III *Aged photosmog phase:*
Deficiency of NO_2 and reactive hydrocarbons, deceleration of ozone formation below the heterogeneous destruction rate.

These phases can be clearly discerned in Fig. 1.31, which shows smog-chamber measurements in a single-hydrocarbon-NO_x system (Akimoto et al. 1980). The analytical method was multipath FTIR spectrometry, which is capable of detecting a large number of reactants and products simultaneously. The smog-chamber product propyleneglycol 1,2-dinitrate (PGDN) was identified by its IR spectrum in this study for the first time.

When ambient air with relatively low levels of pollutants is irradiated in a smog chamber, phases I and II are often not very well separated (Becker et al. 1978, Bruckmann et al. 1980). However, when the light intensity in the smog chamber is not constant throughout the experiment, but simulates the natural diurnal profile, all three phases can be discerned, even in relatively clean ambient air irradiations (Henrich et al. 1982).

Perhaps the most widely noted smog-chamber experiments, which had lasting consequences, were undertaken in the 1960's by the US Bureau of Mines. Car ex-

haust in air was irradiated through a Pyrex window in an aluminium box of
2.8 m³ volume. The aim of the experiments was to establish empirical correlations
between initial NO_x and nonmethane hydrocarbon (NMHC) concentrations, and
ozone yields (Dimitriades 1970). The investigation led to the concept of ozone
isopleths, which will be discussed in more detail in Sect. 1.4.2.3.

1.4.2.2 Chemical Models

A chemical model of the atmosphere can be constructed by arranging all atmospheric constituents: trace gases, atoms, free radicals, and photons, in the first
row and column of a square matrix, as shown in an oversimplified example:

	O_2	NO	NO_2	C_2H_4	O_3	O	OH	$h\nu$
O_2	–	–	–	–	–	O_3	–	–
NO		NO_2	–	–	NO_2	–	HONO	–
NO_2			–	–	NO_3	NO	HNO_3	NO+O
C_2H_4				–	formald. $+H_2COO$	addn.	addn.	–
O_3					–	–	HO_2	O(^{1}D)
O						–	–	–
OH							–	–
$h\nu$								–

(addn. = addition product)

Each matrix element is filled in with the product(s) of the reaction between two
species, one in the first row, one in the first column, of the matrix. The matrix
is symmetric, and only one half of the off-diagonal elements need be considered.
When the reactants do not react, the matrix element is empty, indicated by a minus sign. A minus sign can also be written to a good approximation if the reaction
makes a negligible contribution to the conversion of both reactants. The scheme
is complete when all non-negligible products of the retained reactions (those without a minus sign) are themselves elements of the first column and row of the matrix. This is obviously not the case in our example (new products are framed), and
the matrix must be expanded accordingly. When a self-consistent matrix of reactants and products is finally obtained, the chemical model is considered to be
complete.

Each reaction of a chemical model defines time derivatives of reactants and
products, as outlined in Section 1.4.1. Adding up for each species X_i its time derivatives with respect to all the reactions in which it participates, either as a product or as a reactant,

$$\frac{d(X_i)}{dt} = \sum_{l,m} k_{lm}(X_l)(X_m) - \sum_j k_{ij}(X_i)(X_j),$$

yields a set of simultaneous first-order differential equations, one for each species
concentration, which can be solved by suitable numerical methods (e. g., Chance

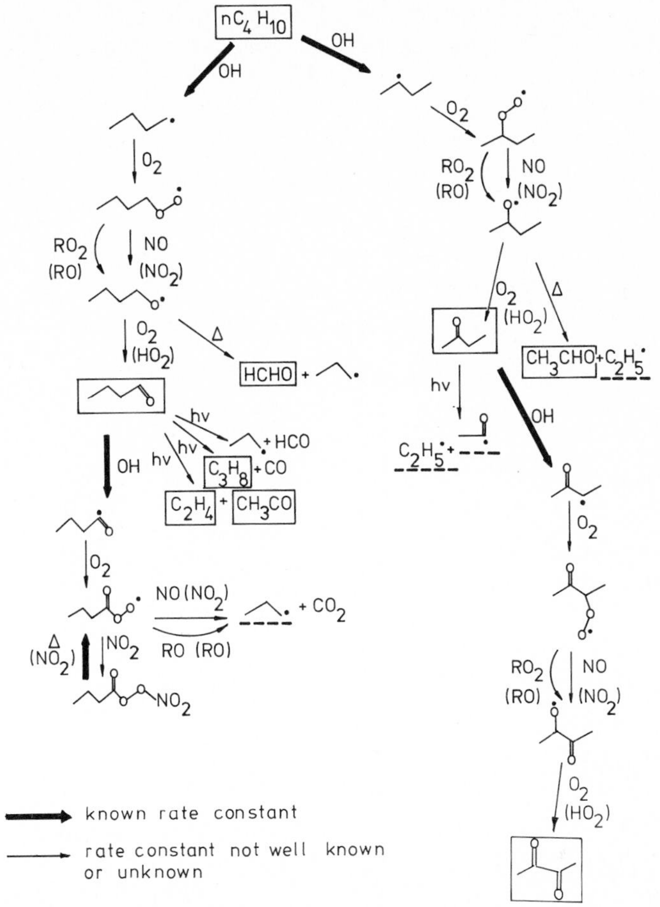

Fig. 1.32. OH radical induced oxidation of n-butane under atmospheric conditions. *Fat arrows* indicate reactions with known rate constants. All other rate constants are not well known or unknown. Observed reaction products are framed

et al. 1977), if the rate constants of the reactions and the initial state of the system are known.

In practice, rate constants of many reactions, particularly of organic radical reactions, are not known to the necessary degree of accuracy, or are entirely unknown. The number of rate constants needed for the reactions of individual organic molecules present in polluted air is already very large. The number of unknown rate constants increases tremendously when the secondary reactions of the organic radical fragments obtained in the first step are also taken into account. This is illustrated by the postulated reaction scheme of the OH radical induced atmospheric oxidation of n-butane in Fig. 1.32: Only 5 out of 27 rate constants in the scheme are known (fat arrows).

Clearly, an "exact" numerical simulation of chemical change in the polluted atmosphere, based on a "complete" reaction scheme, is a forbidding task, although the number of available rate constants has increased considerably in recent years. In current chemical models, the known rate constants are combined with reasonable estimates of unknowns, assuming that the calculated concentration-time profiles of important oxidants like NO_2, O_3 and others are relatively insensitive to these estimates.

The confusing number of reactions which must be included in any fairly self-consistent model, even for a single-hydrocarbon-NO_x smog-chamber experiment (Sakamaki et al. 1982), can be organized into few classes of reactions which have certain common properties or functions:

A *Photochemical sources of atoms and radicals*, e.g.:

$$NO_2 \quad + h\nu \rightarrow NO + O$$

$$H_2CO \quad + h\nu \rightarrow H \quad + HCO \quad (besides \ H_2 + CO)$$

$$O_3 \qquad + h\nu \rightarrow O_2 \quad + O(^1D) \quad (besides \ O_2 + O(^3P))$$

$$HONO + h\nu \rightarrow OH + NO$$

$$higher \ aldehydes \ + h\nu \rightarrow HCO + R$$

B *Inorganic reactions*, e.g.:

$$O \quad + O_2(+M) \quad \rightarrow O_3(+M)$$

$$NO + O_3 \qquad \rightarrow NO_2 + O_2$$

$$HO_2 + NO \qquad \rightarrow OH + NO_2$$

$$OH \quad + NO_2(+M) \rightarrow HNO_3(+M)$$

$$HO_2 + HO_2 \qquad \rightarrow H_2O_2 + O_2$$

C *Reactions of inorganic radicals and molecules with hydrocarbons, aldehydes, etc.*, e.g.:

$$OH + alkane \quad \rightarrow H_2O + R \quad (several \ isomers)$$

$$OH + olefins \quad \rightarrow addition \ (2 \ or \ more \ isomers)$$

$$OH + aromatics \rightarrow addition \ and \ abstraction$$

$$O_3 \ + olefins \quad \rightarrow aldehyde + RCHOO$$

$$OH + aldehyde \ \rightarrow H_2O + RCO$$

D *Secondary reactions of the organic radical products of C with molecular oxygen, decomposition reactions, rearrangements*, e.g.:

$$R + O_2 \quad \rightarrow RO_2$$

$$RO + O_2 \rightarrow aldehyde + HO_2$$

$$RO \qquad \rightarrow rearrangement, \ decomposition \ (chain \ branching)$$

E *Reactions of organic radicals with NO and NO_2*, e.g.:

$$RO_2 \quad + NO \rightarrow RO + NO_2$$

$$RO \quad + NO_2 \rightarrow RONO_2$$

$$RCOO_2 + NO \ \rightarrow RCO_2 + NO_2 \rightarrow R + CO_2 + NO_2$$

$$RCOO_2 + NO_2 \rightarrow RCOO_2NO_2 \ (PAN)$$

(Other groupings are feasible; e. g., into (a) true radical sources (= photodissoci-ations), (b) radical chain reactions which generate HO_2 and RO_2, (c) chain branching reactions which increase the number of radicals, (d) chain termination by radical annihilation, e. g., $OH + NO_2 \rightarrow HNO_3$).

The photochemical reactions of *group A* provide the driving force of free radi-cal chemistry in the polluted atmosphere. In their absence (e. g., during the night), tropospheric chemistry is reduced to a few ozone reactions and heterogeneous processes like dry deposition and rainout.

The inorganic reactions in *group B* have been studied in particular detail. However, the fact that the rate constant of the important reaction

$$HO_2 + NO \rightarrow OH + NO_2$$

had to be revised by a factor of 40 not too long ago (Howard and Evenson 1977) makes it clear that even these reactions require continued attention.

A continuously increasing number of rate constants of OH and O_3 reactions with organic compounds (*group C*) is becoming available. Unfortunately, much less is known about the products of these reactions (Kan et al. 1981, Ghosh et al. 1982, Adeniji et al. 1982, Atkinson et al. 1981, 1982).

Very few rate constants of organic radical reactions in *group D* are known, ex-cept for the simplest members of the organic radical family. Indirect information on rearrangements and decomposition reactions of larger organic radicals has been extracted from product analyses (Martinez et al. 1981).

Rate constants of the type $RO_2 + NO$ (*group E*) are more difficult to measure than their inorganic counterpart, $HO_2 + NO$. The fact that organic peroxinitrates may exist in dynamic equilibrium with their radical fragments was first appre-hended in studies of the thermal monomolecular decomposition of peroxiacetyl nitrate (PAN) (Cox and Roffey 1977, Hendry and Kenley 1977).

In the belief that the removal of precursors and the formation of oxidants in a fixed volume of air can be modeled by numerical integration of the model-de-pendent simultaneous differential equations, if the *essential* steps of the mecha-nism and the boundary conditions are correctly formulated, two basic types of chemical models have evolved: The detailed mechanisms and the so-called "lumped" mechanisms.

Detailed Mechanisms

The first detailed mechanism was advanced by Niki et al. (1972) in a computer modeling exercise of smog-chamber results in a NO/NO_2/propene/humid air sys-tem. It was exemplary for all future mechanisms which were composed of true el-ementary steps. Detailed chemical mechanisms are too time-consuming on a com-puter for most applications. Furthermore, the wealth of input data required for a detailed model which considers a large number of organic molecules individ-ually is often not available in practical situations.

A land-mark in the development of detailed mechanisms is the work of De-merjian et al. (1974). The authors present a detailed discussion of available methods for estimating rate constants of organic radical reactions for which no experimental data were available. Isaksen et al. (1978 b) and Hov et al. (1978 a, b) have coupled detailed chemical models into simple transport models. This work was extended in England by R. G. Derwent and Hov (1979, 1980), making use, as much as possible, of product analyses and other laboratory results pertaining to the breakdown patterns of individual hydrocarbons under atmospheric conditions.

Detailed reaction mechanisms are valuable diagnostic tools for the scientist, e. g., in sensitivity studies of the impact of uncertainties in rate constants, photolysis frequencies, and other important input data on calculated observables (concentration-time profiles of precursors and products) in smog chambers and in the field.

Lumped Mechanisms

The development of lumped mechanisms dates back to the late 1960's. They rely on the hypothesis that a detailed mechanism can be condensed into a small number of representative (or symbolic) reactions which mimic the rates of NO-to-NO_2 conversion, NO_2 removal, and ozone formation, which are of primary concern in air pollution control strategies. Some generalized processes which must be mimicked by suitably lumped reactions are the following:

1. Chain initiation by photolysis of light absorbing molecules;
2. reaction of OH with a hydrocarbon, formation of peroxy radicals;
3. oxidation of NO by peroxy radicals, regeneration of OH as chain carrier;
4. generation of photon absorbers;
5. removal of NO_2 and radicals by recombination;
6. ozone formation by NO_2 photolysis;
7. ozone destruction by reaction with NO.

Most of the rate constants in simple lumped mechanisms are fitting parameters which must not be compared with known rate constants of true elementary reactions.

Lumped mechanisms are devised for inclusion in detailed transport models, which impose limitations on the number of stable species subject to eddy diffusion, for reasons of computer time. An instructive discussion of the capabilities and limitations of transport models in simulations of oxidant formation has been presented by Demerjian (1977). The simple mechanism of Eschenroeder and Martinez (1972) and the early lumped mechanisms of Hecht et al. (1974) were optimized for urban atmospheres, particularly for Los Angeles. The Eschenroeder-Martinez mechanism is reproduced in Table 1.25. Most of its reactions are highly symbolic, with no counterpart in reality. The rate constants were obtained by fitting NO, NO_2, and O_3 profiles of smog-chamber experiments. The mechanism has no diagnostic value and fails when used outside the fitted range. The same holds true, to a lesser extent, for the Seinfeld mechanism and its various stages of development (references in Falls et al. 1979). In a more recent version of the mechanism, care was taken to use true elementary reactions and evaluated rate constants for the inorganic part, while the hydrocarbons and aldehydes were lumped symbolically into alkanes (ALK), olefins (OLE), aromatics (ARO), and higher aldehydes (RCHO). Formaldehyde and ethene are treated individually. The mechanism is stoichiometrically imbalanced, at least in its published version. It is noteworthy that the latest version of the mechanism bears striking similarities

Table 1.25. Lumped mechanism of Eschenroeder and Martinez (1972). The reaction sequence $HO + CO \rightarrow H + CO_2$, $H + O_2 + M \rightarrow HO_2 + M$ was discussed, but not included in the original mechanism

	Reaction	Fitted rate constant
1.	$h\nu + NO_2 \rightarrow NO + O$	0.4 min^{-1}
2.	$O + O_2 + M \rightarrow O_3 + M$	$1.32 \times 10^{-5} \text{ ppm}^{-2} \text{ min}^{-1}$
3.	$O_3 + NO \rightarrow NO_2 + O_2$	$40 \text{ ppm}^{-1} \text{min}^{-1}$
4.	$O + HC \rightarrow 2 RO_2$	$6100 \text{ ppm}^{-1} \text{ min}^{-1}$
5.	$OH + HC \rightarrow 2 RO_2$	$80 \text{ ppm}^{-1} \text{ min}^{-1}$
6.	$RO_2 + NO \rightarrow NO_2 + 0.5 OH$	$1500 \text{ ppm}^{-1} \text{ min}^{-1}$
7.	$RO_2 + NO_2 \rightarrow PAN$	$6 \text{ ppm}^{-1} \text{ min}^{-1}$
8.	$OH + NO \rightarrow HONO$	$10 \text{ ppm}^{-1} \text{ min}^{-1}$
9.	$OH + NO_2 \rightarrow HNO_3$	$30 \text{ ppm}^{-1} \text{ min}^{-1}$
10.	$O_3 + HC \rightarrow RO_2$	$0.0125 \text{ ppm}^{-1} \text{ min}^{-1}$
11.	$(H_2O +) NO + NO_2 \rightarrow 2 HONO$	$0.01 \text{ ppm}^{-1} \text{ min}^{-1}$
12.	$h\nu + HONO \rightarrow OH + NO$	0.001 min^{-1}

to the Carbon Bond Mechanism of Whitten et al. (1980) which is, however, more simple. An earlier version of the Carbon Bond Mechanism, in which some of the rate constants of elementary reactions were quite wrong, was used as submodel in a transport model which successfully simulated the formation of ozone in the Los Angeles Basin.

The merits and limitation of the Carbon Bond Mechanism, particularly its ability to model the effect of hydrocarbon composition on ozone formation at constant and variable light intensity, have been investigated on behalf of the Umweltbundesamt (Schurath et al. 1982, 1983).

It has become apparent in recent years that the oxidant problem is by no means confined to the severely polluted urban areas. The time-scale of oxidant formation and preservation in an air mass can amount to several days, during which time the precursors and oxidants are transported hundreds of kilometers. Lumped models which have been optimized for urban conditions are not suited for modeling formation and transport of oxidants in aging air masses (Demerjian 1977). Detailed mechanisms are more appropriate, but can only be used in conjunction with extremely simple moving box models (Isaksen et al. 1978b).

1.4.2.3. Isopleth Models

It was mentioned in Section 1.4.2.1 that the US Bureau of Mines undertook smog-chamber experiments in the 1960's to evaluate the relative impact of hydrocarbons and NO_x on oxidant formation (more precisely: ozone formation) in polluted air (Dimitriades 1970). Irradiations of variable amounts of car exhaust and additional NO_x ($= NO + NO_2$) in purified air were carried out at constant light intensity in an aluminium chamber of 2.8 m^3 volume. The ozone maxima were determined as function of the initial NO_x mixing ratio, and the initial nonmethane hydrocarbon (NMHC) mixing ratio in ppm C. When the ozone maxima were plotted as function of the independent variables $(NO_x)_o$ and $(NMHC)_o$ in a three-dimensional rectangular coordinate system, the experimental data stretched out an approximately smooth vaulted surface in space. When lines of equal ozone maxima on the surface were projected into the $(NO_x)_o$–$(NMHC)_o$ plane, a system of bent lines was obtained which were termed ozone isopleths.

The perceptible isopleth concept, which can readily be applied to smog-chamber results under a wide variety of conditions, has since been adopted by numerous authors (Becker et al. 1978, Akimoto et al. 1979a, Sakamaki et al. 1980, 1982, Bruckmann et al. 1980, Henrich et al. 1982).

In order to meet the Federal Air Quality Standard for Ozone, Dodge (1977) and others have devised a control strategy on the basis of ozone isopleths. Application of the strategy requires the computer program OZIPP (= Ozone Isopleth Plotting Package), which is disseminated by the USEPA (Whitten et al. 1978). An essential part of the program is a chemical model of photosmog reactions (a Seinfeld-type lumped model, Hecht et al. 1974). The model was validated by Dodge (1977) by comparison with the smog chamber results of Dimitriades (1970).

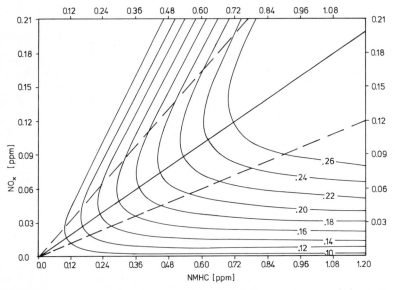

Fig. 1.33. City-specific ozone isopleths computed with OZIPP. The following upwind concentrations (concentrations above the mixing layer) were specified: NMHC 0.05 (0.005) ppm C; NO_x 0.01 (0.005) ppm; O_3 0.05 (0.08) ppm. Emissions in the model stopped after 0800 h, when the air mass was assumed to be transported out of the city area. Default values were used for all other input data. *Straight line* through the origin is drawn for a fixed NMHC:NO_x ratio of 6, which is typical for Bonn (Henrich et al. 1982). *Intermittent lines* delimit the approximate range of observed NMHC:NO_x ratios in Bonn. *Intersection point* of the straight line with the 0.20 ppm ozone isopleth suggests that a NMHC reduction at constant NO_x should be more efficient in Bonn than the same relative reduction in NO_x emissions at constant NMHC. This conclusion is correct only if (a) the chemical model in OZIPP fits the precursor reactions in the city plume of Bonn, (b) the boundary conditions for OZIPP were chosen correctly, (c) the ozone maximum is measured in an air mass which moved away from the urban source area at 0800 h. EPA emphasizes that OZIPP cannot predict absolute ozone concentrations correctly, its validity being confined to predicting *relative* efficiencies of NMHC and NO_x control measures with respect to reducing annual ozone maxima

Ozone isopleths are constructed by OZIPP on the basis of numerical integrations of the optimized chemical model. In order to apply the model to a specified urban area, a number of city-specific input data are required, such as the light-intensity profile, upwind precursor and oxidant concentrations, and the emission pattern of the city. Transport and dispersion are simulated in a simplified manner by a moving box, which drifts across the city and further downwind at a specified wind speed. The height of the box can also be varied, mimicking the diurnal variation in atmospheric mixing height (= dilution of the mixing layer with clean air from aloft).

A set of computed city-specific ozone isopleths is plotted in Fig. 1.33. The co-ordinates NMHC and NO_x denote approximate early morning concentrations. More important than absolute concentrations is the city-specific *emission ratio* NMHC : NO_x, which defines a straight line through the origin of the plot. The intersection point of the straight line with the ozone isopleth corresponding to the actually observed ozone maximum in the city defines the starting point of potential control measures. The effect of *relative* reductions of NMHC and/or NO_x emissions can be readily read off the diagram.

Control strategies based on OZIPP are subject to a number of limitations. As the authors of the model (Whitten et al. 1978) have pointed out themselves, OZIPP can only be expected to work within and some short distance downwind of a city. The important problem of rural ozone is beyond the scope of OZIPP. It should also be kept in mind that the lumped chemical model in OZIPP was validated only for a limited concentration range of NMHC and NO_x, and need not be valid outside this range. Comparison with a detailed chemical model has in fact revealed a number of short-comings of the lumped mechanism in OZIPP (Hov and Derwent 1980). Very recently (Dunker et al. 1984) ozone isopleths have been calculated with four currently used chemical mechanisms, employing the same meteorological conditions and representation of the pollutant mix for all mechanisms. The four mechanisms were found to differ substantially in their predictions of maximum hourly average ozone.

1.4.3 Effects of Meteorological Parameters on the Formation of Oxidants

Elevated ozone concentrations at ground level may be due either to intrusions of stratospheric air or to photochemical reactions of anthropogenically emitted precursors. Stratospheric intrusions, which may be caused by vertical circulation within a high reaching cumulonimbus cloud or by the sinking of stratospheric air at a cold front (Johnson and Viezee 1981) result, in the first case, in short (½ h duration) and locally limited ozone peak values of several hundred ppb (Attmannspacher 1976), and in the second case, in elevated ozone concentrations distributed over several hours or days in the meso-scale range (Derwent et al. 1978). Airborne measurements by Johnson and Viezee (1981) revealed that stratospheric ozone may often enter the upper troposphere in the spring and autumn, whereby concentrations of over 200 $\mu g \, m^{-3}$ (100 ppb) have been registered at an altitude of approximately 5 km. At altitudes below 1,000 m above sea level, however, the O_3 content usually decreased to below 80 $\mu g \, m^{-3}$. In a few spring-time episodes, a direct ozone intrusion from the stratosphere down to ground level occurred behind a cold front.

Clearly distinguishable from these natural episodes are such conditions under which ozone is formed and transported as a result of anthropogenic emissions of precursors. Based on case studies, Guicherit and van Dop (1977) reveal the for-mation and occurrence of high ozone concentrations in large regions of Western Europe in the area of quasi-stationary, warm high pressure areas. Similar reports have been compiled from results in the USA (Altshuller 1978), Canada (Mukam-mal et al. 1982) and Norway (Schjoldager 1979). Altshuller comments especially on the varying frequency of the occurrence of such large-scale weather situations in different years, which results in respective fluctuations in the frequency of el-evated ozone concentrations.

Muschalik (1980) studies the relationship between ozone concentration and the large-scale weather classes (Großwetterlagen, after Hess and Brezowsky 1977) to describe the air pressure distribution over Europe, for several locations in the Federal Republic of Germany. It was found that the following weather classes oc-curred in connection with high ozone concentrations.

High pressure over Central Europa and high pressure over the Arctic Ocean and Iceland with anticyclonal isobaric curvature over Central Europe. These con-ditions are characterized by stable layering, high radiation, high temperatures, and poor mixing.

During such situations, elevated ozone concentrations occur in an area of sev-eral hundred kilometers. The anthropogenic influence is especially well detected leeward of congested areas or intense sources of precursors, since concentration peaks several tens of kilometers wide are superimposed on the background ozone (Fricke 1980).

Elevated ozone concentrations also occur under conditions of high pressure over Scandinavia and Finland anticyclonal, high pressure bridge over Central Eu-rope, northeast anticyclonal situation and high pressure over the North Sea and Fennoscandia cyclonal, although not so often and not as high as in the first two large-scale weather situations. Other situations are associated with precipitation and more intense mixing and, therefore, do not result in such high ozone concen-trations (Muschalik 1980).

The relationship between ozone concentration and meteorological parameters was also examined with various statistical methods. Analyses for the Federal Re-public of Germany have been presented by Muschalik (1980) and by Bruckmann and Langensiepen (1981).

Muschalik determined at several locations that a significant positive correla-tion exists between the ozone concentration at 1400 to 1430 h Central European Time and the global radiation, the duration of sunshine, and the temperature. A negative correlation was found between the ozone concentration and the air velocity at the 850 mbar level (approx. 1,500 m above sea level). She also found a significant negative correlation between the ozone concentration and the hu-midity after exclusion of the influence of temperature. After exclusion of the influence of radiation, only the temperature was found to have a significant effect on the ozone concentration.

Bruckmann and Langensiepen (1981) related ozone data from the summer months (May to September) of the years 1976–1979 from stations in Essen, Cologne, and Bonn with the daily temperature maximum, the square of the temperature maximum, the duration of sunshine and the air velocity near the ground to calculate multiple regression equations. Just as had been found in the principal component analysis, which had also been carried out, the temperature was proved to be a significant parameter influencing the ozone concentration, followed by the duration of sunshine and the wind velocity. The latter demonstrated only a weak (negative) correlation with ozone, since their values at ground level at 0900 h are not adequately representative, neither in time nor space. The multiple correlation coefficients were between 0.80 and 0.86; i.e., the variance of the ozone maxima, which can only be explained by the above meteorological parameters, was between 64 and 74%. This agrees with the results of Bruntz et al. (1974), who found a multiple correlation coefficient of 0.84 for the same parameters in New York.

Stewart et al. (1976) correlated ozone measurements in London with the maximum temperature, radiation, and relative humidity, using the logarithms of these parameters. The multiple correlation coefficient, which was calculated as 0.62, was improved to 0.71 by inclusion of the ozone concentration from the previous day. As is the case with Bruckmann and Langensiepen (1981), high ozone values are underestimated, since the regression equation cannot distinguish between ozone which has been formed locally and that which has been transported from other regions. If these data are to be used for the development of a statistical model for the prediction of ozone concentrations, especially for the prognosis of transgressions over certain threshold values, further parameters will need to be considered.

After preliminary work by Tiao et al. (1976), a predictive model with over 30 variables – mainly from the free atmosphere up to the 500 mbar level – was developed for the Los Angeles Basin (Aron and Aron 1978). The best correlation with the daily ozone maximum was with temperature at the 900 and 850 mbar level and the ozone maximum of the previous day. Lin (1982) developed a statistical model based on discriminant analysis especially to predict transgressions of the smog warning limits of 200 and 350 ppb in the Los Angeles Basin. Depending on the location, it yields correct prognoses in 65 to 88% of the cases for the transgression of the 200 ppb threshold, and 51 to 80% of the cases for the transgression of the 350 ppb warning threshold.

A model has also been designed for the St. Louis area (Karl 1979), using ozone data from 25 stations from April to October of the years 1975–1976. The air velocity and the temperature in the 850 mbar level were found to be important data for the 48-h prediction. For the 24-h prediction they were found to be the temperature in the 850 and 1,000 mbar level, the potential temperature and the relative humidity in the boundary layer and the ozone maximum of the previous day.

Wolff and Lioy (1978) developed a model for the northeastern USA using a stepwise regression analysis in which they attained a correlation coefficient of 0.96. This was achieved with the inclusion of the ozone and temperature maxima upwind from the previous day, the temperature maximum on the actual day and the mean air velocity up to 1,000 m above ground level, as well as the NO_x and hydrocarbon emissions along the trajectory.

It should be noted that all such models allow no statements on cause and effect relationships. They only hold true for the location for which they were developed, and even then, only for unchanged conditions.

1.4.4 Dispersion and Transport Modeling

If one whishes to find a quantitative relationship between precursor emissions and ambient air concentrations of oxidants, chemical reactions, as well as transport and dispersion of the precursors, intermediates, and products must be taken into account at the same time. Chemical reactions have been presented separately in Section 1.4.2; the influence of meteorological parameters on oxidant formation was discussed in Section 1.4.3.

1.4.4.1 Empirical Approach: Maximum Ozone Isopleths

When a large number of daily ozone maxima are plotted versus the corresponding early morning concentrations of NO_x and NMHC in a certain area, a cloud of data points is obtained. The upper envelope of all observations, preferably from many years, delimits maximum ozone concentrations which are unlikely to be exceeded in that area. Such "maximum ozone isopleths" also yield information on the relative impact of NO_x and NMHC emissions on ozone formation. It is an advantage of the "maximum ozone isopleth" concept that area-specific factors, such as the emission pattern, orography, and small-scale meteorology, are fully taken into account. It is a disadvantage of the concept that the isopleths are valid only for a specific area. Furthermore, the number of measuring stations and observations needed to construct statistically significant "maximum ozone isopleths" is rather large.

1.4.4.2 Wind Tunnel Modeling

Formation and transport of oxidants have not yet been studied in wind tunnels, with the sole exception of recent work by Builtjes (1981), who studied the dispersion and oxidation of NO from a point source in ozone-containing air in a wind tunnel. It seems impossible to simulate the complex chemical reactions leading to oxidant formation in the polluted sunlit atmosphere in a wind tunnel. This is because the chemical and photochemical reactions involved in oxidant formation and destruction in the atmosphere cannot be held in realistic proportions, and at the same time be accelerated (e. g., by increasing the precursor concentrations and the light intensity by some appropriate factors), in order to match the much shorter time-scale available for oxidant formation in a wind tunnel.

1.4.4.3 Meteorological Dispersion Models

In most of these models the rate of change in the concentrations c_i of species i is described by the mass conservation equation in the following simplified form:

$$\overbrace{\frac{\delta \bar{c}_i}{\delta t} + \bar{u}\frac{\delta \bar{c}_i}{\delta x} + \bar{v}\frac{\delta \bar{c}_i}{\delta y} + \bar{w}\frac{\delta \bar{c}_i}{\delta z}}^{\text{Transport}}$$

$$= \underbrace{\frac{\delta}{\delta x}\left(K_h\frac{\delta \bar{c}_i}{\delta x}\right) + \frac{\delta}{\delta y}\left(K_h\frac{\delta \bar{c}_i}{\delta y}\right) + \frac{\delta}{\delta z}\left(K_z\frac{\delta \bar{c}_i}{\delta z}\right)}_{\text{Turbulent diffusion}} + R_i + D_i$$

$\bar{c}_i$ = mean concentration of species i at some point in space;
$\bar{u}, \bar{v}, \bar{w}$ = x-, y-, and z-components of the mean wind vector;
K_h, K_z = horizontal and vertical eddy diffusion coefficients;
R_i = net rate of change of concentration $\bar{c}_i$ due to chemical and photochemical reactions;
D_i = net rate of change of concentration $\bar{c}_i$ due to physical processes (wet and dry deposition, emission).

The mass conservation equations, one for each species i, are coupled partial differential equations which must be solved simultaneously. The coupling is due to the chemical reaction term, R_i, which also contains concentrations $\bar{c}_j$ of species $j \neq i$.

The stationary solution of an extremely simplified version of the mass conservation equation is known as the Gaussian plume model for point sources. It neglects the chemical coupling terms R_i, and is thus only valid for chemically inert pollutants. Gaussian models cannot be used for oxidants which are subject to strong chemical coupling.

Integration of the mass conservation equation for reactive pollutants ($R_i \neq 0$) is only feasible by numerical methods. However, an unrestricted integration of the mass conservation equation is beyond the capabilities of present-day computers. Therefore several simplified schemes have been devised for practical situations. In the simplest of all possible schemes, the concentration gradients and transport terms of species i are assumed to be zero inside a large volume ("box") of air. The result is a purely chemical model which can be integrated by suitable numerical methods (cf. Sect. 1.4.2.2). Important offspring of the so-called box model are the Eulerian or multibox models. Lagrangian or trajectory models assume that the well-mixed box is not fixed in space, but moves along a trajectory traced out by the mean wind. Eulerian and Lagrangian elements have been amalgamated in the particle-in-a-box method (Sklarew et al. 1972, Seinfeld et al. 1972).

Each particular modeling scheme yields a set of differential equations which must be integrated on a computer. This is usually done by a method of small differences (Reynolds 1977). For example, the purely chemical rate R_i of a species i is given by (cf. Sect. 1.4.2.2).

$$R_i = \sum_{l,m} k_{lm} \cdot c_l \cdot c_m - \sum_j k_{ij} \cdot c_i \cdot c_j.$$

Unless the concentrations c_l, c_m, c_i, c_j vary too rapidly, the chemical rates R_i may be held constant over suitable time intervals Δt (Röth 1981), during which the overall changes in the concentrations c_i, including transport and turbulent diffusion, are evaluated numerically. Computer time is saved by introducing the steady-state approximation $R_i = 0$ for atoms, radicals, and possibly other short-lived species (Hesstvedt et al. 1978). Further savings in computer time are possible by lumping of compounds (cf. "lumped mechanisms" in Sect. 1.4.2.2), particularly of hydrocarbons, into a few symbolic species (Reynolds 1977; Eschenroeder and Martinez 1972; Schere and Demerjian 1978).

In order to calculate concentration changes by transport and turbulent diffusion, both the wind vector (u,v,w) and the eddy diffusion coefficients K_h and K_z are required. These are often not available, and thus must be evaluated beforehand by means of a meteorological submodel which, however, requires other specific input data (Tangermann-Dlugi and Fiedler 1979).

Emission rates of the precursors which are essential for modeling photochemical pollution may be deduced from emission inventories. However, the emission rates thus obtained are annual averages which may be inadequate for calculations of specific summer episodes.

Some facets of the most widely used modeling schemes are briefly discussed in the following sections.

1.4.4.4 Box Models

The model of a well-mixed box-shaped volume of air of height H is an acceptable approximation for an urban atmosphere devoid of prominent point sources. The spatial resolution of the model varies with the size of the box. No eddy diffusion coefficients are required, and the concentrations c_i of the species i are assumed to vary according to an extremely simple mass conservation equation:

$$\frac{dc_i}{dt} = \frac{(c_i' - c_i)}{H}\frac{dH}{dt} - (c_i - c_i') \cdot \frac{\bar{u}}{L} + R_i;$$

$H = $ height of the box ($=$depth of the mixing layer which is allowed to vary as function of time);
$L = $ length of the box in wind direction;
$\bar{u} = $ mean wind speed;
$c_i' = $ concentration of species i outside the box, upwind and aloft.

The utter simplicity of the box model allows for a rather detailed treatment of atmospheric chemistry in the chemical coupling terms R_i.

A box model has been fairly successfully applied to the Los Angeles Basin by Hanna (1973). Venkatram (1978), who validated a similar model against field measurements, concludes that box models are of rather limited value for modeling oxidants, in contrast to Schere and Demerjian (1978), who recommend that box models should be used more extensively for that purpose. A box model has also been tested in the Federal Republic of Germany, in a study of oxidant formation in the Cologne/Bonn area (Stern and Scherer 1980; Scherer and Stern

1982). Atmospheric chemistry was simulated in the model by means of EKMA, which is recommended by EPA for ozone isopleth calculations (cf. Sect. 1.4.2.3).

Derwent and Hov (1980) assumed a box size of 450 km × 360 km × H in a simulation of photochemical smog formation in Southern England. Three hundred chemical reactions plus dry deposition were included in a 7-day simulation of O_3, PAN, HNO_3, and other pollutant concentrations.

1.4.4.5 Eulerian Models

A Eulerian or grid model of the atmosphere above an area is obtained by subsecting the mixing layer into a grid of rectangular boxes, several layers on top of each other. The mass conservation equation has to be solved for each box separately, as outlined in Section 1.4.4.3. Transport of species between boxes is effected by the wind field and by eddy diffusion.

Grid models are well suited for calculating the concentration fields of precursors and oxidants in some detail, provided that the resolution of the available input parameters needed for the calculation (wind field, source distribution, eddy diffusion coefficients as function of altitude) is commensurate with the horizontal and vertical grid size. The grid models reported in the literature (Reynolds et al. 1973; Roth et al. 1974; Reynolds et al. 1974; McCracken et al. 1978; Schiavone and Graedel 1981) differ mainly by grid size, meteorological submodel of mixing layer dynamics, and chemical reaction mechanism. At present one of the most flexible grid models is the SAI model of Reynolds et al. (1976). It has recently been used in the Netherlands (Builtjes 1980). The SAI model requires rather detailed input parameters, e. g., a detailed knowledge of the wind field and source distribution.

1.4.4.6 Lagrangian Models

Trajectory or Lagrange models consider a vertical column of air of specified dimensions. The air column moves along a trajectory across the area of interest. Inside the column, the mass conservation equation has to be solved, taking into account sources and sinks at the ground, chemical reactions, and vertical (but no horizontal) transport by eddy diffusion, while horizontal transport by the wind field is eliminated by definition (Eschenroeder and Martinez 1972; Wayne et al. 1973; Hov et al. 1978a). It is one of the disadvantages of trajectory models that they do not yield horizontal distributions of trace species in an area. The number and detail of input parameters needed for trajectory models is less than for grid models.

Tilden and Seinfeld (1982) investigated the sensitivity of a photochemical trajectory model for ozone to uncertainties in input parameters, allowing for variations of $\pm 50\%$ in each parameter. They found that the calculated mixing ratios of O_3, NO_x, and NMHC depended only weakly on vertical diffusion coefficients, deposition velocities, and relative humidity, while uncertainties in light intensity, initial conditions, and source strengths had a stronger impact on the calculated results.

1.4.4.7 Modeling Ozone Formation in Power Plant Plumes

Recently, ozone formation in long-range transported power plant plumes has become a matter of concern. Ozone in aged plumes is taken as an indicator of photochemical reactivity, which results in the formation of H_2SO_4 and HNO_3 via OH radical reactions in the gas phase. In order to study photochemistry in an expanding plume, Bottenheim and Strausz (1982) have modified a Gaussian plume model (neglecting reflection of the plume at the ground) in the following way: the plume is subdivided into concentric elliptical rings. The mass conservation equation of reactive species, which is based on a detailed chemical reaction mechanism and includes time-dependent eddy parameters, is solved numerically for each elliptical ring, assuming that turbulent transport is effective only between adjacent rings. In a similar model of Meagher and Luria (1982), eddy diffusion coefficients for turbulent transport between adjacent rings were derived from σ-parameters after Pasquill, and chemistry was modeled by the Carbon Bond Mechanism (cf. Sect. 1.4.2.2).

Chapter 1.5   Surveillance of Ambient Air Quality

The relationship between the ambient air concentrations of precursors and oxidants, the analysis of long-term trends in oxidant concentrations, and the prognosis of short-term episodes can be established only after measuring these trace constituents over sufficiently long periods of time. The choice of suitable analytical techniques, the planning of measuring sites and measuring periods, and methods of data evaluation, will be discussed in this section.

1.5.1 Analytical Techniques

This subsection deals with routine procedures for ambient air measurements of precursor and oxidant concentrations at justifiable expense of staff and technical equipment. Furthermore, some analytical techniques for detailed field investigations, or for random sampling, will be briefly described. These methods are often in the stage of research and development, or find only limited application as routine methods for other practical reasons. Unfortunately, standardized analytical methods for most of the intermediates and some of the final products of photochemical reactions in the atmosphere are not yet available.

1.5.1.1 Determination of Precursors in Ambient Air

1.5.1.1.1 Nitrogen Oxides

NO is most widely measured continuously by means of chemiluminescence analyzers which are based on the chemiluminescent reaction of NO with excess ozone (Fontijn et al. 1970). Ozone is admixed to ambient air which is pumped through a light-tight reaction chamber in front of a red-sensitive photomultiplier. Some of the resulting NO_2 is formed in an electronically excited state, which may be either quenched by collisions, or – depending on pressure – emit a photon in the red to near infrared spectral region. At constant reduced pressure, the light intensity is proportional to the NO mixing ratio in the air. The method can be modified to measure total $NO_x = NO + NO_2$ and, by taking the difference of two signals, NO and NO_2 separately. For the purpose, NO_2 is reduced to NO by passing the air through a thermal or catalytic converter. Commercial NO/NO_2 analyzers use either a single reaction chamber which measures NO and NO_x intermittently (see Fig. 1.34 and VDI-Richtlinie 2453, part 6), or two photomultipliers and reaction

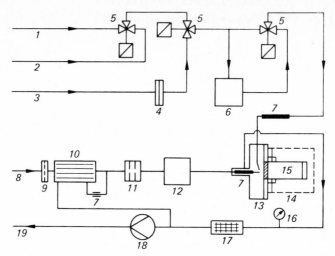

Fig. 1.34. Gas flow diagram of a chemiluminescence analyzer for NO_x; single channel method. (VDI-Richtlinie 2453, part 6). *1* zero gas input; *2* calibration gas input; *3* air sample input; *4* dust filter; *5* magnetic valve; *6* converter; *7* capillary; *8* indoor air input; *9* dust filter; *10* permeation dryer; *11* humidity indicator; *12* ozone generator; *13* reaction chamber; *14* detector cell; *15* photomultiplier; *16* pressure meter; *17* activated charcoal filter; *18* vacuum pump; *19* exhaust gas exit

chambers, only one of which is preceded by a converter. The detection limit of commercial instruments for NO and NO_2 is typically 3 to 5 ppb. The relative standard deviation is 1 to 2%. Some of the thermal converters oxidize NH_3 to NO. The resulting interference is of no practical importance, since the $NO_x : NH_3$ ratio is substantially higher than 10:1 under normal atmospheric conditions. Positive interference by other convertible nitrogen compounds (PAN, HONO, HNO_3), or negative interference by strong quenchers of the chemiluminescence, may become more serious under extreme conditions, but do not pose a problem in routine applications.

NO_2 in ambient air can also be determined by continuous absorption in buffered KI solution: NO_2 releases molecular iodine, which is measured by continuous coulometry in a suitable electrochemical cell. Other oxidizing (e. g., O_3, PAN) and reducing (e. g., SO_2, H_2S) compounds cause interference, and must be removed from the sampled air by means of scrubbers, which cannot be achieved without difficulty. The Beckmann coulometric NO_2 analyzer has a detection limit of 4 ppb. NO can also be detected after oxidation, e. g., by ozone.

The well-known Saltzman-Method (Saltzman 1954; VDI-Richtlinie 2453, part 1) has also been adapted for automated continuous NO_2 measurements in ambient air. NO_2 is absorbed in a solution of sulfanilic acid and α-naphthylamine. The resulting violet azo dye is determined by colorimetry. The detection limit of commercial analyzers (WACO Ltd., Precision Scientific Ltd., Technicon) amounts to several ppb.

Long-path optical absorption in the UV, which has been used to measure NO_2 and several other pollutants in the atmosphere in special investigations (see Chap. 1.3.2.2.3), is not suited for routine measurements. However, correlation spectrometry (Barringer Research) has been applied for remote sensing of NO_2 column densities in power plant plumes (e. g., Beilke et al. 1982).

1.5.1.1.2 Hydrocarbons

Quasi-continuous hydrocarbon measurements are usually carried out by automatic flame ionization detection. When used without separation column, the flame ionization detector (FID) measures total hydrocarbons, including methane in ppm-C units since, under suitable conditions, the FID signal is proportional to the number of carbon atoms contained in the air sample in the form of hydrocarbons. The sum of the nonmethane hydrocarbons (NMHC) can be measured after separation from methane on a pre-column (VDI-Richtlinie 3483, parts 1–3). In practical applications care must be taken to maintain equal sensitivities for different hydrocarbons on a C-atom basis, and to eliminate water vapor effects. Automatic separation and analysis of individual alkanes up to C_8, alkenes up to C_5, and aromatic hydrocarbons up to C_9, may be achieved with more refined GC-analyzers (e. g., Siemens U 180), using cryogenic pre-concentration columns, as shown schematically in Fig. 1.35. The detection limit is in the lower ppb range. Recently, GC-MS separation, identification, and determination of atmospheric hydrocarbons has become feasible with extreme sensitivity.

Atmospheric hydrocarbons can also be enriched on activated charcoal, followed by liquid extraction and analysis of the eluate by thin layer chromatogra-

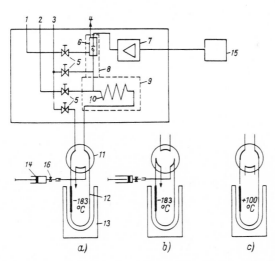

Fig. 1.35 a–c. Schematic diagram of a gas chromatograph for the determination of hydrocarbons. (VDI-Richtlinie 3782, part 2). **a** intake valve positioned for "flushing"; **b** intake valve positioned for "concentration"; **c** intake valve positioned for "intake". *1* detector flush gas input; *2* fuel gas input (H_2); *3* carrier gas input (N_2); *4* detector exhaust; *5* regulating valve for constant gas flow; *6* flame ionization detector; *7* electrometer amplifier; *8* detector thermostat; *9* column thermostat; *10* separation column; *11* six-way intake valve for sample; *12* precolumn for concentration; *13* cooling/heating unit; *14* gas-tight syringe; *15* registration unit; *16* micro valve

phy (VDI-Richtlinie 3482, part 4). The method is particularly suited for high molecular weight compounds. Nondispersive infrared (NDIR) detection of hydrocarbons is not sensitive enough for ambient measurements.

1.5.1.1.3 Aldehydes

The sum concentration of aliphatic aldehydes in ambient air may be determined by absorption in a solution of 3-methyl-benzothiazolinonhydrazon hy-
drochloride in the presence of $FeCl_3$, and subsequent photometry of the resulting
blue coloration. The method also detects aromatic aldehydes to some extent
(Sawicki et al. 1961). Specific colorimetric detection of formaldehyde is achieved
by absorption of the compound in a sulfuric acid solution of chromotropic acid
(Altshuller et al. 1961).

Another specific colorimetric technique for the determination of formaldehyde in air uses a solution of pararosaniline. SO_2, which causes negative interference, must be removed by absorption in tetrachloromercurate solution (VDI-
Richtlinie 3484, part 1). Detection limits of less than 0.1 ppb formaldehyde in air
have recently been achieved by Schmidt and Lowe (1982), and Neitzert and Seiler
(1981), who used an aqueous solution of 2,4-dinitrophenyl hydrazine and special
trapping techniques. The hydrazone can be extracted and determined by HPLC.
HPLC separation of the hydrazones is also suitable for the detection of individual
aliphatic and aromatic aldehydes in polluted air (Grosjean 1982).

1.5.1.2 Determination of Oxidants

Routine methods for oxidant measurements in ambient air are available only
for NO_2 (see Sect. 1.5.1.1.1), ozone, and PAN. The latter two will be discussed
in the following subsections. Recent attempts at measuring atmospheric concentrations of gaseous H_2O_2 have been dealt with in Chapter 1.3.2.2.3. Several other
oxidants (organic peroxides, peracids, nitrates) have been detected in absorption
by long-path Fourier transform infrared spectrometry, mostly in smog-chamber
studies, but cannot normally be detected in the atmosphere.

1.5.1.2.1 Ozone

The most important instrumental technique for measuring ozone in ambient
air is based on the chemiluminescent reaction of ozone with an excess of ethene,
which is well suited for continuous detection. Chemiluminescence is due to formaldehyde in an electronically excited state. Emission occurs in the visible and
near-UV spectral region, where extremely sensitive photomultipliers are available. The detection limit of commercial chemiluminescence analyzers for ozone
is less than 1 ppb. No interference by other atmospheric constituents has been reported.

Another chemiluminescence technique for ozone, which does not require a
gaseous reactant, is based on the chemiluminescent surface reaction of ozone with

rhodamine B adsorbed on silica gel (Regener 1964). The poor stability of the solid reactant has been overcome by adding gallic acid (Guicherit 1971). The strong negative interference of water vapor could be eliminated by hydrophobization of the solid surface with a silicone resin coating (Hodgeson 1972). The detection limit of the method is below 1 ppb.

Chemiluminescence analyzers have to be calibrated at regular intervals, using either a standardized KI method (VDI-Richtlinie 2468, part 1), or the UV photometric technique (VDI-Richtlinie 2468, part 6).

The UV photometric method, which measures ozone by optical absorption of monochromatic mercury lamp emission at 254 nm, has also been adapted for field measurements with a commercial instrument (Dasibi). The detection limit can be as low as 1 ppb.

A wet chemical technique for the determination of ozone is based on the decoloration of indigosulfonic acid by ozone. The detection limit is approximately 5 ppb. Interference by NO_2 is not very serious (VDI-Richtlinie 2468, part 5).

The most widely used wet chemical method for ozone measurements in the laboratory and in very clean air uses a buffered aqueous potassium iodide solution which releases molecular iodine in stoichiometric amounts. The iodine may be measured either colorimetrically or coulometrically (VDI-Richtlinie 2468, part 3). The stoichiometry of the reaction is strongly dependent on the pH of the solution. The method suffers from positive interference by other oxidants (PAN, NO_2, H_2O_2), and negative interference by reducing compounds (particularly SO_2). These interferences cannot be satisfactorily eliminated by pre-filters. The KI method is therefore said to measure "net" oxidant, which is often much less than the true ozone concentration in polluted air.

1.5.1.2.2 Peroxyacetyl Nitrate (PAN)

Peroxyacetyl nitrate (PAN) and peroxypropionyl nitrate (PPN) in ambient air can be determined after gas chromatographic separation with an electron capture detector, which is highly specific for these compounds and for chlorinated hydrocarbons. Air samples of 1–5 ml are directly injected on a suitable short column (VDI-Richtlinie 2468, part 7), either manually with a gas-tight syringe, or with an automatic sampling system. Contact of the samples with warm or metallic surfaces must be avoided to prevent thermal decomposition of the unstable compounds (see Chap. 1.2.2.2). Retention times at ambient or slightly above ambient column temperatures are in the order of 2–5 min. The detection limit is about 0.1 ppb.

Calibration methods based on pure PAN are not feasible, since the pure liquid compound is highly explosive. Gaseous PAN is prepared by photolysis of ethyl nitrite vapor in oxygen, or by other suitable methods (see Chap. 1.2.2.2). Several hundred ppm PAN in nitrogen can be determined by IR spectrometry using published absorption coefficients (Bruckmann and Willner 1983). The sample gas must be further diluted to be suitable for calibrating the GC-ECD, which has a limited range of linear response (VDI-Richtlinie 2468, part 8).

Very recently, cryotrapping techniques for field and airborne sampling of PAN have been developed, as well as novel calibration techniques which extend to much lower concentrations (Glavas and Schurath 1983; Schurath et al. 1984; Meyrahn et al. 1984).

1.5.1.2.3 Nitric Acid, Nitrous Acid, and Aerosol Nitrate

The sum concentration of gaseous nitric acid (HNO_3) and particulate nitrate can be measured colorimetrically after absorption in aqueous solution. Gas diffusion denuders have been used to selectively remove gaseous HNO_3, while collecting aerosol nitrate on a filter. Shaw et al. (1982) have extracted the nitrate collected on the filter with benzene, and determined the resulting nitrobenzene by GC-ECD. Filters impregnated with NaCl collect both gaseous nitric acid and aerosol nitrate. Gaseous nitric acid may thus be measured by difference, using two identical impregnated filters in parallel, one without and the other with HNO_3 denuder (Forrest et al. 1982).

Some NO_2 converters which are used in conjunction with chemiluminescence analyzers for NO_x have been found to convert nitric acid (Joseph and Spicer 1978). Several authors have used this fact to measure HNO_3 by difference, using a nylon filter cartridge to remove gaseous HNO_3 selectively on one channel. Kelly (1979) employed hot glass beads at 350° to convert gaseous HNO_3, and $FeSO_4$ to convert NO_2. An intercomparison of the chemiluminescence, infrared, filter, and diffusion denuder techniques for nitric acid detection under field conditions has been reported by Spicer et al. (1982).

Although several laboratory studies of gaseous nitrous acid (HONO) have been carried out, employing thermal NO_x converters, selective scrubbers, and chemiluminescence analyzers to measure HONO by difference (Cox and Derwent 1976, 1977; Sanhueza et al. 1984), very few field measurements of the compound have been reported. The only available data (see Chap. 1.3.2.2.3 for references) were obtained by long-path differential UV absorption.

1.5.2 Monitoring Policy

1.5.2.1 Key Oxidants and Other Pollutants Which Should Be Monitored

The term key oxidant applies to a compound which, by its chemical properties, ambient concentrations, and environmental effects, is highly representative of the entire class of photochemical oxidants, as defined in Chapter 1.2. It is required that a key oxidant can be measured with high reliability on a routine basis.

To date, the most abundant photochemical pollutant, ozone, is commonly regarded as key oxidant. It is highly correlated with PAN in the atmosphere (see e. g., Bos et al. 1978), and is also largely held responsible for environmental effects of photochemical oxidants. Furthermore, automated equipment for continuous monitoring of ozone is readily available.

PAN is known to cause specific plant injury. It is therefore recommended, under certain conditions, to measure this compound also (see Sect. 1.5.2.2).

In order to elucidate in a quantitative way the interdependence between oxidant formation and precursor concentrations in the atmosphere, important precursors, particularly NO, NO_2, and C_2–C_8 hydrocarbons (alkanes, alkenes, aromatic compounds) should be measured continuously. Continuous measurements of aldehydes, ketones, and a few other reactive organics would be desirable, but cannot be carried out on a routine basis with commercially available equipment.

The identification of specific precursor sources by ambient air measurements is feasible only on the basis of detailed precursor analyses. Measurements by FID of the total hydrocarbon load including methane are of little diagnostic value. Even bulk monitoring of the nonmethane hydrocarbons (NMHC) only, although preferable to total hydrocarbon monitoring, cannot convey a reliable measure of the oxidant-forming potential.

1.5.2.2 Monitoring Sites and Monitoring Periods

Because of their *regional* occurrence, measurements of oxidants and their precursors should be as representative of a larger area as possible. Monitoring in the vicinity of prominent precursor sources, such as heavy traffic roads, and at sites which are subject to unusual micro-scale meteorology, should thus be avoided.

Ozone measurements in street-canyons, near highways, and downwind of power plants are falsified by the rapid reaction of ozone with NO. Heterogeneous ozone losses on solid surfaces can also cause unduly low ozone measurements which are not representative of the area, if the ozone monitor is improperly positioned, either too close to the ground, or next to tall buildings. In summary, monitoring sites should be selected at an adequate distance from strong precursor sources, the wind field in luff of the monitors should be unimpeded by tall buildings etc., and both ozone and precursors should be measured at equal heights.

For special diagnostic purposes, vertical concentration profiles of oxidants and precursors are needed up to the top of the mixing layer.

The assessment of oxidant formation on a regional scale requires data from several representative monitoring sites.

Monitoring of ozone can be restricted to the time period May–September. Outside this range, the probability of high ozone levels is negligible.

In view of recent speculations that increasing oxidant levels ensuing from medium- to long-range transport of oxidants and precursors from the source areas might enhance forest decline due to "acid rain" in remote rural areas, long-term measurements should be carried out in these areas.

1.5.2.3 Data Evaluation

Data evaluation should be aimed at two major issues: (1) establishment of long-term trends, and (2) development and testing of short-term prognosis schemes. Half-hour mean values of ozone, NO, NO_2, and individual hydrocarbon concentrations are suitable for statistical treatment. Shorter averaging times are not required in stationary monitoring programs. As distinct from other harmful pollutants, annual averages of ozone and other oxidant concentrations are not very meaningful, since the formation of photochemical oxidants occurs predominantly during times of intense solar irradiation, mainly in the 5-month period between May and September. This is the recommended monitoring period for both, oxidants and precursors, from which the following key data should be extracted:

1. The maximum half-hour mean values between May and September,
2. a high percentile value (e. g., the 95 percentile) of the half-hour mean values,
3. daily means of oxidant concentrations, calculated from the half-hour mean values between 0800 and 1800 h local time.

Chapter 1.6 Abatement Strategies

Since abatement strategies for secondary pollutants must necessarily be based on precursor emission control measures, quantitative correlations have to be established (1) between the ambient concentrations of oxidants and various precursors, and (2) between ambient levels and source strengths of the precursors. Emission control measures can then be taken against hydrocarbons and/or nitrogen oxides, depending on the established correlations.

An established method for obtaining the required relationships uses the isopleth concept, which correlates field or smog-chamber measurements of initial nitrogen oxides and nonmethane hydrocarbon concentrations with observed ozone maxima, as outlined in Chapter 1.4.2.3.

The correlation between source strengths and ambient concentrations of precursors have been discussed in Chapter 1.4.4.

To date, both smog-chamber experiments and model calculations have been found to provide insufficient data for quantitative recommendations, i. e., stating in absolute numbers the reductions in NO_x and/or NMHC emissions required for specified reductions in oxidant levels. In spite of this fact, numerous abatement schemes have been proposed in the literature from which recommendations for precursor emission reductions may be derived:

1. Assumption of linear correlation between ozone maxima
and NMHC emissions

This scheme, which requires one-sided nonmethane hydrocarbon reductions only, has proven useful in the heavily polluted basins of California. It fails under differing climatic conditions, particularly in suburban and rural areas.

2. Statistical correlation of ozone maxima, precursor concentrations,
and meteorological parameters

This method provides a nonlinear relationship between the highest tolerable ozone maximum and the pertinent reduction of hydrocarbon emissions, in percent. Nitrogen oxide emissions are not considered. The correlation is derived from early morning hydrocarbon concentrations in the atmosphere (0600 to 0900 h), and is valied only for the respective city (Appendix J Method, US EPA, US Federal Register 1971). The method predicts larger required reductions in hydrocarbon emissions than does scheme 1.

Abatement strategies based on statistical analysis of ozone and precursor concentrations can be further refined by inclusion of meteorological observables (temperature maximum, solar irradiation, wind velocity) in the analysis (Georgii and Muschalik 1980). The prognostic capabilities of statistical models including meteorological observables are only valid for the respective area (Wolff and Lioy 1978), and require input data from long-term measurements extending over several years.

3. Correlations between ozone maxima, hydrocarbons, and nitrogen oxides

Such correlations are usually visualized in the form of isopleth diagrams. In this way ambient NMHC and NO_x concentrations from 0600–0900 h can be correlated with the highest hourly mean of the ozone concentration (see US EPA 1977 b and 1981; Dodge 1977). Kinosian (1982) proposed a linear analytical expression relating the ozone maximum with the geometric mean of the precursor concentrations, which is, however, valid only within certain concentration limits.

Numerous ozone isopleth diagrams based on smog-chamber experiments and/or kinetic model calculations, have been published (Dodge 1977; Dimitriades 1972; Bruckmann et al. 1980; Guicherit et al. 1978; Yoshida et al. 1977; Becker et al. 1978; Akimoto et al. 1979 a; Dimitriades 1977; Post 1979; Carter et al. 1982; Löbel 1979 a; Sakamaki et al. 1982). The agreement is rather poor. Improvements of the chemical kinetic models on which isopleth calculations are based have been proposed by Harris (1982) and Jones et al. (1983).

Although it is known that compounds other than ozone make important contributions to the environmental effects of photochemical smog, few isopleths are available for other oxidants (D. F. Miller and Joseph 1976; Löbel 1978, 1979 b). Such isopleths, which consider HNO_3, aldehydes, PAN, aerosols, and various other compounds, show at least in a qualitative way that the reduction of nitrogen oxides concentration is not necessarily less efficient than hydrocarbon reduction in abating environmental effects of photochemical oxidants.

Conclusions

Current simulation models are still burdened with so many uncertainties that they do not form a safe basis for quantitative abatement strategies. It appears, however, that reductions in precursor emissions (NO_x, NMHC, or both) will *always* result in some reduction of the large field oxidant load of the atmosphere, even if the beneficial effects cannot be observed in the near field, or might be even negative, as has been postulated for one-sided NO_x reductions on the basis of model calculations and ozone isopleth schemes for the near field.

In order to alleviate decision-making in oxidant abatement efforts, further progress should be encouraged in physicochemical dispersion modeling. This requires in practice that more detailed meteorological and emission data are made available as input data. In parallel, continuous monitoring stations should be installed in areas which are susceptible to oxidant pollution, in order to elucidate the correlations between precursor concentrations, meteorological parameters, and oxidant levels. Such investigations will establish a firm basis for area-specific control measures in the future.

Chapter 1.7 Summary and Final Conclusions

1. It has clearly been shown that the formation of *anthropogenic* ozone together with other photochemical oxidant components occur in Western Europe during the summer months, May to September. In the Federal Republic of Germany from which the data were more extensively analyzed, a threshold of 200 µg m^{-3} ozone near ground is exceeded quite frequently in many areas (Table 1.11). In regions which are more heavily polluted with nitrogen oxides and hydrocarbons, as for example, Cologne, Frankfurt, Mannheim, or Karlsruhe, ½-h mean values of ozone in excess of 400 µg m^{-3} are occasionally measured (Table 1.11, Figs. 1.7 and 1.9). The annual frequency of exceeding high thresholds exhibits a strong correlation with certain large scale weather situations (Figs. 1.12 and 1.13, see also Chap. 1.4.3). To date, high ozone maxima, with ½- and 3-h mean values, have most frequently been observed in 1976. The highest ½-h value for ozone at ground level ever measured in Germany was 664 µg m^{-3} on the 23rd of June 1976 at the monitoring station Mannheim (Table 1.11 and Fig. 1.7). Such local peak values often occur in conjunction with widespread and persistent (several hours to days) high ozone concentration fields. The available data reveal that an industrialized area such as the Upper Rhine Valley is subject to particularly high ozone levels due to its climatic conditions. Further, it becomes evident that high peak values of ozone are restricted to densely populated urban areas. However, the ozone concentrations averaged over longer times – annual means or summer means – increase in peripheral areas, particularly in the upper range of medium altitude mountains (Fig. 1.10).

2. On the basis of the presently known data, it is recommended to measure ozone at few, but properly selected sites, instead of increasing the number of monitoring stations. Airborne measurements should be carried out more regularly under various weather conditions in order to follow the longer-ranging oxidants and precursor distributions.

3. In view of the annual and diurnal fluctuations of the ground level ozone concentrations, future air quality standards should preferentially refer to ozone concentrations from May to September between 0800 and 1800 h local time.

4. Observance of an ozone threshold value of 250 µg m^{-3} (½ h mean) is presumably not attainable by employing regional emission reducing measures only, since ozone levels in this order are determined by long-range formation and transport processes also (Chap. 1.3.2.2.3). Regional and local peak values in excess of 250 µg m^{-3}, however, may be abated by such measures. A reduction of the large-scale ozone load is certainly impossible without a coordinated emission reduction on a European scale.

5. It is not possible at the present time to make a general prognosis pertaining to the effect of control measures of individual classes of precursors on the abatement of oxidants. The few, but instructive, airborne measurements in several polluted areas of the Federal Republic of Germany demonstrate that elevated oxidant concentrations mostly occur downwind of sources which emit reactive hydrocarbons. Therefore, the reduction of reactive hydrocarbon emissions appears to be most promising for suppression of small-scale elevated ozone levels. A reduction of nitrogen oxides emissions is, on the other hand, probably more effective for larger areas and on longer time-scales on which the natural and the less reactive hydrocarbons also contribute to the formation of oxidants. If the reduction of hydrocarbon emissions alone is pursued, only a shift of the oxidant load further away from the precursor source areas might result.

6. It should also be of concern that in connection with an increased formation of photochemical oxidants, the production of sulfuric acid and nitric acid from the sulfur and nitrogen oxides, respectively, is accelerated. Hydrocarbons are also rapidly transformed into organic acids during build-up of photooxidants.

Chapter 1.8 Appendix

Nomenclature of chemical compounds and mixtures

PAN: Peroxyacetyl Nitrate ($CH_3CO_3NO_2$)
Alkanes: Methane (CH_4), Ethane (C_2H_6), Propane (C_3H_8) etc.
Alkenes: ($=$Olefins) Ethene (C_2H_4), Propene (C_3H_6) etc.
Alkines: Ethine (C_2H_2) ($=$Acethylene) etc.

The expression *total hydrocarbons* refers to the sum of compounds consisting of the elements carbon and hydrogen

$$C_nH_m : \text{THC (Total Hydrocarbons)}.$$

The *nonmethane hydrocarbons* (NMHC) denote total hydrocarbons less methane which, to a first approximation, represent the reactive fraction of total hydrocarbons.

Table 1.26. Conversion factors of concentration units: ($mg\,m^{-3}$) to (ppm) or ($\mu g\,m^{-3}$) to (ppb) at 20 °C; see Chapter 1.3.2

Substance		$c_{ppm} = A \cdot c_{mg/m^3}$ $c_{ppb} = A \cdot c_{\mu g/m^3}$	$c_{mg/m^3} = B \cdot c_{ppm}$ $c_{\mu g/m^3} = B \cdot c_{ppb}$
Formula	Molar mass	A	B
Methane: CH_4	16.0	1.50	0.67
Ethine: C_2H_2	26.0	0.93	1.08
Ethene: C_2H_4	28.1	0.86	1.17
Nitrogen Monoxide: NO	30.0	0.80	1.25
Formaldehyde: H_2CO	30.0	0.80	1.25
Ethane: C_2H_6	30.1	0.80	1.25
Methanol: CH_3OH	32.0	0.75	1.33
Hydrogen Peroxide: H_2O_2	34.0	0.71	1.41
Propene: C_3H_6	42.1	0.57	1.75
Acetaldehyde: CH_3CHO	44.1	0.55	1.83
Propane: C_3H_8	44.1	0.55	1.83
Nitrogen Dioxide: NO_2	46.0	0.52	1.91
Ethanol: C_2H_5OH	46.1	0.52	1.92
Ozone: O_3	48.0	0.50	2.00
Butane: C_4H_{10}	58.1	0.41	2.42
Nitric Acid: HNO_3	63.0	0.38	2.62
Benzene: C_6H_6	78.1	0.31	3.25
Toluene: C_7H_8	92.1	0.26	3.83
PAN: $CH_3CO_3NO_2$	121.1	0.20	5.02

Concentration Units

Concentrations are given in volume (molar) mixing ratios ppm, ppb, and ppt:

$$1 \text{ ppm} = 1{,}000 \text{ ppb} = 1{,}000{,}000 \text{ ppt}$$

or in mass concentration units mg m^{-3} and µg m^{-3}:

$$1 \text{ mg m}^{-3} = 1{,}000 \text{ µg m}^{-3}.$$

Volume mixing ratios can be converted to mass concentration units, or vice versa, by means of Table 1.26.

References

Adeniji SA, Kerr JA, Williams MR (1981) Rate constants for ozone-alkene reactions under atmospheric conditions. Int J Chem Kinet 13:209–217

Ahrens D (1983) Klima am Südlichen Oberrhein. Regionalverband Südl. Oberrhein, No 11, Freiburg, pp 61–71

Akimoto H, Sakamaki F, Hoshino M, Inoue G, Okuda M (1979a) Photochemical ozone formation in propylene – nitrogen oxide – dry air system. Environ Sci Technol 13:53–58

Akimoto H, Hoshino M, Inoue G, Sakamaki F, Washida N, Okuda M (1979b) Design and characterization of the evacuable and bakable photochemical smog chamber. Environ Sci Technol 13:471–475

Akimoto H, Bandow H, Sakamaki F, Inoue G, Hoshino M, Okuda M (1980) Photooxidation of the propylene-NO$_x$-air system studied by long-path Fourier transform infrared spectrometry. Environ Sci Technol 14:172–179

Altshuller AP (1978) Association of oxidant episodes with warm stagnating anticyclones. J Air Pollut Control Assoc 28:152–155

Altshuller AP, Miller DL, Sleva SF (1961) Determination of formaldehyde in gas mixtures by the chromotropic acid method. Anal Chem 33:621–625

Apling AJ, Sullivan EJ, Williams ML, Ball DJ, Eggleton AEJ, Hampton L, Waller RE (1977) Ozone concentrations in South-East England during the summer of 1976. Nature (London) 269:569–573

Aron RH, Aron IM (1978) Statistical forecasting models, II. Oxidant concentrations in the Los Angeles Basin. J Air Pollut Control Assoc 28:684–688

Atkins DHF, Cox RA, Eggleton AEJ (1972) Photochemical ozone and sulphuric acid formation in the atmosphere over Southern England. Nature (London) 235:372–376

Atkinson R, Aschmann SM, Winer AM, Pitts JN (1981) Rate constants for the gas phase reactions of O$_3$ with a series of carbonyls at 296 K. Int J Chem Kinet 13:1133–1142

Atkinson R, Winer AM, Pitts JN (1982) Rate constants for the gas phase reactions of O$_3$ with the natural hydrocarbons isoprene and α- and β-pinene. Atmos Environ 16:1017–1020

Atkinson R, Aschmann SA, Winer AM, Pitts JN (1984) Kinetics of the gas-phase reactions of NO$_3$ radicals with a series of dialkenes, cycloalkenes, and monoterpenes at 295 ± 1 K. Environ Sci Technol 18:370–375

Attmannspacher W (1976) Über Extremwerte des natürlichen und anthropogenen bodennahen Ozons. Meteorol Rundsch 29:33–38

Attmannspacher W (1977) Extrem hohe Konzentrationen natürlichen Ozons auf dem Hohenpeißenberg, aktuelle Ergebnisse. VDI-Berichte 270: Ozon und Begleitsubstanzen im photochemischen Smog. VDI, Düsseldorf, pp 71–74

Attmannspacher W (1981) 200 Jahre meteorologische Beobachtungen auf dem Hohenpeißenberg, 1781–1981. Ber Dtsch Wetterdienst 155:1–84

Attmannspacher W (1983) Änderungen der lebenswichtigen Ozonschicht der Atmosphäre aufgrund langjähriger Ozonmessungen am Meteorologischen Observatorium Hohenpeißenberg. Symposium „Probleme der Umwelt- und Medizinmeteorologie im Gebirge", Rauris, Austria, Sept 1983

Attmannspacher W, Hartmannsgruber R (1973) On extremely high values of ozone near the ground. Pure Appl Geophys 106–108:1091–1096

Attmannspacher W, Hartmannsgruber R, Ziegler B (1980) Normales Verhalten des bodennahen Ozons am Hohenpeißenberg und kurzzeitiger Einbruch anthropogen verursachten Ozons. Photochemische Luftverunreinigungen in der Bundesrepublik Deutschland. Tagung des Umweltbundesamtes, Okt. 1979. VDI, Düsseldorf, pp 229–238

Bahe FC, Illner H, Marx WN, Schurath U, Röth P (1979a) Messungen der von Veränderungen der Ozonschicht stark abhängigen kurzwelligen Sonnenstrahlung. BPT-Ber 4/79, Ges Strahlen- Umweltforsch, München, Bereich Projektträgerschaften

Bahe FC, Marx WN, Schurath U, Röth EP (1979b) Determination of the absolute photolysis rate of ozone by sunlight, $O_3 + h\nu \rightarrow O(^1D) + O_2(^1\Delta_g)$, at ground level. Atmos Environ 13:1515–1522

Bahe FC, Schurath U, Becker KH (1980) The frequency of NO_2 photolysis at ground level, as recorded by a continuous actinometer. Atmos Environ 14:711–718

Bass AM, Glasgow LC, Miller C, Jesson JP, Filkin DP (1980a) Temperature dependent absorption cross sections for formaldehyde (CH_2O): The effect of formaldehyde on stratospheric chlorine chemistry. Planet Space Sci 28:675–679

Bass AM, Glasgow LC, Miller C, Jesson JP, Filkin DP (1980b) Temperature dependent absorption cross sections for formaldehyde (CH_2O): The effect of formaldehyde on stratospheric chlorine chemistry. Proc NATO Adv Study Inst Atmos Ozone: Its Variation and Human Influences. Rep No FAA-EE-80-20, pp 467–477

Baulch DL, Cox RA, Hampson RF, Kerr JA, Troe J, Watson RT (1980) Evaluated kinetic and photochemical data for atmospheric chemistry. J Phys Chem Ref Data 9:295–471

Baulch DL, Cox RA, Crutzen PJ, Hampson RF, Kerr JA, Troe J, Watson RT (1982) Evaluated kinetic and photochemical data for atmospheric chemistry. Suppl I J Phys Chem Ref Data 11:327–496

Becker KH, Schurath U (1973) Entsteht photochemischer „Smog" in der Bundesrepublik Deutschland? Umschau 73:310–311

Becker KH, Schurath U (1975) Der Einfluß von Stickstoffoxiden auf atmosphärische Oxidationsprozesse. Staub-Reinhalt Luft 35:156–161

Becker KH, Löbel J, Schurath U (1978) Simulationsexperimente zur photochemischen Luftverschmutzung. Staub-Reinhalt Luft 38:278–283

Becker KH, Schurath U, Georgii HW, Deimel M (1979) Untersuchungen über Smogbildung, insbesondere über die Ausbildung von Oxidantien als Folge der Luftverunreinigung in der Bundesrepublik Deutschland. Forschungsber 79-10402502/03/04 Umweltbundesamt, Berlin

Becker KH, Löbel J, Schurath U (1983) Bildung, Transport und Kontrolle von Photooxidantien. In: Umweltbundesamt (ed) Luftqualitätskriterien für photochemische Oxidantien. Schmidt, Berlin, pp 3–132

Beilke S, Markusch H, Jost D (1982) Measurements of NO-oxidation in power plant plumes by correlation spectroscopy. See Versino B, Ott H (eds) (1982), pp 448–459

Bénarie M, Benec'hi A, Chuong BT, Menard T (1979) Etude de la pollution sur site rural. Mesure de l'ozone et des oxydes d'azote à Vert-Le-Petit. Pollut Atmos 81:44–52

Bos R, Goudena EJG, Guicherit R, Hoogeveen A, de Vreede JAF (1978) Atmospheric precursors and oxidants concentrations in the Netherlands. TNO-Rep Photochem Smogform Netherlands, S'-Gravenhage 1978, pp 20–60

Bottenheim JW, Strausz OP (1982) Modelling study of a chemically reactive power plant plume. Atmos Environ 16:85–97

Božičević Z, Klasinc L, Cvitaš T, Güsten H (1976a) Photochemische Ozonbildung in der unteren Atmosphäre über der Stadt Zagreb. Staub-Reinhalt Luft 36:363–366

Božičević Z, Butković V, Klasinc L (1976b) Foto-smog u Zagrebu. Kem Ind 25:333–337

Božičević Z, Butković V, Cvitaš T, Klasinc L (1978) Statistička obrada podataka fotokemijskog zagadenja u Zagrebu. Kem Ind 27:177–181

Bravo H, Magaña A, Magaña R (1979) The actual air pollution situation in Mexico City. Staub-Reinhalt Luft 39:427–428

Bruckmann P, Eynck P (1979) Analyse der Bildung von Photooxidantien an der Meßstelle Essen-Süd. Schriftenr Landesanst Immissionssch Landes NW H 49:19–28

Bruckmann P, Eynck P (1980a) Messung von Photooxidantien (Ozon und PAN) im Rhein-Ruhr-Gebiet. Photochemische Luftverunreinigungen in der Bundesrepublik Deutschland. Tagung des Umweltbundesamtes, Okt. 1979. VDI, Düsseldorf, pp 92–107

Bruckmann P, Eynck P (1980b) Aufbau und Eigenschaften einer Reaktionskammer zum Studium photochemischer Reaktionen in der Atmosphäre. Schriftenr Landesanst Immissionssch Landes NW H 50:23–30

Bruckmann P, Langensiepen EW (1981) Untersuchungen über Zusammenhänge zwischen Ozonkonzentration und meteorologischen Parametern im Rhein-Ruhr-Gebiet. Staub-Reinhalt Luft 41:79–85

Bruckmann P, Mülder W (1979) Die Messung von Peroxiacetylnitrat (PAN) in der Außenluft – Verfahren und erste Ergebnisse. Schriftenr Landesanst Immissionssch Landes NW H 47:30–41

Bruckmann P, Buck M, Eynck P (1980) Modelluntersuchungen über den Zusammenhang zwischen Vorläufer- und Photooxidantienkonzentrationen. Staub-Reinhalt Luft 40:412–417

Bruckmann P, Ellermann K, Ixfeld H (1982) Die Ozonbelastung im Köln-Bonner Raum in den Meßjahren 1977–80. Schriftenr Landesanst Immissionssch Landes NW H 55:14–20

Bruckmann PW, Willner H (1983) Infrared spectrosopic study of peroxyacetyl nitrate (PAN) and its decomposition products. Environ Sci Technol 17:352–357

Bruntz SM, Cleveland WS, Graedel TE, Kleiner B, Warner JL (1974) Ozone concentrations in New Jersey and New York: Statistical association with related variables. Science 186:257–258

Builtjes PJH (1980) Application of a photochemical dispersion model to the Netherlands and its surroundings. Proc 11th NATO-CCMS Int Tech Meet, Amsterdam 1980

Builtjes PJH (1981) Some remarks considering the modeling of a chemically reacting plume in a wind tunnel. Report „Ausbreitung von Schadstoffen in der Atmosphäre. Physikalisch-Chemische Modelle", VDI-Kommission Reinhalt Luft, 1981

Bundesminister des Innern (1984) Umwelt 104, p 36–37

Butković V, Cvitaš T, Gotovac V, Klasinc L (1983) Variation of tropospheric ozone concentrations in selected areas of Croatia. Proc VIth World Congr Air Qual, May 1983, Paris, 3:175–180

Calvert JG (1976 a) Hydrocarbon involvement in photochemical smog formation in Los Angeles atmosphere. Environ Sci Technol 10:256–262

Calvert JG (1976 b) Test of the theory of ozone generation in Los Angeles atmosphere. Environ Sci Technol 10:248–256

Calvert JG (1981) The homogeneous chemistry of formaldehyde generation and destruction within the atmosphere. Proc NATO Adv Study Inst Atmos Ozone: Its Variation and Human Influences. FAA-EE-80-20:153–190

Calvert JG, Pitts JN (1966) Photochemistry. Wiley, New York

Campbell MJ, Sheppard JC, Au BF (1979) Measurement of hydroxyl concentration in surface air by monitoring CO oxidation. Geophys Res Lett 6:175–178

Carter WPL, Atkinson R, Winer AM, Pitts JN (1981) Evidence for chamber-dependent radical sources: Impact on kinetic computer models for air pollution. Int J Chem Kinet 13:735–740

Carter WPL, Winer AM, Pitts JN (1982 a) Effects of kinetic mechanisms and hydrocarbon composition on oxidant-precursor relationships predicted by the EKMA isopleth technique. Atmos Environ 16:113–120

Carter WPL, Atkinson R, Winer AM, Pitts JN (1982 b) Experimental investigation of chamber dependent radical sources. Int J Chem Kinet 14:1071–1103

Chamberlain AC (1960) Aspects of the deposition of radioactive and other gases and particles. Int J Air Pollut 3:63–88

Chameides WL, Davis DD (1982) The free radical chemistry of cloud droplets and its impact upon the composition of rain. J Geophys Res 87:4863–4877

Chameides WL, Tan A (1981) The two-dimensional diagnostic model for tropospheric OH: An uncertainty analysis. J Geophys Res 86:5209–5223

Chameides WL, Stedman DH, Dickerson RR, Rush DW, Cicerone RJ (1977) NO_x Production in lightning. J Atmos Sci 34:143–149

Chance EM, Curtis AR, Jones IP, Kirby CR (1977) FACSIMILE: A computer program for flow and chemistry simulation and general initial value problems. AERE-R 8775, HM Stationery Office

Cleemput O van, El-Sebaay AS, Baert L (1982) Production of gaseous hydrocarbons in soil. See Versino B, Ott H (eds) (1982), pp 349–355

Cofer III WR (1982) Methane and nonmethane hydrocarbon concentrations in the North and South Atlantic marine boundary layer. J Geophys Res 87:7201–7205

Cox RA, Derwent RG (1976) The ultra-violet absorption spectrum of gaseous nitrous acid. J Photochem 6:23–34

Cox RA, Roffey MJ (1977) Thermal decomposition of peroxyacetylnitrate in the presence of nitric oxide. Environ Sci Technol 11:900–906

Cox RA, Eggleton AEJ, Derwent RG, Lovelock JE, Back DH (1975) Long-range transport of photochemical ozone in North-Western Europe. Nature (London) 255:118–121

Cox RA, Patrick KF, Chant SA (1981) Mechanism of atmospheric photooxidation of organic compounds. Reactions of alkoxy radicals in oxidation of n-butane and simple ketones. Environ Sci Technol 15:587–592

Crutzen PJ, Fishman J (1977) Average concentration of OH in the troposphere, and the budget of CH_4, CO, H_2, and CH_3CCl_3. Geophys Res Lett 4:321–324

Cvitaš T, Klasinc L (1979) „Trenutni snimak" zagradenja zraka u Kvarnerskom zaljevu. Zast Atmos 15:13–16

Cvitaš T, Güsten H, Klasinc L (1979) Statistical association of the photochemical ozone concentrations in the lower atmosphere of Zagreb with meteorological variables. Staub-Reinhalt Luft 39:92–95

Danielsen ES (1968) Stratospheric-tropospheric exchange based on radioactivity, ozone and potential vorticity. J Atmos Sci 25:502–518

Davenport JE (1982) Parameters for ozone photolysis as a function of temperature at 280–330 nm. FAA-EE-80-44R, May 1982

Deimel M (1982a) Bericht FE-Vorhaben „Einbezug chemischer Umwandlungsprozesse in die Ausbreitungsrechnung", Teil I: Messung von Oxidantien und Vorläufern im Kölner Raum 1979–81, Amt Umweltsch Stadt Köln

Deimel M (1982b) Luftuntersuchungen im Raum Köln 1979–80. Amt Umweltsch Stadt Köln H 1

Deimel M (1979) Ozon- und Stickoxidbelastung im Köln-Bonner Raum. See Becker et al. (eds) (1979) Sect III, pp 1–25

Demerjian KL (1977) Photochemical air quality simulation modeling: Current status and future prospects. Proc Int Conf Photochem Oxidant Pollut and Its Control, EPA-600/3-77-001b, NTIS, pp 777–794

Demerjian KL, Kerr JA, Calvert JG (1974) The mechanism of photochemical smog formation. Adv Environ Sci Technol 4:1–261

Derwent RG, Stewart HNM (1973) Elevated ozone levels in the air of Central London. Nature (London) 241:342–343

Derwent RG, Eggleton AEJ, Williams ML, Bell CA (1978) Elevated ozone levels from natural sources. Atmos Environ 12:2173–2177

Derwent RG, Hov Ø (1979) Computer modeling studies of photochemical air pollution formation in North West Europe. Environ Med Sci Div, AERE Harwell, Great Britain

Derwent RG, Hov Ø (1980) Computer modeling studies of the impact of vehicle exhaust emission controls on photochemical air pollution in the United Kingdom. Environ Sci Technol 14:1360–1366

Deutscher Wetterdienst, Meteorologisches Observatorium Hamburg (1976–1983) Ergebnisse von Strahlungsmessungen in der Bundesrepublik Deutschland sowie von speziellen Meßreihen am Meteorologischen Observatorium Hamburg, No 1–8

Deutscher Wetterdienst, Meteorologisches Observatorium Hohenpeißenberg. Sonderbeobachtungen des Meteorologischen Observatoriums Hohenpeißenberg. Ergebnisse der aerologischen und bodennahen Ozonmessungen. Two regular issues per year

Deželjin S, Gotovac V, Cvitaš T (1981) Ozon u splitskom zraku. Kem Ind 30:57–61

Dickerson RR (1980) Direct measurements of ozone and nitrogen dioxide photolysis rates in the atmosphere. Coop thesis No 58, Univ Mich Natl Cent Atmos Res, NCAR/CT-58

Dickerson RR, Stedman DH, Delany AC (1982) Direct measurements of ozone and nitrogen dioxide photolysis rates in the troposphere. J Geophys Res 87:4933–4946

Dimitriades B (1970) On the function of hydrocarbon and nitrogen oxides in photochemical smog formation. US Bur Mines Rep Invest RI 7433, Sept 1970

Dimitriades B (1972) Effects of hydrocarbon and nitrogen oxides on photochemical smog formation. Environ Sci Technol 6:253–260

Dimitriades B (1977) Oxidant control strategies. Part I. Urban oxidant control strategy derived from existing smog chamber data. Environ Sci Technol 11:80–88

Dodge MC (1977) Combined use of modeling techniques and smog chamber data to derive ozone-precursor relationships. Proc Int Conf Photochem Oxidant Pollut and Its Control, EPA-600/3-77-001b, pp 861–889

Doman M, Jessel U (1977) Über neue Ozon-Meßergebnisse auf der Nordseeinsel Sylt. VDI-Berichte 270, Ozon und Begleitsubstanzen im photochemischen Smog. VDI, Düsseldorf, pp 69–70

Dop H van, Guicherit R, Lanting RW (1977) Some measurements of the vertical distribution of ozone in the atmospheric boundary layer. Atmos Environ 11:65–71

Duce RA (1978) Speculations on the budget of particulate and vapor phase non-methane organic carbon in the global troposphere. Pure Appl Geophys 116:244–273

Dulson W (1978) Organisch-chemische Fremdstoffe in atmosphärischer Luft. Schriftenr Ver Wasser-Boden- Lufthyg 47:47–53

Dulson W (1979) Kohlenwasserstoff-Belastung des Köln-Bonner Raumes. See Becker et al. (eds) (1979) Sect IV, pp 1–18

Dulson W (1981) Messungen von Kohlenwasserstoff-Immissionen in Köln. Ber Amt Umweltsch Stadt Köln, H 1:1–88

Dunker AM, Sundarshan Kumar, Berzins PH (1984) A comparison of chemical mechanisms used in atmospheric models. Atmos Environ 18:311–321

Duuren H van, Römer FG, Diederen HSMA, Guicherit R, van ten Hout KD (1982) Measurements by aeroplane of the distribution of ozone and primary air pollutants. See Versino B, Ott H (eds) (1982), pp 460–468

Edinger JG (1973) Vertical distribution of photochemical smog in Los Angeles Basin. Environ Sci Technol 7:247–252

Ehhalt DH, Drummond JW (1982) The tropospheric cycle of NO_x. In: Georgii HW, Jaeschke W (eds) Chemistry of the unpolluted and the polluted troposphere. Reidel, Dordrecht, pp 219–251

Ehmert A (1949) Über den Ozongehalt der unteren Atmosphäre bei winterlichem Hochdruckwetter nach Messungen. Ber Dtsch Wetterdienst US-Zone No 11:63–66

Ehmert A (1952) Gleichzeitige Messungen des Ozongehaltes erdnaher Luft an mehreren Stationen mit einem einfachen Verfahren. J Atmos Terrest Phys 2:189–195

Emonds H (1954) Das Bonner Stadtklima. Ber Geogr Inst Univ Bonn, pp 1–64

EPA/EPRI (1980) Contrib Proc Joint Symp Stationary Combustion NO_x Control, Oct 6–9, Denver, vol I, EPA-600/7-79-050

Eschenroeder AQ, Martinez JR (1972) Concepts and applications of photochemical smog models. Adv Chem Ser, vol 113:101–168. Am Chem Soc, Washington DC

Fabian P, Junge CH (1970) Global rate of ozone destruction on the earth's surface. Arch Meteorol Geophys Bioklimatol Ser A 19:161–172

Fabian P, Pruchniewicz PG, Zand A (1971) Transport- und Austauschvorgänge in der Atmosphäre und ihre Erforschung mit Spurenstoffen. Naturwissenschaften 58:541–549

Fabry C, Buission H (1921) Etude de l'extrémité ultraviolette du spectre solaire. J Phys Radium Ser 6, 2:226

Falls AH, McRae GJ, Seinfeld JH (1979) Sensitivity and uncertainty of reaction mechanisms for photochemical air pollution. Int J Chem Kinet 11:1137–1162

Fett W (1970) Zum Nachweis des Stadteinflusses auf den Ozongehalt der Luft mittels seiner Windgeschwindigkeitsabhängigkeit. Schriftenr Ver Wasser-Boden-Lufthyg. Berlin-Dahlem H 33:117–128

Fishman J, Seiler W (1983) Correlative nature of ozone and carbon monoxide in the troposphere: Implications for the tropospheric ozone budget. J Geophys Res 88:3662–3670

Fishman J, Solomon S, Crutzen PJ (1979) Observational and theoretical evidence in support of a significant in-situ photochemical source of tropospheric ozone. Tellus 31:432–446

Fontijn A, Sabadell AJ, Ronco RJ (1970) Homogeneous chemiluminescent measurement of nitric oxide with ozone. Implications for continuous selective monitoring of gaseous air pollutants. Anal Chem 42:575–579

Forrest J, Spandau DJ, Tanner RL, Newman L (1982) Determination of atmospheric nitrate and nitric acid employing a diffusion denuder with a filter pack. Atmos Environ 16:1473–1485

Fricke W (1980) Die Bildung und Verteilung von anthropogenem Ozon in der unteren Troposphäre. Ber Inst Meteorol Geophys Univ Frankfurt No 44:1–133, Dec 1980

Fricke W, Rudolf W (1977) Ozonkonzentrationen in Luv und Lee von Ballungsgebieten auf der Fluglinie München–Rotterdam. Staub-Reinhalt Luft 37:341–345

Frohne HCH, Schneider W (1977) Vergleich der Kohlenwasserstoff-Immissionen im industrienahen und -fernen Bereich. VDI-Berichte 270, Ozon und Begleitsubstanzen im photochemischen Smog. VDI, Düsseldorf, pp 83–91

Gaffney JS, Fajer R, Senum GI (1984) An improved procedure for high purity gaseous peroxyacetyl nitrate production: use of a heavy lipid solvent. Atmos Environ 18:215–218

Galbally IE (1971) Ozone profiles and ozone fluxes in the atmospheric surface layer. Q J R Meteorol Soc 97:18–29

Garland JA, Derwent RG (1979) Destruction at the ground and diurnal cycle of concentration of ozone and other gases. Q J R Meteorol Soc 105:169–183

Garland JA, Penkett SA (1976) Absorption of peroxyacetyl nitrate and ozone by natural surfaces. Atmos Environ 10:1127–1131

Gay BW, Arnts RR (1977) The chemistry of naturally emitted hydrocarbons. EPA 600/3-77-001, pp 745–751

Georgii HW (1982) The atmospheric sulfur budget. In: Georgii HW, Jaeschke W (eds) Chemistry of the unpolluted and the polluted troposphere. Reidel, Dordrecht, pp 295–324

Georgii HW, Muschalik B (1980) Der Einfluß meteorologischer Parameter auf die bodennahe Oxidantienverteilung. Forschungsprojekt 10402504 Umweltbundesamt, Berlin

Georgii HW, Fricke W, Rudolf W, Deimel M, Becker KH, Schurath U (1977) Bildung und Transport von Photooxidantien im Raum Bonn–Köln und Frankfurt/M. VDI-Berichte 270: Ozon und Begleitsubstanzen im photochemischen Smog. VDI, Düsseldorf, pp 19–24

Gerold F, Brieda F, Heidenfels F, Treusch P (1980) Emissionsfaktoren für Luftverunreinigungen – Feuerungs- und Aufbereitungsanlagen sowie Lagerung und Umschlag fester und flüssiger Stoffe. Forschungsber No 77/10402704. Umweltbundesamt Materialien 2/80. Schmidt, Berlin

Ghosh PN, Rakoczi F, Bauder A, Guenthard HH (1982) Product analysis of chemical reactions relevant to atmospheric chemistry. See Versino B, Ott H (eds) (1982), pp 149–154

Giovanelli G, Fortezza F, Minguzzi L, Stochi V, Wandini W (1982) Photochemical ozone transport in an industrial costal area. See Versino B, Ott H (eds) (1982), pp 492–501

Glavas S, Schurath U (1983) Concentration of peroxyacetylnitrate (PAN) for mobile field measurement in tropospheric air. Chim Chron New Ser 12:89–97

Gloria HR, Bradburn G, Reinisch RF, Pitts JN, Behar JV, Zafonte L (1974) Airborne survey of major air basins in California. J Air Pollut Control Assoc 24:645–652

Grennfelt P (1977) Ozone episodes on the Swedish West Coast. Proc Int Conf Photochem Oxidant Pollut and Its Control, EPA-600/3-77-001a, pp 329–377

Grennfelt P, Samuelsson U, Nielsen T, Thomsen EL (1982) The presence of PAN in long-range transported polluted air masses. See Versino B, Ott H (eds) (1982), pp 619–624

Grosjean D (1982) Formaldehyde and other carbonyls in Los Angeles ambient air. Environ Sci Technol 16:254–262

Grosjean D, Kochy Fung, Collins J, Harrison J, Breitung E (1984) Portable generator for on-site calibration of peroxyacetyl nitrate analyzers. Anal Chem 56:569–573

Guicherit R (1971) Ozone analysis by chemiluminescence measurement. Z Anal Chem 256:177–182

Guicherit R (1973) Photochemical smog formation in the Netherlands. Congr No 00320, IG-TNO Publ No 459

Guicherit R (1975) Fotochemische smogforming in nederland. Rep No G 646, TNO Res Inst Environ Hyg 104:1–104

Guichert R, Dop H van (1977) Photochemical production of ozone in Western Europe (1971–1975) and its relation to meteorology. Atmos Environ 11:145–155

Guicherit R, Jeltes R, Lindqvist F (1972) Determination of the ozone concentration in outdoor air near Delft, the Netherlands. Environ Pollut 3:91–110

Guicherit R, Blokzijl PJ, Plasse CJ (1978) Some notes on the abatement of photochemical ozone production. In: Photochemical smog formation in the Netherlands. TNO Rep, pp 187–201

Ham J van (1978) Objections to the use of polyvinyl fluoride in smog chamber experiments. Chemosphere 4:315–318

Hameed S, Stewart RW (1983) Latitudinal variation of tropospheric ozone in a photochemical model. J Geophys Res 88:5153–5162

Hampson RF (1980) Chemical kinetic and photochemical data sheets for atmospheric reactions. Rep No FAA-EE-80-17, April 1980

Hampson RF, Garvin D (1978) Reaction rate and photochemical data for atmospheric chemistry – 1977. NBS Spec Publ 513, May 1978

Hampson RF, Braun W, Brown RW, Garvin D, Herron JT, Huie RE, Curylo MJ, Laufer AH, McKinley JD, Okabe H, Scheer MD, Tsang W, Stedman DH (1973) Survey of photochemical and rate data for twenty-eight reactions of interest in atmospheric chemistry. J Phys Chem Ref Data 2:267–312

Handbook of Chemistry and Physics (1974) 55th edn. CRC-Press, Cleveland

Hanna S (1973) A simple dispersion model for the analysis of chemically reactive pollutants. Atmos Environ 7:803–817

Hanst PL, Wong NW, Bragin J (1982) A long-path infra-red study of Los Angeles smog. Atmos Environ 5:969–981

Harris GW, Carter WPL, Winer AM, Pitts JN, Platt U, Perner D (1982) Observations of nitrous acid in the Los Angeles atmosphere and implications for prediction of ozone-precursor relationship. Environ Sci Technol 16:414–419

Hartley WN (1881) On the absorption of solar rays by atmospheric ozone. J Chem Soc 39:111–128

Hecht TA, Seinfeld JH, Dodge MC (1974) Further development of generalized kinetic mechanism for photochemical smog. Environ Sci Technol 8:327–339

Heikes BG, Lazarus AL, Kok GL, Kunen SM, Gandrud BW, Githin SN, Sperry PD (1982) Evidence for aqueous phase hydrogen peroxide synthesis in the troposphere. J Geophys Res 87:3045–3051

Helas G, Warneck P (1981) Background NO_x mixing ratios in air masses over the North Atlantic Ocean. J Geophys Res 86:7283–7290

Hendry DG, Kenley RA (1977) Generation of peroxy radicals from peroxy nitrates (RO_2NO_2). Decomposition of peroxyacetyl nitrates. J Am Chem Soc 99:3198–3199

Henrich K, Lippmann H, Schurath U, Wendler W (1982) Ozone formation in a smog chamber at constant and variable light intensity. See Versino B, Ott H (eds) (1982), pp 218–227

Hess P, Brezowsky H (1977) Katalog der Großwetterlagen (1881–1976). Ber Dtsch Wetterdienst No 113

Hesstvedt E, Hov Ø, Isaksen ISA (1978) Quasi-steady-state approximation in air pollution modeling: Comparison of two numerical schemes for oxidant prediction. Int J Chem Kinet 10:971–994

Hill RD, Rinker RG, Wilson HG (1980) Atmospheric nitrogen fixation by lightning. J Atmos Sci 37:179–192

Hodgeson JA (1972) Review of analytical methods for atmospheric oxidants measurements. Int J Environ Anal Chem 1:233–241

Holdren MW, Spicer CW (1984) Field compatible calibration procedure for peroxyacetyl nitrate. Environ Sci Technol 18:113–116

Holdren MW, Westberg HH, Zimmerman PR (1979) Analysis of monoterpene hydrocarbons in rural atmospheres. J Geophys Res 84:5083–5088

Hov Ø (1984) Modelling of the long-range transport of peroxyacetylnitrate to Scandinavia. J Atmos Sci 1:187–202

Hov Ø, Derwent RG (1980) Sensitivity studies of the effects of model formulation on the evaluation of control strategies for photochemical air pollution formation in the United Kingdom. Manuscript 1980

Hov Ø, Hesstvedt E, Isaksen ISA (1978 a) Long-range transport of tropospheric ozone. Nature (London) 273:341–344

Hov Ø, Isaksen ISA, Hesstvedt E (1978 b) Diurnal variations of ozone and other pollutants in an urban area. Atmos Environ 12:2469–2476

Hov Ø, Schjoldager J, Wathne BM (1983) Measurement and modeling of the concentrations of terpenes in coniferous forest air. J Geophys Res 88:10679–10688

Howard CJ, Evenson KM (1977) Kinetics of the reactions of HO_2 with NO. Geophys Res Lett 4:437–440

Hübler G, Ehhalt DH, Pätz HW, Perner D, Platt U, Schröder T, Tönnissen A (1982) Determination of the ground level OH concentration by long path laser absorption technique. See Versino B, Ott H (eds) (1982), pp 2–9

Hyde R, Hawke GS (1977) The transport of photochemical smog across the Sydney Bay. Proc Int Conf Photochem Oxidant Pollut and Its Control, EPA-600/3-77-001a, pp 285–297

IG-TNO (1971) Analyse van de smogsituation in de Randstad Holland. Werkrapport G 500. IG-TNO, Delft, pp 1–135

Isaksen ISA, Hesstvedt E, Hov Ø (1978 a) A chemical model for urban plumes: Test for ozone and particulate sulfur formation in St. Louis urban plume. Atmos Environ 12:599–604

Isaksen ISA, Hov Ø, Hesstvedt E (1978 b) Ozone generation over rural areas. Environ Sci Technol 12:1279–1284

Ixfeld H (1977) Weiträumige Immissionsbelastung des Ballungsgebietes Rhein-Ruhr durch Oxidantien-Vorläufer. VDI-Berichte 270: Ozon und Begleitsubstanzen im photochemischen Smog. VDI, Düsseldorf, pp 63–68

Jander K (1979) Prüfung und Bewertung von Meßverfahren zur Bestimmung smogbildender Komponenten im Abgas ölgefeuerter Großkessel. Inst Wasser- Boden- Lufthyg Bundesgesundheitsamt, Berlin

Jeffries H, Fox D, Kamens R (1976 a) Photochemical conversion of NO to NO_2 by hydrocarbons in an outdoor smog chamber. J Air Pollut Control Assoc 26:480–484

Jeffries H, Fox D, Kamens R (1976 b) Outdoor smog chamber studies: Light effects relative to indoor chambers. Environ Sci Technol 10:1006–1011

Jet Propulsion Laboratory (1981) Chemical kinetic and photochemical data for use in stratospheric modelling. Evaluation No. 4: NASA Panel for Data Evaluation. JPL Publ 81–3, Jan 1981

Johnson WB, Viezee W (1981) Stratospheric ozone in the lower troposphere. I. Presentation and interpretation of aircraft measurements. Atmos Environ 15:1309–1323

Jones KH, Ruch RB, Barone JB, Walsh JF, Karpovich RA (1983) The rationale and need to consider an alternative to EKMA. J Air Pollut Control Assoc 33:330–332

Joseph DW, Spicer CW (1978) Chemiluminescence method for atmospheric monitoring of nitric acid and nitrogen oxides. Anal Chem 50:1400–1403

Jost D (1970) Survey of the distribution of trace substances in pure and polluted atmospheres. Pure Appl Chem 24:643–654

Junge CH (1962) Global ozone budget and exchange between stratosphere and troposphere. Tellus 14:363–377

Junge CH (1978) Der natürliche Kreislauf der Gase. Ber Bunsenges 82:1128–1132

Kamens RM, Jeffries HE, Gery MW, Wiener RW, Sexton KG, Howe GB (1981) The impact of α-pinene on urban smog formation: Outdoor smog chamber study. Atmos Environ 15:969–981

Kan CS, Fu Su, Calvert JG, Shaw JH (1981) Mechanism of the ozone-ethene reaction in dilute N_2/O_2 mixtures near 1-atm pressure. J Phys Chem 85:2359–2363

Kanter HJ, Reiter R, Pötzl K (1979) Untersuchungen über die Häufigkeit und Ursache hoher Ozonkonzentrationen unter Reinluftbedingungen. Forschungsber FE-Vorhaben 78-10402800, Umweltbundesamt, Berlin

Kanter HJ, Reiter R, Munzert KH (1982) Untersuchungen zur Frage der photochemischen Produktion von Ozon in Reinluftgebieten und ihrer vertikalen Verteilung. Forschungsber FE-Vorhaben 10402800, Umweltbundesamt, Berlin

Karl TR (1979) Potential application of model output statistics (MOS) to forecasts of surface ozone concentrations. J Appl Meteorol 18:254–265

Karlinsky (1874) Zwanzigjährige Ozonbeobachtungen in Krakau. Z Oesterr Ges Meteorol 9:94–95

Kelly TJ, Stedman DH, Kok GL (1979) Measurement of H_2O_2 and HNO_3 in rural air. Geophys Res Lett 6:375–378

Kessler C, Perner D, Platt U (1981) Spectroscopic measurements of nitrous acid and formaldehyde – Implications for urban photochemistry. Presented at the 2nd meeting of Working Party 2 On Mechanisms of Oxidantion Processes in the Atmosphere. COST 61 a bis, Löwen 1981

Kessler C, Perner D, Platt U (1982) Spectroscopic measurements of nitrous acid and formaldehyde – Implications for urban photochemistry. See Versino B, Ott H (eds) (1982), pp 393–400

Kinosian JR (1982) Ozone-precursor relationship from EKMA-diagrams. Environ Sci Technol 16:880–883

Kley D, Drummond JW, McFarland M, Liu SC (1981) Tropospheric profiles of NO_x. J Geophys Res 86C:3153–3161

Knoeppel H, Versino B, Schlitt H, Peil A, Schauenburg H, Vissers H (1980) Organics in air. Sampling and identification. Proc 1st Eur Symp Phys Chem Behav Atmos Pollut, EEC/COST 61 a bis, Ispra, Oct 1979, pp 25–40

Kok GL, Holler TP, Lopez MB, Nachtrieb AH, Yuan M (1978 a) Chemiluminescent method for determination of hydrogen peroxide in the ambient atmosphere. Environ Sci Technol 12:1072–1076

Kok GL, Darnall KR, Winer AM, Pitts JN, Gay BW (1978 b) Ambient air measurement of hydrogen peroxide in the California South Coast Air Basin. Environ Sci Technol 12:1077–1080

Kosak-Channing LF, Helz GR (1983) Solubility of ozone in aqueous solutions of 0–0.6 M ionic strength at 5–30 °C. Environ Sci Technol 17:145–149

Kravetz TM, Martin SW, Mendenhall GD (1980) Synthesis of peroxyacetyl and peroxyaryl nitrates. Complexation of peroxyacetyl nitrate with benzene. Environ Sci Technol 14:1262–1264

Lahmann E (1969) Ozon in städtischer Luft. Umschau 21:693–694

Lahmann E, Westphal J, Damaschke K, Lübke M (1968) Kontinuierliche Ozonmessungen in einer verkehrsreichen Straße. Gesund-Ing 89:144–147

Laird AR, Miksad RW, Middleton P (1982) A simple model for urban ozone impact predictions. J Air Pollut Control Assoc 32:1221–1225

Lee YN, Schwartz SE (1981) Reaction kinetics of nitrogen dioxide with liquid water at low partial pressure. J Phys Chem 85:840–848

Leighton PA (1961) Photochemistry of air pollution. Academic Press, London New York

Levine JS, Rogowski RS, Gregory GL, Howell WE, Fishman J (1981) Simultaneous measurements of NO_x, NO, and O_3, and O_3 production in a laboratory discharge: Atmospheric implications. Geophys Res Lett 8:357–360

Lin G-Y (1982) Oxidant prediction by discriminant analysis in the South Coast Air Basin of California. Atmos Environ 16:135–143

Liu SC, Kley D, McFarland M, Mahlman JD, Levy II H (1980) On the origin of tropospheric ozone. J Geophys Res 85:7546–7552

Löbel J (1978) Reduzierungsstrategien zur photochemischen Luftverschmutzung. Staub-Reinhalt Luft 38:284–285

Löbel J (1979a) Simulation of photochemical smog in a 425 l smog chamber. Proc Photochemische Luftverunreinigungen in der Bundesrepublik Deutschland. Tagung des Umweltbundesamtes, Oct 1979, VDI, Düsseldorf, pp 13–45

Löbel J (1979b) Formation of sulfuric acid and sulfates in ambient air – A smog chamber study. Proc Int Symp Sulphur Emiss Environ. London 8–10 May 1979. Soc Chem Industry, London SW1X8PS

Löbel J, Wipprecht V, Schurath U (1980) Messungen von Peroxiacetylnitrat (PAN) in Außenluft. Staub-Reinhalt Luft 40:243–244

Logan JA (1983) Nitrogen oxides in the troposphere: global and regional budgets. J Geophys Res 88:10785–10807

Logan JA, Prather MJ, Wofsy SC, McElroy MB (1981) Tropospheric chemistry: A global perspective. J Geophys Res 86:7210–7254

Lonneman WA, Seila RL, Bufalini JJ (1978) Ambient air hydrocarbon concentrations in Florida. Environ Sci Technol 12:459–463

Lonneman WA, Bufalini JJ, Kuntz RL, Meeks SA (1981) Contamination from fluorocarbon films. Environ Sci Technol 15:99-103

Lowe DC, Schmidt U (1983) Formaldehyde (HCHO) Measurements in the nonurban atmosphere. J Geophys Res 88:10844–10858

Lowe DC, Schmidt U, Ehhalt DH (1980) A new technique for measuring tropospheric formaldehyde (CH_2O). Geophys Res Lett 10:825–828

Lowe DC, Schmidt U, Ehhalt DH (1981a) The tropospheric distribution of formaldehyde. Ber Kernforschungsanl Jülich, No 1756

Lowe DC, Schmidt U, Ehhalt DH, Frischkorn CGB, Nürnberg HW (1981b) Determination of formaldehyde in clean air. Environ Sci Technol 15:819–823

Martinez RI, Herron JT, Huie RE (1981) The mechanism of ozone-alkene reactions in the gas phase. A mass spectrometric study of the reactions of eight linear and branched-chain alkenes. J Am Chem Soc 103:3807–3820

Marx W, Monkhouse PB, Schurath U (1983) Kinetik und Intensität photochemischer Reaktionsschritte in der Atmosphäre. Abschlußber FKW 20, Ges Strahlen-Umweltforsch, München

McCracken M, Wuebbles D, Walton J, Duewer W, Grant K (1978) The Livermore regional air quality model: Concept and development. J Appl Meteorol 17:254–272

McFarland M, Kley D, Drummond JW, Schmeltekopf AL, Winkler PH (1979) Nitric oxide measurements in the equatorial pacific region. Geophys Res Lett 6:605–608

Meagher JF, Luria M (1982) Model calculations of the chemical processes occuring in the plume of a coal-fired power plant. Atmos Environ 16:183–195

Meyrahn H, Hahn J, Helas G, Warneck P, Penkett S (1984) Cryogenic sampling and analysis of peroxyacetyl nitrate in the atmosphere. See Versino B, Angeletti G (eds) (1984), pp 38–43

Michetschläger C, Köhler G, Strauss R (1978) Oxidantienmessung in München. Umwelt 2:90–94

Miller DF (1981) Modeling of SO_2 oxidation in smog. EPA-600/S3-81-040, Project Summary

Miller DF, Joseph DW (1976) Smog chamber studies on photochemical aerosol-precursor relationship. EPA-600/3-76-080

Miller S (1983) Reviewing health effects of pollutants. Environ Sci Technol 17:128A–130A

Mizogushi I, Makino K, Kudou S, Mikami R (1977) On the relationship of subjective symptoms to photochemical oxidant. EPA-600/3-77-001a. Proc Int Conf Photochem Oxidant Pollut and Its Control, pp 477–495

Mukammal EI, Neumann HH, Gillespie TJ (1982) Meteorological conditions associated with ozone in Southwestern Ontario, Canada. Atmos Environ 16:2095–2106

Muschalik B (1980) Der Einfluß meteorologischer Parameter auf die bodennahe Oxidantienverteilung. Im Auftrag des Umweltbundesamtes, March 1980, Universitätsinst Meteorol Geophys, Frankfurt aM

Muschalik B (1979) Der Einfluß meteorologischer Parameter auf die bodennahe Oxidantienverteilung. See K. H. Becker et al. (eds) (1979), Sect VI, pp 1–32

Nassar J, Goldbach J (1977) Zeitliche Verläufe von reaktiven Kohlenwasserstoff-Immissionen. VDI-Berichte 270, Ozon und Begleitsubstanzen im photochemischen Smog. VDI, Düsseldorf, pp 93–99

Nederbragt GW, Horst A van der, Duijn J van (1965) Rapid ozone determination near an accelerator. Nature (London) 206:87

Neitzert V, Seiler W (1981) Measurement of formaldehyde in clean air. Geophys Res Lett 8:79–82

Nelson PF, Quigley SM (1984) The hydrocarbon composition of exhaust emitted from gasoline fuelled vehicles. Atmos Environ 18:79–87

Neuber E (1982) Messungen von Vorläufer- und Oxidantienkonzentrationen in der verunreinigten Troposphäre. Forschungsprojekt 104-02551/03 im Auftrag des Umweltbundesamtes. Final Report, Berlin

Neuber E (1983) Untersuchungen zur Oxidantienbildung in der verunreinigten Troposphäre. Dissertation, Johann-Wolfgang-Goethe-Univ Frankfurt

Neuber E, Georgii H-W, Müller J (1981) Verteilung leichter Kohlenwasserstoffe an Meßstellen unterschiedlicher Luftqualität. Staub-Reinhalt Luft 41:91–97

Neuber E, Georgii H-W, Müller J (1982) Aircraft measurement of oxidants and precursors in the plumes of heavy-industrialized areas. See Versino B, Ott H (eds) (1982), pp 469–481

Nicolet M (1975) Stratospheric ozone: An introduction to its study. Rev Geophys Space Phys 13:593–636

Nicolet M (1978) Etude des réactions chimiques de l'ozone dans la stratosphère. Inst R Meteorol Belg, Brussels 1978

Nieboer H, Ham J van (1976) Peroxyacetyl nitrate (PAN) in relation to ozone and some meteorological parameters at Delft in the Netherlands. Atmos Environ 10:115–120

Nielsen T, Samuelsson U, Grennfelt P, Thomsen EL (1981) Peroxyacetyl nitrate in long-range transported polluted air. Nature (London) 293:553–555

Nielsen T, Hansen AM, Thomsen EL (1982) A convenient method for preparation of pure standards of peroxiacetyl nitrate for atmospheric analysis. Atmos Environ 16:2447–2450

Niki H, Daby EE, Weinstock B (1972) Mechanisms of smog reactions. Adv Chem Ser, vol 113:16–57. Am Chem Soc, Washington DC

Novak I, Sabljić A (1981) „Trenutni snimak" zagradenja zraka u Kvarnerskom zaljevu 1979. Kem Ind 30:5–8

Obländer W, Siegel D (1977) Ergebnisse der Außenluftmessung von Ozon und Kohlenwasserstoffen in Baden-Württemberg. VDI-Berichte 270: Ozon und Begleitsubstanzen im photochemischen Smog. VDI, Düsseldorf, pp 55–61

Ozone Chemistry and Technology (1959) Adv Chem Ser, vol 21. Am Chem Soc Washington DC

Paetzold HK, Regener E (1957) Ozon in der Atmosphäre. In: Flügge S (ed) Handbuch der Physik, vol XLV-III. Springer, Berlin Heidelberg New York, pp 370–426

Paffrath D, Funk H, Körner G (1983) Bodennahe Ozonmessungen in Oberpfaffenhofen vom August 1974 bis Februar 1981. Ber IB-553-7-83 Inst Phys Atmos, Oberpfaffenhofen, DFVLR

Penkett SA, Sandalls FJ, Lovelock JE (1975) Observations of peroxyacetyl nitrate (PAN) in the air in Southern England. Atmos Environ 9:139–141

Perner D, Hübler G (1982) Experimental detection of OH in the troposphere. In: Georggi HW, Jaeschke W (eds) Chemistry of the unpolluted and polluted troposphere. Reidel, Dordrecht, pp 267–294

Perner D, Ehhalt DH, Pätz HW, Platt U, Röth EP, Volz A (1976) OH-radicals in the lower troposphere. Geophys Res Lett 3:466–468

Perros P, Toupance G (1984) Production et transfert d'ozone sur le bassin de Fos-Berre. See Versino B, Angeletti G (eds) (1984), pp 596–602

Peterson JT (1976) Calculated actinic fluxes (290–700 nm) for air pollution photochemistry applications. EPA-600/4-76-025

Platt U (1981) The diurnal variation of tropospheric nitrate radicals (NO_3). Presented at the 2nd meeting of Working Group 2 on Mechanisms of Oxidation Processes in the Atmosphere, COST 61 a bis, Löwen 1981

Platt U, Perner D (1980) Direct measurements of atmospheric CH_2O, HNO_2, O_3, NO_2, and SO_2 by differential optical absorption in the near UV. J Geophys Res 85:7453–7358

Platt U, Perner D, Pätz W (1979) Simultaneous measurements of atmospheric CH_2O, O_3, and NO_2 by differential optical absorption. J Geophys Res 84:6329–6335

Platt U, Perner D, Winer AM, Harris GW, Pitts JN (1980a) Detection of NO_3 in the polluted troposphere by differential optical absorption. Geophys Res Lett 7:89–92

Platt U, Perner D, Harris GW, Winer AM, Pitts JN (1980b) Observations of nitrous acid in an urban atmosphere by differential optical absorption. Nature (London) 285:312

Platt UF, Winer AM, Biermann HW, Atkinson R, Pitts JN (1984) Measurement of nitrate radical concentrations in continental air. Environ Sci Technol 18:365–369

Post K (1979) Precursor distributions, ozone formation and control strategy options for Sydney. Atmos Environ 13:783–790

Post K (1981) Ozone formation and the spatial distribution of precursor emissions in Sydney. Atmos Environ 15:743–747

Post K, Bilger RW (1978) Ozone-precursor relationships in the Sydney Airshed. Atmos Environ 12:1857–1865

Prestel MA (1874) Die periodische Veränderung des Ozongehaltes der Luft im Laufe des Jahres. Z Oesterr Ges Meteorol 9:166

Pruchniewicz PG (1970) Über ein Ozon-Registriergerät und Untersuchung der zeitlichen und räumlichen Variation des troposphärischen Ozons auf der Nordhalbkugel der Erde. Mitt Max-Planck-Inst Aeronomie, No 42 (S). Springer, Berlin Heidelberg New York, pp 1–70

Rasmussen RA (1981) A review of the natural hydrocarbon issue. In: Bulfani JJ, Arnts RR (eds) Atmospheric biogenic hydrocarbons, vol I. Ann Arbor Sci, pp 3–12

Ratsch HC, Tingley DT (1977) Emission rates for biogenic NO_x. Emissions inventory factor workshop, Sept 13–15, 1977, Vol 2

Regener VH (1938) Neue Messungen der vertikalen Verteilung des Ozons in der Atmosphäre. Z Phys 109:642–670

Regener VH (1964) Measuring of atmospheric ozone with the chemiluminescent method. J Geophys Res 69:3795–3800

Reiter RE (1975) Stratospheric-tropospheric exchange processes. Rev Geophys Space Phys 13:459–474

Reiter R, Kanter HJ (1982a) Reasons for seasonal and daily variations of CO_2 and O_3 at 0.7, 1.8 and 3.0 km altitude recorded since 1977. See Versino B, Ott H (eds) (1982), pp 551–560

Reiter R, Kanter HJ (1982b) Time behavior of CO_2 and O_3 in the lower troposphere based on recordings from neighboring mountain stations between 0.7 and 3.0 km ASL including the effects of meteorological parameters. Arch Meteorol Geophys Bioklimatol Ser B 30:191–225

Reiter R, Munzert K, Sladkovic R, Pötzl K, Kanter HJ (1983) Basiserarbeitung zum Problem „Waldschäden" im bayerischen Nordalpenraum. Ber 8272-623-27768 Fraunhofer Inst Atmos Umweltforsch, Garmisch-Partenkirchen

Reynolds SD (1977) The Systems Applications, Inc. Urban airshed model: An overview of recent developmental work. Proc Int Conf Photochem Oxidant Pollut and Its Control, EPA-600/3-77-001b, pp 795–802

Reynolds SD, Roth PM, Seinfeld JH (1973) Mathematical modeling of photochemical air pollution. I. Formulation of the model. Atmos Environ 7:1033–1061

Reynolds SD, Liu MK, Hecht TA, Roth PM, Seinfeld JH (1974) Mathematical modeling of photochemical air pollution, III. Evaluation of the model. Atmos Environ 8:563–596

Reynolds SD, Ames J, Hecht TA, Meyer JP, Whitney DC (1976) Continued research in mesoscale air pollution simulation modeling, vol II SAI/EF-75-25. Systems Applications, San Rafael, Calif, and EPA 600/4-76-016d

Roberts JM, Fehsenfeld FC, Albritton DL, Sievers RE (1983) Measurement of monoterpene hydrocarbons at Niwot Ridge, Colorado. J Geophys Res 88:10667–10678

Röth P (1981) Vorschlag für die Berücksichtigung chemischer Vorgänge in einem dreidimensionalen Ausbreitungsmodell. Report „Ausbreitung von Schadstoffen in der Atmosphäre. Physikalisch-chemische Modelle", VDI-Kommission Reinhalt Luft, Düsseldorf

Roth PM, Roberts PJW, Lui MK, Reynolds SD, Seinfeld JH (1974) Mathematical modeling of photochemical air pollution. II. A model and inventory of pollutant emissions. Atmos Environ 8:97–130

Rudolph J, Ehhalt DH, Gravenhorst G (1980) Recent measurements of light hydrocarbons in remote areas. Proc 1st Eur Symp Phys-Chem Behav Atmos Pollut, EEC/COST 61 a bis, Ispra, Oct 1979, pp 41–51

Sakamaki F, Akimoto H, Okuda M (1980) Water vapor effect on photochemical ozone formation in a propylene-NO_x-air system. Environ Sci Technol 14:985–989

Sakamaki F, Okuda M, Akimoto H, Yamasaki H (1982) Computer modeling study of photochemical ozone formation in the propene-nitrogen oxides – dry air system. Generalized maximum ozone isopleths. Environ Sci Technol 16:45–52

Sakamaki F, Hatakeyama S, Akimoto H (1983) Formation of nitrous acid and nitric oxide in the heterogeneous dark reaction of nitrogen dioxide and water vapor in a smog chamber. Int J Chem Kinet 15:1013–1029

Saltzman BE (1954) Colorimetric microdetermination of NO_2 in the atmosphere. Anal Chem 26:1949–1955

Sanhueza E, Plum CN, Pitts JN (1984) Positive interference of nitrous acid in the determination of gaseous HNO_3 by the NO_x chemiluminescence-nylon cartridge method: Applications to measurements of ppb levels of HONO in air. Atmos Environ 18:1029–1031

Sawicki E, Hauser TR, Stanley TW, Elbert W (1961) 3-methyl-2-benzothiazolone hydrazone test. Sensitive new methods for detection, rapid estimation, and determination of aliphatic aldehydes. Anal Chem 33:93–96

Schere KL, Demerjian KL (1978) A photochemical box model for urban air quality simulation. Proc 4th Joint Conf Sensing Environ Pollut, New Orleans 1977. Am Chem Soc, Washington DC

Scherer B, Stern R (1982) Analysis of a photochemical smog episode and preparation of the meteorological input data for a three dimensional air quality dispersion model. See Versino B, Ott H (eds) (1982), pp 561–571

Schiavone JA, Graedel TE (1981) 2-D Studies of the kinetic photochemistry of the urban troposphere. I. Air stagnation conditions. Atmos Environ 15:163–176

Schjoldager J (1979) Observation of high ozone concentrations in Oslo, Norway, during the summer of 1977. Atmos Environ 13:1689–1696

Schjoldager J, Stige L (1980) Malinger av ozon i nedre Telemark, Oslo og Oslofjorden sommeren 1979. Norw Inst Air Res, Rep No 5/80, Ref 21179

Schjoldager J, Sivertsen B, Hanssen JE (1978) On the occurrence of photochemical oxidants at high latitudes. Atmos Environ 12:2461–2467

Schjoldager J, Grennfelt P, Saltbones J (1983) Photochemical oxidants in North-West Europe 1976–79, a pilot study. Paper presented at the EPA-OECD international conference on long range transport models for photochemical oxidants and their precursors. Research Triangle Park, NC/USA, 12–14 April 1983

Schmidt U, Lowe DC (1982) Vertical profiles of formaldehyde in the troposphere. See Versino B, Ott H (eds) (1982), pp 377–386

Schönbein CF (1844) Über die Erzeugung des Ozons auf chemischem Wege. Basel 1844

Schubert B, Schmidt U, Ehhalt DH (1984) Sampling and analysis of acetaldehyde in tropospheric air. See Versino B, Angeletti G (eds) (1984), pp 44–52

Schurath U (1979a) Physikalisch-chemische Grundlagen der Photosmog-Bildung unter Berücksichtigung von Smogkammer-Messungen. See Becker et al. (eds) (1979), Sect VII, pp 1–232

Schurath U (1979b) Fallstudien: Photosmog-Episoden im Köln-Bonner Raum. See Becker et al. (eds) (1979), Sect IX, pp 1–28

Schurath U, Wipprecht V (1980) Reactions of peroxiacyl radicals. Proc 1st Eur Symp Phys-Chem Behav Atmos Pollut, EEC/COST 61 a bis, Ispra, Oct 1979, pp 157–166

Schurath U, Wiese A, Becker KH (1976) Ein Chemilumineszenzanalysator für ungesättigte Kohlenwasserstoffe in der Außenluft. Staub-Reinhalt Luft 36:379–385

Schurath U, Marx WN, Monkhouse PB (1981) Field measurements of photolysis frequencies in the atmosphere. 10th Int Conf Photochem, Crete/Greece, Sept 1981. J Photochem 17:140

Schurath U, Goeth N, Henrich K, Lippmann H (1982) Untersuchungen zur Entwicklung eines chemischen Reaktionsmodells atmosphärischer Spurengasumsetzungen. Teil II: Analyse des chemischen Reaktionsmechanismus im „SAI Airshed Model 1978". Forschungsber 81-10402511/02, Umweltbundesamt, Berlin 1982

Schurath U, Henrich K, Lippmann H, Wendler W (1983) Untersuchungen zur Entwicklung eines chemischen Reaktionsmodells atmosphärischer Spurengasumsetzungen. Teil I: Durchführung und Analyse der Smogkammermessungen. Beschreibung des chemischen Modells. Forschungsber 81-10402511/02, Umweltbundesamt, Berlin 1983

Schurath U, Kortmann U, Glavas S (1984) Properties, formation, and detection of peroxyacetyl nitrate. See Versino B, Angeletti G (eds) (1984), pp 27–37

Seiler W, Fishman J (1981) The distribution of carbon monoxide and ozone in the free troposphere. J Geophys Res 86:7255–7265

Seinfeld JH, Reynolds SD, Roth PM (1972) Simulation of urban air pollution. Adv Chem Ser, vol 113:58–100. Am Chem Soc, Washington DC

Shaw RW, Stevens RK, Bowermaster J, Tesch JW, Tew E (1982) Measurements of atmospheric nitrate and nitric acid: The denuder difference experiment. Atmos Environ 16:845–853

Singh HB, Hanst PL (1981) Peroxyacetyl nitrate (PAN) in the unpolluted atmosphere: An important reservoir for nitrogen oxides. Geophys Res Lett 8:941–944

Singh HB, Salas LJ (1982) Measurement of selected light hydrocarbons over the Pacific Ocean: Latitudinal and seasonal variations. Geophys Res Lett 9:842–845

Singh HB, Salas LJ (1983) Peroxyacetyl nitrate in the free troposphere. Nature (London) 302:326–328

Singh HB, Ludwig FL, Johnson WB (1978) Tropospheric ozone: Concentrations and variabilities in clean remote atmospheres. Atmos Environ 12:2185–2196

Sklarew RC, Fabrick AJ, Prager JE (1972) Mathematical modelling of photochemical smog using the PICK method. J Air Pollut Control Assoc 22:865–869

Spicer CW (1977) Photochemical atmospheric pollutants derived from nitrogen oxides. Atmos Environ 11:1089–1095

Spicer CW, Joseph DW, Stickels PHD, Ward GF (1979) Ozone sources and transport in the Northeastern United States. Environ Sci Technol 13:975–985

Spicer CW, Howes JE, Bishop TA, Arnold LH, Stevens RK (1982) Nitric acid measurement methods: An intercomparison. Atmos Environ 16:1487–1500

Stangl H, Lohse C, Payrissat M, Versino B, Nicollin B, Ottbrini G, Rau H (1980) Discussion of field measurements of pollutants made at Ispra and other sites of the Po Valley. Proc 1st Eur Symp Phys-Chem Behav Atmos Pollut, EEC/COST 61 a bis, Ispra, Oct 1979, pp 472–478

Stephens ER (1969) The formation, reactions, and properties of peroxyacetyl nitrates in photochemical air pollution. In: Pitts JN, Metcalf RL (eds) Adv Environ Sci, vol I. Wiley, New York, pp 119–146

Stern R, Scherer B (1980) An application of the empirical kinetic modeling approach (EKMA) to the Cologne area. Proc 11th NATO-CCMS Int Tech Meet, Amsterdam 1980

Stewart HNM, Sullivan EJ, Williams ML (1976) Ozone levels in Central London. Nature (London) 263:582–584

Stockwell WR, Calvert JG (1978) The near ultraviolet absorption spectrum of gaseous HONO and N_2O_3. J Photochem 8:193–203

Strauss R (1980) Ergebnisse von Messungen der Stickoxid-, Kohlenwasserstoff- und Ozonimmissionen in bayerischen Städten. Photochemische Luftverunreinigungen in der Bundesrepublik Deutschland. Tagung des Umweltbundesamtes, Oct 1979, VDI, Düsseldorf, pp 156–167

Tangermann-Dlugi G, Fiedler F (1979) Numerisches Simulationsmodell ALAG. Manuskr Meteorol Inst Univ Karlsruhe, 1979

Taylor WD, Allston TD, Moscato MJ, Fazekas GB, Kozlowski R, Takacs GA (1980) Atmospheric photodissociation lifetimes of nitromethane, methyl nitrite, and methyl nitrate. Int J Chem Kinet 12:213–240

Ten Brink HM, Kelly TJ, Lee YN, Schwartz SE (1984) Attempted measurements of gaseous H_2O_2 in the ambient atmosphere. See Versino B, Angeletti G (eds) (1984), pp 20–26

Tiao GC, Phadke MS, Box GEP (1976) Some empirical models for the Los Angeles photochemical smog data. J Air Pollut Control Assoc 26:485–490

Tiefenau H, Fabian P (1972) The specific ozone destruction at the ocean surface and its dependence on horizontal wind velocity from profile measurements. Arch Meteorol Geophys Bioklimatol Ser A 21:399–412

Tilden JW, Seinfeld JH (1982) Sensitivity analysis of a mathematical model for photochemical air pollution. Atmos Environ 16:1357–1364

TNO (1978) Photochemical smogformation in the Netherlands. Guicherit R (ed). TNO's-Gravenhage

Troe J (1979) Predictive possibilities of unimolecular rate theory. J Phys Chem 83:114–126

Tuazon EC, Winer AM, Pitts JN (1981) Trace pollutant concentrations in a multiday smog episode in the California South Coast Air Basin by long path length Fourier transform infrared spectroscopy. Environ Sci Technol 15:1232–1237

Tuazon EC, Atkinson R, Plum CN, Winer AM, Pitts JN (1983) The reaction of gas phase N_2O_5 with water vapor. Geophys Res Lett 10:953–956

TÜV Rheinland e V (1980) Das Emissionsverhalten von Personenkraftwagen in der Bundesrepublik Deutschland im Bezugsjahr 1980. Schmidt, Berlin

Turco RT (1975) Photodissociation rates in the atmosphere below 100 km. Geophys Surv 2:153–192

Umweltbundesamt (1981) Luftreinhaltung '81. Entwicklung – Stand – Tendenzen. Schmidt, Berlin

Uno I, Wakamatsu S, Suzuki M, Ogawa Y (1984) Three-dimensional behavior of photochemical pollutants covering the Tokyo metropolitan area. Atmos Environ 18:751–761

U.S.E.P.A. (1977a) A survey of sulfate, nitrate, and acid aerosol emissions and their control. EPA-600/7-77-041, Research Triangle Park

U.S.E.P.A. (1977b) Uses, limitations, and technical basis of procedures for quantifying relationships between photochemical oxidants and precursors. US-EPA-450/2-77-021a

U.S.E.P.A. (1981) Guideline for use of city-specific EKMA in preparing ozone SIPs. USEPA-450/4-80-027

US Federal Register (1971) 36:15491

VDI-Richtlinien: A series of technical notes and recommendations. VDI (ed) VDI, Düsseldorf

Venkatram A (1978) An examination of box models for air quality simulation. Atmos Environ 12:2243–2249

Versino B, Ott H (eds) (1982) Phys Chem Behav Atoms Pollut. Proc 2nd Eur Symp, Varese/Italy, 29 Sept–1 Oct 1981, Reidel, Dordrecht

Versino B, Angeletti G (eds) (1984) Phys Chem Behav Atmos Pollut. Proc 3rd Eur Symp, Varese/Italy, 10–12 April 1984, Reidel, Dordrecht

VGB (1982) Kraftwerksemissionen und saure Niederschläge. VGB-TW 302, Essen 1982

Viezee W, Johnson WB, Singh HB (1983) Stratospheric ozone in the low lower troposphere – II. Assessment of downward flux and ground-level impact. Atmos Environ 17:1979–1993

Volz A (1979) Messung von atmosphärischem ^{14}CO: Eine Methode zur Bestimmung der troposphärischen OH-Radikalkonzentration. Dissertation, Universität Bonn, 1979

Volz A, Ehhalt DH, Derwent RD (1981) Seasonal and latitudinal variation of ^{14}CO and the tropospheric concentration of OH radicals. J Geophys Res 86:5163–5171

Wakamatsu S, Ogawa Y, Murano K, Goi K, Abutamoto Y (1983) Aircraft survey of the secondary photochemical pollutants covering the Tokyo metropolitan area. Atmos Environ 17:827–835

Walker JF (1975) Formaldehyde, 3rd edn. Krieger, Humbington, New York

Wang CC, Davis LI (1974) Measurement of hydroxyl concentrations in air using a tunable UV-laser beam. Phys Rev Lett 32:349–352

Warmbt W (1979) Ergebnisse langjähriger Messungen des bodennahen Ozons in der DDR. Meteorol 29:24–31

Warmbt W (1981) Results of long-term measurements of near-surface ozone in the GDR. A contribution to actual problems on tropospheric ozone. Proc KAPG Symp Invest Atmos Ozone, Tblissi, USSR, pp 242–248, Nov 23–27

Wayne LG, Kokin A, Weisburd MI (1973) Controlled evaluation of the reactive environmental simulation model (REM). USEPA-R4-73-013a, vol I

Whelpdale DM (1982) Wet and dry deposition. In: Georgii HW, Jaeschke W (eds) Chemistry of the unpolluted and the polluted troposphere. Reidel, Dordrecht, pp 375–391

Whitten GZ, Hogo H, Dodge MC (1978) User's manual for kinetic model and ozone isopleth plotting package. EPA-600/8-78-014a

Whitten ZG, Hogo H, Killus JP (1980) The carbon-bond mechanism: A condensed kinetic mechanism for photochemical smog. Environ Sci Technol 14:690–700

Winkler P (1980) Kritische Bemerkungen zur Frage photochemischer Ozonproduktion aus meteorologischer Sicht. Photochemische Luftverunreinigungen in der Bundesrepublik Deutschland, Tagung des Umweltbundesamtes, Oct 1979. VDI, Düsseldorf, pp 197–207

Wisse JA, Velds CA (1970) Preliminary discussion on some oxidant measurements at Vlaadeningen, the Netherlands. Atmos Environ 4:79–85

Wofsy SC (1976) Interactions of CH_4 and CO in the Earth's atmosphere. Annu Rev Earth Planet Sci 4:442–469

Wolff GT, Lioy PJ (1978) An empirical model for forecasting maximum daily ozone levels in the Northeastern US. J Air Pollut Control Assoc 28:1034–1038

Yokouchi Y, Okaniwa M, Ambe Y, Fuwa K (1983) Seasonal variation of monoterpenes in the atmosphere of a pine forest. Atmos Environ 17:743–750

Yoshida T, Akita M, Takeyama S (1977) Simulation of photochemical reactions in smog chambers. Proc 4th Int Clean Air Congr, Tokio, pp 501–503

Yoshizumi K, Aoki K, Nouchi I, Okita T, Kobayashi T, Kamakura S, Tajima M (1984) Measurements of the concentration in rainwater and of the Henry's law constant of hydrogen peroxide. Atmos Environ 18:395–401

Zafiriou OC, McFarland M (1981) Nitric oxide from nitrite photolysis in the Central Equatorial Pacific. J Geophys Res 86:3173–3182

Zika RG, Saltzman ES (1982) Interaction of ozone and hydrogen peroxide in water: Implications for analysis of H_2O_2 in air. Geophys Res Lett 9:231–234

Zika RG, Saltzman ES, Chameides WL, Davies DD (1982) H_2O_2 levels in rainwater collected in South Florida and the Bahama Islands. J Geophys Res 87:5015–5017

Zimmerman PR (1979) Testing of hydrocarbon emissions from vegetation, leaf litter and aquatic surfaces, and development of a methodology from compiling biogenic emission inventories. EPA-450/4-79-004

Zimmerman PR, Chatfield RB, Fishman J, Crutzen PJ, Hanst PL (1978) Estimates on the production of CO and H_2. From the oxidation of hydrocarbon emissions from vegetation. Geophys Res Lett 5:679–682

Zittel (1874) Über den Ozongehalt der Wüstenluft. Z Oesterr Ges Meteorol 1:9

Part 2

Effects of Photochemical Oxidants on Plants

ROBERT GUDERIAN

Institut für Angewandte Botanik der Universität Essen

DAVID T. TINGEY

U.S. Environmental Protection Agency, Corvallis, Oregon

RUDOLF RABE

Rheinisch-Westfälischer Technischer Überwachungsverein, Essen

Chapter 2.1 Introduction

Although numerous secondary components occur in the photochemical smog complex (Part 1, this vol.), only the most important phytotoxic components, ozone (O_3) and peroxyacetyl nitrate (PAN) are discussed. Phytotoxic concentrations of ozone are more widely distributed and generally occur in higher ambient concentrations than do those for PAN. However, on volume (molecular) basis, PAN is more phytotoxic. PAN is the first compound of a homologous series whose phytotoxicity increases with molecular weight in the following sequence: peroxyacetyl nitrate (PAN) < peroxypropionyl nitrate (PPN) < peroxyisobutyryl nitrate (PisoBN) < peroxybutyryl nitrate (PBN) < peroxybenzoyl nitrate (PBzN). Based on current knowledge, none of these compounds, except PAN, occurs at phytotoxic concentrations in the atmosphere (Mudd 1975, EPA 1978a). The possible practical importance of the co-occurrence of PAN and ozone (O. C. Taylor 1969, Temple 1982) will be discussed, as will the influence of the photochemical smog complex.

The basis for material selection, its organization and presentation, was influenced by the relationships among ambient exposures, environmental factors, plant responses (Fig. 2.1), and the desire to provide a perspective for practical environmental protection measures, in addition to reviewing available literature for students and researchers.

The reaction of a plant to a given air pollutant depends on the ambient exposure, the amount of pollutant diffusing into the leaves, and the plant's autonomous and environmentally modified resistance. The degree of resistance is controlled primarily by the plant's genetic complement (Chap. 2.4.3.2) and its developmental stage at the time of exposure (Chap. 2.4.3.1) and, secondarily, by the modifying influence of various external factors (Chap. 2.4.2). Depending on the intensity, nature, and frequency of the photochemical oxidant exposure (Chap. 2.4.1), the plant response is influenced by given environmental conditions such as climate (Chap. 2.4.2.1), soil (Chap. 2.4.2.2), and biotic factors (Chap. 2.2.5). The response can range from acute, chronic (Chap. 2.3.1.1), or subtle (Chap. 2.3.1.2) injury in an individual plant to changes in plant communities, their composition and structure (Chap. 2.2.5). Injury to the ecological and economical functions of the affected plants can lead to an additional reduction in their ideal values and finally to gene erosion through the loss of species (Chap. 2.3.3). A range of experimental and survey methods are available to determine the various air-pollution effects at the individual organizational levels of the ecosystem (Chap. 2.3.4).

Studies of air-pollution effects on vegetation are not an end in themselves. Consequently, it is not sufficient to describe only the ambient exposure and resul-

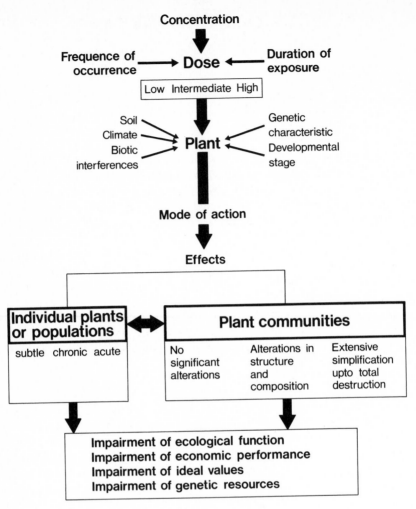

Fig. 2.1. Reactions of plants as individuals, populations, and communities to air pollutants, depending on various response-influencing factors

tant effects; a basis for remedial measures should be given. The implementation of these measures assumes that one has information about the type, intensity, spatial distribution, and cause of the pollution-induced injury. Presentations in Chapters 2.2 and 2.3 should help answer these questions, and also the extent that biological indicators can be used to detect and evaluate the risks arising from photochemical oxidants. The uses of, and possible changes in, plant cultural practices as a way to reduce effects of air pollutants on plants are discussed in Chapter 2.4.4.

The decisive factor in protecting vegetation is to reduce the ambient pollutant exposure. A significant basis for establishing air quality criteria is an accurate knowledge of the ambient exposure characteristics and the resulting responses.

Data developed under various limiting conditions and from different countries were used to evaluate and interpret the quantitatively determined relationships between ambient exposure and effect, and also to derive the maximal acceptable ozone concentrations for the protection of vegetation (Chap. 2.6).

Photochemical oxidants constitute a complex mixture of secondary pollutants, many of which are biologically active and may co-occur with primary air pollutants. Therefore, the effects of the single gases and also pollutant combinations are discussed (Chap. 2.5). The evidence that ozone may be a primary cause of the forest injury now occurring in Central Europe is reviewed (Chap. 2.5.5). These topics, integrated with Chapter 2.2, Mode of Action, provide a comprehensive analysis and description of pollution uptake and its effects on plants. A unique concept used to evaluate the results concerning the mode of action for ozone and PAN provides a significant basis for the recognition and evaluation of their effects in relation to practical environmental protection measures and for the formulation of additional research topics.

Chapter 2.2 Mode of Action

The various plant responses to photochemical oxidants at various levels of biological organization are based on a sequence of biochemical and physiological events terminating in injury (Fig. 2.2). From this perspective, Tingey and G. E. Taylor (1982) proposed a conceptual model to summarize plant responses which included pollutant uptake (leaf conductance), perturbation, homeostasis, and injury which can be equated with or lead into damage (see Chap. 2.3). This concept is summarized as a basis for discussing the mechanisms of ozone and PAN actions on vegetation.

Pollutant uptake is governed by leaf conductance (conductance = 1/diffusive resistance) which regulates pollutant movement from the ambient air into cellular perturbation sites. The gas phase movement of ozone into the leaf depends on the concentration gradient between the ambient air and sorptive sites within the leaf and the ease of mass transfer along the diffusive path. Following the sorption of pollutants onto the wet cell surfaces, their movement is controlled by liquid phase reactions, including diffusion and reaction with scavenging systems.

Perturbation is the initial phytotoxic event and results from ozone-induced alterations in cellular function or structure. The initial phytotoxicity is caused by the reactions of ozone or its derivatives with macromolecules that are active in cellular functions. The primary site for ozone reactions is cellular membranes causing changes in selective permeability.

Homeostasis is the recovery process that occurs following perturbations. The cells may respond with either repair or compensatory mechanisms.

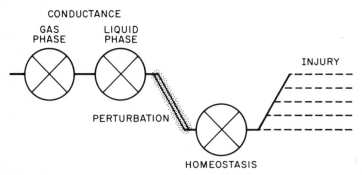

Fig. 2.2. Conceptual model for the biochemical and physiological processes controlling plant response to ozone. The major processes controlling the response are leaf conductance (gas and liquid phases) and homeostasis. (Tingey and G. E. Taylor 1982)

Injury is caused by pollutant absorption and its subsequent reaction at active sites within the cell. Injury is the result of the inability to repair, or compensate for cellular dysfunctions. The primary role of membranes in metabolism and their reactivity with ozone suggests that metabolic repercussions will occur throughout the cell following membrane disruption.

2.2.1 Pollutant Uptake

Photochemical oxidants are absorbed directly from the atmosphere in contrast to some other air pollutants, for example, acidic gases and heavy metals which are absorbed both by the aboveground plant parts and the roots following deposition into the soil. Uptake is limited essentially to the chlorophyll-containing organs; thus practically all the absorption occurs within the leaves. Gas absorption processes in leaves and plant canopies are described in the following sections.

2.2.1.1 Pollutant Uptake into Leaves

Photochemical oxidant injury results from biochemical and physiological reactions occurring within the cells of the leaf interior. The rate and amount of ozone and PAN that diffuse from the ambient air to reactive sites within the leaf are determinants of injury. Pollutant flux is a function of the chemical and physical properties along the gas-to-liquid diffusive pathway. Physical structures may restrict pollutant flux; chemical reactions may scavenge ozone or PAN molecules within the gas and liquid phases.

2.2.1.1.1 Gas Phase Conductance

The absorption of gases, such as ozone or PAN, is the consequence of the chemical potential gradient between the bulk air (C_a) and the leaf interior (C_i). Flux is proportional to the pathway conductance (g) and inversely related to the resistances (R) to mass transfer along the diffusive pathway as the pollutant moves through the boundary layer (R_a), stomata and intercellular spaces (R_s), and the liquid phases (R_r). These relations have been used in many modeling studies of gas uptake (e. g., J. H. Bennett et al. 1973, Ting and Dugger 1968, Tingey and G. E. Taylor 1982) and are illustrated in Fig. 2.3. Total pollutant flux (J_{total}) is the sum of the flux to the leaf surface ($J_{surface}$) and the leaf interior ($J_{internal}$). Although fluxes to the leaf surface and leaf interior may both be large, cellular perturbations result only from gases diffusing into the leaves ($J_{internal}$). Because both the surface and internal fluxes may change during exposure, the ambient dose (concentration × duration of exposure) to which the plant is exposed is not necessarily a reliable estimate of the cellular dose that the plant actually receives.

The flux of the gases to the leaf surface is controlled by the boundary layer resistance (R_a), which is a function of leaf orientation, size, shape, epidermal char-

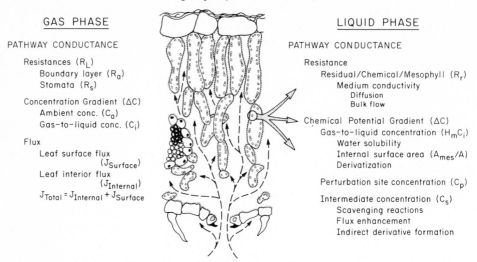

DETERMINANTS OF OZONE FLUX (J)
$$J = \Delta C / (R_a + R_s + R_r) \quad \text{or} \quad J = \Delta C \cdot g$$

GAS PHASE

PATHWAY CONDUCTANCE

Resistances (R_L)
  Boundary layer (R_a)
  Stomata (R_s)

Concentration Gradient (ΔC)
  Ambient conc. (C_a)
  Gas-to-liquid conc. (C_i)

Flux
  Leaf surface flux
        ($J_{Surface}$)
  Leaf interior flux
        ($J_{Internal}$)
$J_{Total} = J_{Internal} + J_{Surface}$

LIQUID PHASE

PATHWAY CONDUCTANCE

Resistance
  Residual/Chemical/Mesophyll (R_r)
    Medium conductivity
      Diffusion
      Bulk flow

Chemical Potential Gradient (ΔC)
  Gas-to-liquid concentration ($H_m C_i$)
    Water solubility
    Internal surface area (A_{mes}/A)
    Derivatization

Perturbation site concentration (C_p)

Intermediate concentration (C_s)
  Scavenging reactions
  Flux enhancement
  Indirect derivative formation

Fig. 2.3. The relationship of pollutant flux, chemical potential gradient, and leaf resistances or conductances in the gas and liquid phases. (Tingey and G. E. Taylor 1982)

acteristics, trichomes, and wind speed. The boundary layer thickness decreases with increasing wind speed, and at speeds above 2 m s^{-1} it can be neglected (Hill 1971). However, in exposure chambers with low wind speeds or closed chambers with little or no air exchange, the boundary layer resistance can be appreciable and influence pollutant uptake. For example, ryegrass (*Lolium perenne*) exposed to SO$_2$ at a wind speed of 0.17 m s^{-1} was uninjured; when exposed at a wind speed of 0.42 m s^{-1} it displayed considerable injury (Ashenden and Mansfield 1977). Similar results were obtained by Brennan and Leone (1968) in plant studies using ozone and SO$_2$. A sufficient air exchange and turbulence to minimize the boundary layer and prevent pollution depletion within the boundary layer are important considerations in establishing the comparability of results among experimental studies.

Leaf morphology can have a significant effect on pollutant flux. Elkiey et al. (1979) related differential ozone resistance in petunia cultivars to trichome density, raised and roughened cuticular features, and small epidermal cells. They suggested that these morphological features influenced the amount of pollutant diffusing into the leaf. J. H. Bennett et al. (1973) reported that pubescent leaf surfaces absorbed more ozone that glabrous ones. Studies with white pine clones showed that ozone exposure (0.30 ppm 6 h/day^{-1} for 7 days) did not alter cuticular wax structure (Trimble et al. 1982). Cuticular wax structure was also not correlated with ozone sensitivity in the clones.

In addition to the boundary layer resistance, stomatal resistance also influences pollutant uptake. Stomatal resistance is determined by stomatal number, size, anatomical characteristics of the guard cells, and the size of the stomatal pore. There is general agreement that when the stomata are closed, pollutant uptake is reduced and essentially no plant injury results. The opening of the stomatal pore is controlled by the internal CO_2 concentration and the hydration of the guard cells; it is also influenced by temperature, relative humidity, light, plant nutrition, and water availability. The stomatal opening is controlled by the turgor differential between the guard cells and the adjacent subsidiary cells. When the guard cells have a higher turgor than the epidermal cells, the stomatal pores open; when it is less, the pore opening is reduced. The turgor differential results from the establishment of an osmotic gradient by the accumulation of K ions in the guard cells (Zelitch 1969). The osmotically active K^+ causes an increased water absorption into the guard cells, increasing turgor.

The external and internal factors which directly or indirectly influence the stomatal control mechanisms alter the stomatal aperture, influencing pollutant uptake, as shown in Chapters 2.4.2 and 2.4.3.

Numerous studies have focused on the association between stomatal resistance, ozone uptake, and the resultant foliar injury; most have concluded that stomatal resistance was the principal or sole factor controlling differential foliar injury (Table 2.1). This mechanism has been invoked to explain the differences in plant reactions to ozone as influenced by environmental factors – edaphic and climatic, internal growth factors, and inter- and intraspecific differences in foliar sensitivity. The basic assumption is that pollutant uptake is proportional to stomatal resistance, therefore an increase in stomatal resistance will decrease internal ozone dose, reducing injury. However, experimental verification of this assumption is lacking (Table 2.1). The failure to measure ozone uptake to support this assumption may explain why injury is frequently not related to pollutant uptake.

Table 2.1. Examples of genetic and environmental control of ozone injury in which the mechanism was thought to be stomatal conductance. Ozone flux was not measured. (Tingey and G. E. Taylor 1982)

Genetic variation	Environmental variation
Interspecific	Edaphic
Pinus spp. (Evans and Miller 1972)	Potassium (Leone 1976, Noland
Multiple spp. (Thorne and Hanson 1972)	and Kozlowski 1979)
Intraspecific	Water (Markowski and Grzesiak 1974)
Acer rubrum (Townsend and Dochinger 1974)	Salinity (Hoffman et al. 1973, 1975)
Nicotiana tabacum (Ting and Dugger 1971,	Nitrogen (MacDowall 1965a)
Dean 1972, Rich and Turner 1972,	Climatic
Turner et al. 1972)	Light (Dunning and Heck 1973)
Allium cepa (Engle and Gabelman 1966)	Humidity (Otto and Daines 1969)
Phaseolus vulgaris (Thorne and Hanson 1976,	Atmospheric
Butler and Tibbits 1979b)	Ozone dose (MacDowall 1965a,
Developmental	Hill and Littlefield 1969,
Hypericum spp. (Ledbetter et al. 1959)	Runeckles and Rosen 1977)
Nicotiana tabacum (Glater et al. 1962)	Abscisic acid (Fletcher et al. 1972)
	Other pollutants (Jacobson and
	Colavito 1976)

Table 2.2. Examples of variable plant responses to ozone that are not associated with differential gas phase conductance. (Tingey and G. E. Taylor 1982)

Variation source	Flux measurement		Reference
	Water vapor	Ozone	
Genetic			
Interspecific			
Cucumis sativus	X		Beckerson and Hofstra (1979a)
Raphanus sativus			
Glycine max			
Multiple Species		X	Townsend (1974)
Intraspecific			
Vitis labrusca	X		Rosen et al. (1978)
Glycine max		X	G. E. Taylor et al. (1982)
Developmental			
Phaseolus vulgaris	X		Evans and Ting (1974)
Nicotiana tabacum		X	Leuning et al. (1979b)
Gossypium hirsutum	X		Ting and Dugger (1968)
Glycine max		X	G. E. Taylor et al. (1982)
Environmental			
Biotic			
Nicotiana tabacum virus	X		Brennan (1975)
Glycine max virus	X		Vargo et al. (1978)
Atmospheric			
Ethylene diurea chemical	X		Carnahan et al. (1978)
Phaseolus vulgaris			

However, other studies have not reported an association between stomatal resistance and foliar injury (Table 2.2). In some of these studies, ozone uptake was inferred from water flux data, but in others ozone flux was measured. It appears unwarranted to establish a priori a relationship between stomatal resistance and foliar injury.

The data relating ozone uptake to foliar injury are also contradictory. Ozone uptake was greater in young than old foliage of Camellia (Thorne and Hanson 1972). Differences in total ozone flux were related to leaf-age response differences in tobacco (Craker and Starbuck 1972). However, there was no consistent pattern between ozone uptake and the sensitivity of nine shade tree species to ozone (Townsend 1974). Ozone flux for two sensitive species, *Fraxinus americana* and *Quercus alba*, ranged from 398 to 1,058 μg m^{-2} s^{-1}, respectively.

If ozone injury is dependent solely on internal ozone flux, then internal flux should be a reliable predictor of foliar injury. In *Petunia hybrida* cultivars, the sensitive cultivar absorbed significantly more ozone than the resistant one (Elkiey and Ormrod 1980a). However, differential sensitivity in soybean cultivars was not associated with the internal pollutant flux (G. E. Taylor et al. 1982). The sensitive cultivar consistently absorbed less ozone than the resistant cultivar. Ozone uptake per cell has been estimated for several plant species (Craker and Starbuck 1973, Heath 1975, G. E. Taylor et al. 1982). Based upon these studies, injury thresholds appear to be 20 to 40 fmoles per cell.

The stomatal conductance of bean leaves exposed to subinjurious levels of PAN was similar to the controls (Metzler and Pell 1980). Following exposure to injurious levels, stomatal conductance was higher in the dark and lower in the light than in control plants, suggesting that the injury reduced the diurnal amplitude of leaf resistance. Proximity to the stomata was correlated with the appearance of cellular injury.

2.2.1.1.2 Liquid Phase Conductance

The diffusion of ozone from extracellular deposition sites to cellular perturbation sites includes movement through the liquid phase of the cell. Along this pathway it may experience diffusion, and/or bulk flow through hydrated cell walls, membranes, and cytosol, each of which is more viscous than air. Therefore, diffusion will proceed several orders of magnitude more slowly than for equivalent distances in air. Also, the transfer of ozone may be impeded by scavenger reactions that consume ozone. This chemical reactivity in the liquid phase may affect ozone movement, i.e., potential gradient and distribution within the cells. In solution, unsaturated fatty acids, sulfhydryl, and ring-containing compounds are ozone-susceptible. The reaction of ozone with olefinic double bonds produces an aldehyde or ketone and a Criegee zwitterion, which decomposes to produce another aldehyde or ketone and hydrogen peroxide (Tingey and G.E. Taylor 1982). Ozone has also been shown to increase free radicals in plant leaves indirectly or by decomposition in solution (Tingey and G.E. Taylor 1982). Biological systems have evolved scavenging or buffering systems such as superoxide dismutase, catalase, or peroxidases to metabolize free radicals (McCord et al. 1971, Fridovich 1975, 1978). The importance of oxidant-scavenging processes in aerobic metabolism (Fridovich 1975, 1978) suggests that these processes would be important in controlling ozone distribution within the cell. Foliar or root applications of an ethylene diurea (EDU) prevented foliar injury (Carnahan et al. 1978, E.H. Lee and J.H. Bennett 1982) without affecting pollutant uptake (J.H. Bennett et al. 1978). The EDU-induced resistance was associated with higher tissue levels of superoxide dismutase and catalase (E.H. Lee and J.H. Bennett 1982). Young leaves of *Populus euramericana* are more resistant to sulfur dioxide and contain higher superoxide dismutase activity than older leaves (Tanaka and Sugahara 1980). Similarly, the higher ozone resistance of young bean leaves compared to older ones has been associated with higher superoxide dismutase levels (E.H. Lee and J.H. Bennett 1982).

Other scavenging mechanisms have also been suggested. Giese and Christensen (1954) proposed that the cell's first defense against ozone was reactive organic matter on the cell surface coat. The variable ozone response of cell cultures of *Saccharomyces cerevisiae* was related to an array of unspecified intracellular scavenger molecules (Dubeau and Chung 1979). Sulfhydryls (e.g., glutathione), sulfhydryl proteins, cysteine, and small amounts of methionine apparently serve as free-radical scavengers and peroxide decomposers (Tingey and G.E. Taylor 1982).

2.2.1.2 Pollutant Uptake by Plant Canopies

Gas uptake by plant canopies is controlled by the same general processes (concentration gradient, and resistance to mass transfer) that control diffusion into leaves. Only the magnitude of the parameters and some of the physical impediments are changed. Pollutant uptake by alfalfa canopies has been measured in both gas exchange chambers and in field studies (Hill 1971). Given constant environmental conditions, pollutant uptake by alfalfa was proportional to pollutant concentration. However, at higher ozone concentrations, stomatal closure occurred, causing a proportional reduction in ozone uptake.

Using a wind tunnel, Garland and Penkett (1976) estimated a sorption rate of PAN to a grass stand, suggesting a deposition velocity of 0.25 cm s^{-1}. However, this value did not consider the influence of internal and external factors on the pollutant uptake rate.

Treshow and Stewart (1973) measured ozone concentrations of 0.09 ppm above an aspen canopy and 0.01 ppm within the canopy, indicating a significant sorption by the canopy. This concentration differential was maintained even at wind velocities up to 9 m s^{-1}. Using micrometeorological techniques, ozone uptake has been measured for corn and tobacco fields (Leuning et al. 1979a, b, Wesely et al. 1978). Uptake rates ranged from 1.1 to 9.0 nmol cm^{-2} h^{-1}.

2.2.2 Perturbation

Probable sites of ozone-induced perturbations are those highly ordered molecular configurations that are essential to the function of the cell. Experimental results, primarily from ultrastructural and biochemical studies, suggest that the membranes are the primary site of ozone action. Initial ultrastructural changes are diverse, but most indicate membrane dysfunction (see Sect. 2.2.4.1). Tobacco plants exposed to ozone (0.18 ppm) exhibited a loss of turgor within a few minutes after the exposure started, indicating an increase in membrane permeability (Keitel and Arndt 1983). Subsequently, photosynthesis and transpiration were reduced. These observations support the theory that the primary effect of ozone is to increase the permeability of the plasmalemma causing ionic imbalances which are reflected in altered metabolic processes.

Various indicators of membrane permeability have been proposed as measures of ozone perturbations at the cellular level (Heath 1975). Many studies have shown the effects of ozone on plant membranes. For example, ozone exposure to citrus and bean leaves doubled their permeability to radioactive glucose (Dugger and Palmer 1969, Perchorowicz and Ting 1974). Also, there was an increased loss of ^{86}Rb (Evans and Ting 1973); K^{+} loss was increased and uptake was reduced (Heath et al. 1974, Chimiklis and Heath 1975, Heath and Frederick 1979). The effects were reversible following short-term exposures, but continued ozonation caused irreparable injury to membrane components. These data suggest an effect at a limited number of transport sites rather than a general deterioration of the membrane. Alterations in membrane integrity may cause spatial dislocations of membrane components affecting metabolic processes such as photosynthesis. Ozone effects on photosynthesis have been associated with an impairment of membrane-bound systems (Chang and Heggestad 1974, Koiwai and Kisaki 1976). Coulson and Heath (1974) concluded that "... ozone disrupts the normal pathway of energy flow from light-excited chlorophyll into the photoacts by the disruption of the components of the membrane but not a general disintegration of the membrane."

The effects of ozone on various chemical components of membranes have been studied in both artificial systems and intact membranes. The lipid component of cellular membranes is ozone-labile, with the olefinic double bonds of the unsaturated fatty acids being oxidized to form shorter chain saturated carboxylic acids, aldehydes, and a few other reaction products (Teige et al. 1974, Goldstein 1977, Mudd 1982). However, these results from isolated systems have not been confirmed in most vegetation studies. Tomlinson and Rich (1969) observed a reduction in the fatty acid of tobacco leaves following exposure. In contrast, Swanson et al. (1973) found that some fatty acids were decreased but others increased. Mudd (1982) has concluded: "Although the reaction of ozone with unsaturated lipids as a basis for ozone toxicity is intellectually appealing, the body of evidence does not clearly support such a hypothesis".

Ozone can oxidize free amino acids, the most susceptible being cysteine, tryptophane, methionine and, to a lesser extent, histidine and tyrosine (Mudd 1982). Although the oxidation of free amino acids can have harmful consequences, the reaction of ozone with susceptible amino acid residues in various proteins would also have serious metabolic effects for the cell, depending on the protein turnover rate. The sensitivity of various proteins to ozone and identification of the labile amino acid residues have been reviewed (Mudd 1982). Sulfhydryls are also oxidized by ozone to disulfides and sulphonates (Tomlinson and Rich 1970, Mudd 1982). Thiol compounds are oxidized by PAN (Mudd 1982). Sulfhydryl compounds are part of the active site of several enzymes and contribute to the tertiary structure of the various proteins. The effects of ozone and PAN on the protein constituents of membranes will have important consequences for normal membrane function.

2.2.3 Homeostasis

Following a perturbation, living cells attempt to re-establish their normal metabolic levels using repair and/or compensatory processes (Levitt 1980). By definition, repair processes correct the perturbation, eliminating the dysfunction. In compensation, the perturbation and its physiological consequences remain, but the cells react with processes that counter the detrimental effects of the perturbation. Because repair and compensation may range from partial to complete, the degree of injury will vary as well.

Mudd (1980) suggested that resistant plants possess a greater capacity of unspecified mechanisms that repaired cellular injury. In microscope studies of pollution-impacted spruce needles, Sorauer and Ramann (1899) observed the repair of cellular dysfunction. On dicotyledonous plants, an early symptom of ozone injury is the appearance of water-soaked (shiny or oily) areas on the leaf surface which result from an increased membrane permeability and cell contents leaking into the intercellular spaces. In resistant plants this symptom may be temporary, while in sensitive plants it can develop into necrosis. In beans, leaf age variability in ozone susceptibility was attributed to variation in energy pools which governed the repair rate (Dugger et al. 1962b). Ozone elevated potassium efflux in *Chlorella;* however, within 2–3 min after ozonation stopped, the normal efflux rate was re-established (Heath et al. 1974). The time required to re-establish the normal potassium efflux rate decreased with increasing temperature (Chimiklis and Heath 1975).

Ozone increased membrane permeability to glucose and 2-deoxy-glucose (Perchorowicz and Ting 1974), which have been used as indicators of membrane permeability and repair (Sutton and Ting 1977a, b). Continuous light or glucose additions enhanced membrane repair, while low temperatures or darkness retarded it (Sutton and Ting 1977a, b). These observations support the concept that the repair of altered membrane permeability is energy-dependent.

The hexose monophosphate shunt provides reducing power (NADPH) which may be used, in part, to reduce disulfides and peroxides (Mudd and Freeman 1977, Freeman et al. 1979) and supply biosynthetic reductant. Ozone increased the activity of glucose-6-phosphate dehydrogenase in soybean cultivars within 3–6 h following an ozone exposure and it remained elevated for several days (Tingey et al. 1975, 1976a). Sutton and Ting (1977a, b) concluded that an endogenous energy source was "... needed for the re-establishment of the important membrane sulfhydryls and unsaturated lipids oxidized by ozone so that normal transport and permeability are restored to the cell membrane".

2.2.4 Injury

Injury results from the inability to repair or compensate photochemical oxidant-induced perturbations. Biochemically, injury is expressed as alterations in metabolism, including enzyme activities and metabolite pools. These cellular disturbances can also be expressed cumulatively as foliar pathologies, altered carbohydrate allocation, reduced growth and yield, and impacts on plant communities and ecosystems. These ozone responses (i.e., altered metabolism) are the consequences of pollutant-induced perturbations rather than the primary effect of ozone.

Changes in metabolism result from alterations in the structure and function of specific membrane components. The increased membrane permeability is associated with increased water and ion losses and partial loss of cellular compartmentalization, resulting in altered metabolic processes (Tomlinson and Rich 1967, Howell 1974, Tingey et al. 1975). The numerous biochemical and physiological reactions that have been correlated with photochemical oxidant impacts are only partially specific for oxidants. Most reactions may also be induced by other pollutants and various environmental stresses (Levitt 1980).

2.2.4.1 Ultrastructural Changes

The development of visible foliar injury symptoms results from biochemical and physiological changes. These changes include both functional and structural characteristics resulting from disruption of cellular membranes (see Sect. 2.2.2). Foliar injury is the consequence of biochemical and ultrastructural changes which have been established in some detail for injurious oxidant effects (Thomson 1975). The first identifiable ultrastructural changes occur within the chloroplast (Thomson et al. 1966), not the plasmalemma.

Thomson et al. (1966) observed an increased granulation and electron density of the chloroplast stroma following the exposure of bean leaves to ozone (0.6 to 1.0 ppm for 30 min). Crystalline fibril arrays composed of fibers approximately 85 nm in diameter were formed during the exposures, similar to the fibrils induced in bean plants exposed to PAN (Thomson et al. 1965). Similar fibrils were reported by Thomson and Swanson (1972) following treatment of cotton (*Gossypium hirsutum*) leaves with ozone. The crystalline fibril arrays were not observed in tobacco leaves, but indentation of the outer chloroplast membrane such as that reported for bean leaves (Thomson et al. 1966). Membrane indenting can co-occur with the crystalline arrays (Thomson et al. 1974) and can be induced by other pollutants, e.g., NO_2 (Dolzmann and Ullrich 1966, Thomson 1975). De Greef and Verbelen (1973) showed that various stresses such as mineral deficiency, mechanical injury, or water stress induced the crystalloids, with water stress being the chloroplasts from wilted leaves, or from plasmolyzed leaf cells (Wrischer 1973). suspended in a hypotonic medium (Perner 1962, 1963, Shumway et al. 1967), in the chloroplasts from wilted leave, or from plasmolyzed leaf cells (Wrischer 1973). Dehydration could also be the primary cause of photochemical oxidant-induced fibril formation as a consequence of the permeability changes in the plasmalemma and chloroplast membranes following exposure. As stated above, these changes occur in the chloroplasts prior to observable ultrastructural changes in the plasmalemma, which shows modifications only after other fine structure changes. These data led Swanson et al. (1973) to question the hypothesis that the plasmalemma was the primary site of ozone attack. Other data also suggest that the plasmalemma may not be the primary site of ozone reactions. Using data from red blood cells, Mudd (1982) provided evidence to suggest that ozone or a derivative may diffuse through membranes. To explain the observations of structural changes in the chloroplasts prior to changes in the plasmalemma, Tingey and G. E. Taylor (1982) proposed that ozone indirectly induced free radical formation in chloroplasts. Water and ion loss from oxidant-injured cells are not necessarily related to structural changes in the plasmalemma but represent biochemical disruptions of the membrane, as shown in Section 2.2.2. In addition, there is a series of secondary effects that are difficult to explain without membrane injury.

Because the enzymes of photosynthesis are located in the chloroplast, it is obvious that ultrastructural changes in the chloroplast could readily inhibit photosynthesis. The increased electron density in the chloroplast stroma, formation of crystalloids, and rupturing of the outer chloroplast membrane are the first fine structural changes observable from acute oxidant exposure. In morning glory leaves exposed to ozone (0.15 ppm for 8 h) there was a reduction in chloroplast size, disintegration of the thylakoids, and a decrease in ribosomes (Toyama 1976). Plastoglobules and phytoferritin increased in the chloroplasts and there was swelling in the cristae of the mitochrondria. With time, the chloroplast and mito-chondria swelled (T. T. Lee 1968, Swanson et al. 1973). The electron density of the cytoplasm increased, the plasmalemma withdrew from the cell wall and the cellular organelles congregated in the cell center. The vacuole ruptured and the remaining organelles were destroyed; the cell contents collapsed into a thick mass and only the granal membranes and crystalline fibrils were recognizable (Thom-son et al. 1966, 1974, Thomson 1975). There are deviations from this general scheme, depending on plant species, leaf age, and oxidant dose (Mitchell et al. 1979, Pell and Weissberger 1976). An 8-h PAN exposure (0.15 ppm) also induced swelling in the mitochondria (Toyama 1976).

Ultrastructural changes induced by moderate short-term ozone exposures may be reversible (Athanassious 1980): invagination of the chloroplast lamellae, granulation, and increased electron density in the stroma were no longer observ-able several hours following ozone treatment. These responses are typical of water deficiency (Shumway et al. 1967) and are symptomatic of indirect ozone ef-fects on the organelles (Pell and Weissberger 1976). At higher ozone concentra-tions, the cell membranes – the tonoplast, plasmalemma, and endoplasmic reticu-lum – are destroyed, leading to cell death (Athanassious 1980, Pell and Weiss-berger 1976). Petunia pollen exposed to ozone also displayed fine structure changes. The organelles were withdrawn from the plasmalemma and congregated in the cell center, as occurred in mesophyll cells (Harrison and Feder 1974). The biochemical, physiological, and ultrastructural changes in the cell (Chap. 2.2), lead to histological and anatomical injury expressions (Chap. 2.3).

2.2.4.2 Nitrogen Metabolism

Free amino acids increased in the foliage of several plant species following ozone exposure (Tomlinson and Rich 1967, Howell 1970, Ting and Mukerji 1971, Craker and Starbuck 1972, Mumford et al. 1972, Tingey et al. 1973a). An impair-ment in protein synthesis may be inferred from the ozone-induced dissociation of polysomes (Chang 1971a, b). However, effects on foliar protein contents are less clear, either protein decreases, increases, or no effects being reported (Ting and Mukerji 1971, Craker and Starbuck 1972, Beckerson and Hofstra 1972c, Tingey et al. 1973c).

2.2.4.3 Carbohydrate Metabolism

Following the observation that PAN reduced the elongation of oat coleoptiles (Ordin 1962), it was shown that ozone and PAN inhibited the enzymes involved in cell wall biosynthesis, both in vitro and in exposed seedlings (Ordin and Hall 1967, Ordin et al. 1967, 1969, Gordon and Ordin 1972). PAN-exposed tobacco leaves yielded similar results except that the activity of phosphoglucomutase was unaffected and the activity of UDP-glucose-pyrophosphorylase was doubled

(Ordin et al. 1971). This increased enzyme activity was associated with PAN-induced membrane disruptions releasing bound enzymes.

Immediately following ozone exposure, soluble sugars in soybean leaves and pine needles decreased (Tingey et al. 1973 c, Wilkinson and Barnes 1973). Subsequently they increased, frequently in association with foliar injury (J. H. Bennett et al. 1977, Tingey et al. 1973 a, P. R. Miller et al. 1969, Barnes 1972 b, Dugger et al. 1962 a, 1966). The increase in soluble sugars could result from reduced translocation or inhibition of starch synthesis. Acute exposures reportedly reduce starch content (Dugger et al. 1962 a, Youngner and Nudge 1980). The increase in soluble sugars and decrease in starch was also observed following chronic exposures (P. R. Miller et al. 1969, Dugger and Palmer 1969).

2.2.4.4 Stress Metabolism

Ethylene is an important plant metabolite whose production is stimulated by ozone and other stresses (Craker 1971 a, Abeles and Abeles 1972, Tingey et al. 1976 b, Stan et al. 1981). This stress ethylene precedes the appearance of visual injury; it increases proportionally to the applied stress and its production has been associated with many of the ozone response symptoms, including ozone-enhanced leaf abscission and the formation of pigment systems characteristic of ozone injury. The activities of phenylalanine ammonia lyase, polyphenol oxidase, and peroxidases are stimulated by ozone exposure prior to the appearance of visual symptoms, and remain elevated for several days (Tingey et al. 1975, 1976 a, Curtis et al. 1976). These enzymes are involved in phenol biosynthesis and oxidation. An ozone-induced accumulation of phenols has been reported in many plant species and associated with the extent of foliar injury (Howell and Kremer 1973, Howell 1970, Howell 1974, Koukol and Dugger 1967, Tingey et al. 1976 a). Howell and Kremer (1973) identified the pigments formed in ozone-injured leaves as polymerized phenols, proteins, and amino acids. They suggested that ozone induced cellular changes that either allowed more oxygen to enter the cells or oxidative enzymes to react with phenols, forming quinones that polymerized with protein and amino acids, thus yielding the pigments seen in ozone injury.

2.2.4.5 Photosynthesis

Numerous reports have described the effects of ozone and PAN in vivo and in vitro on individual photosynthetic processes. The following discussion illustrates that photosynthesis has numerous sites susceptible to photochemical oxidants.

When isolated chloroplasts were exposed to PAN (0.6 ppm for 0.5 h), oxygen evolution was reduced, but photophosphorylation was unaffected. At higher PAN concentrations, especially when illuminated during exposure, photophosphorylation was inhibited (Dugger et al. 1965). The sulfhydryl reagent N-ethylmalimid inhibited photophosphorylation only in the light (McCarty et al. 1972), suggesting an association between illumination and the presence of reactive sulfhydryl compounds. The redox system of photosystem II was oxidized (Dugger and Ting 1968). Photosystem II also is sensitive to disruption by ozone (Chang and Heggestad 1974, Koiwai and Kisaki 1976). Coulson and Heath (1974, 1975) proposed that ozone and PAN inhibited photosynthesis by different mechanisms. Ozone attacked chloroplast membranes, disrupting the membrane-bound process of photosynthesis. In contrast, PAN diffused into the chloroplast which in the light was alkaline, causing the decomposition of the PAN where its reaction products could attack the sulfhydryl groups of the membranes and ribulose-1,5-biphosphate (RuBP) carboxylase inhibiting photosynthesis. Ozone also reduced the activity of the carboxylase, decreasing CO_2 fixation (Nakamura and Saka 1978). Inhibition of RuBP carboxylase activity occurs shortly after ozone exposure. Pell and Pearson (1983) reported decreases of 36, 68, and 80%, respectively, in RuBP carboxylase activity in the foliage of three alfalfa cultivars 48 h following ozone exposure. These decreases occurred in the absence of visible injury. Coyne and Bingham (1981) found that the photosynthetic rate of field-grown ponderosa pine trees exposed to ambient photochemical oxidants was depressed to a greater extent than could be explained by stomatal closure; they suggested an impairment of excitation or carboxylation reactions. Several enzymatic processes of the Calvin cycle and the C_4-dicarboxylic acid cycle are activated by light and sulfhydryl compounds (Mudd 1975, L. E. Anderson et al. 1974). Photochemical oxidant reactions with these sulfhydryl groups would reduce the secondary reactions of photosynthesis.

Ozone in high concentrations for a short time can destroy chlorophyll, leading to a reduced photosynthetic rate. Chlorophyll loss following ozone exposure has been reported in many species (e. g., Adedipe et al. 1973 a, P. R. Miller et al. 1963, Runeckles and Resh 1975 a, Knudson et al. 1977). Chlorophyll b was more susceptible than chlorophyll a to ozone inactivation (De Koning and Jegier 1968 b, Kühl and Wagner 1970, Nobel 1974); however, Verkroost (1974) found that chlorophyll a was more sensitive. Carotenoids were more susceptible to PAN than chlorophyll (Dugger et al. 1963 a) but chlorophyll a was more susceptible than chlorophyll b (Gross and Dugger 1969).

Carbon dioxide fixation was reduced following short-term exposures to acute-injurious concentrations of ozone (Todd 1958, Todd and Probst 1963, Barnes 1972 a, Furukawa and Kadota 1975). This reduction results from ozone-induced stomatal closure and inhibition of the enzymes of photosynthesis (MacDowall 1965 b, Hill and Littlefield 1969). Numerous studies have shown that the net photosynthetic rate was reduced shortly after the start of the ozone exposure; when the exposures were terminated, the net photosynthetic rate rapidly returned to its original level (Hill and Littlefield 1969, Pell and Brennan 1973). The impact of ozone on photosynthesis was influenced by the season of year and the oxidant resistance of the test plants (Lorenc-Plucińska 1979). Net photosynthesis can be impaired without the development of visible symptoms (Hill and Littlefield 1969, Botkin et al. 1972), however, the appearance of visible symptoms is always associated with a measurable reduction in photosynthesis (Pell and Brennan 1973).

Ponderosa pine trees were exposed to 0.15 ppm ozone for 30 days without the development of visible symptoms, but the net photosynthesis was reduced 10% compared to the controls (P. R. Miller et al. 1969). Similar results were reported for citrus trees grown in field conditions receiving polluted air (Thompson et al. 1967); the photosynthetic rate was reduced in the absence of visible injury when compared to trees grown in filtered air. Under chronic exposures, ozone caused a reduction in soluble sugars and starch (P. R. Miller et al. 1969, Tingey 1974, Tingey et al. 1976a), leading to reduced growth (Jensen 1981a) and yield (see Sect. 2.2.5). Associated with reduced photosynthate production there is an alteration in the partitioning of photosynthate, reducing root growth and root processes which restrict shoot growth (Manning et al. 1971, Tingey 1974, Tingey et al. 1976c, Oshima et al. 1978, Blum and Tingey 1977). Ponderosa pine seedlings exposed to 0.1 ppm ozone 6 h day^{-1} during the growing season had lower root reserves of soluble sugars and starch in autumn (Tingey et al. 1976c), so that the plants had a diminished carbohydrate reserve for initial growth the following spring.

2.2.4.6 Respiration

Respiration was increased following short-term exposures to ozone and PAN. In bean leaves, there was a significant increase in respiration after exposure to 4.0 ppm ozone for 40 min; the increase paralleled the development of visible symptoms and then decreased (Todd 1958). Immediately following a 1 h ozone exposure of tobacco leaves, respiration initially decreased, then increased (Mac-Dowall 1965b). Following long-term exposures at 0.15 to 0.25 ppm for 5 days, the respiration rate of citrus leaves was increased (Dugger and Palmer 1969). The ozone-induced alterations in foliar metabolic processes also influence root reactions. The root respiration of bean plants was reduced following 2 days of exposure (0.15 ppm for 6 h day^{-1} for 2 days) to ozone (Hofstra et al. 1981).

2.2.4.7 Senescence

Particularly following chronic oxidant impacts, there have been numerous associations made between subtle injury reactions and premature senescence, which is characterized by increases in catabolic reactions and decreased anabolic reactions. These catabolic reactions include loss of starch (Dugger and Palmer 1969, Youngner and Nudge 1980), proteins (Craker and Starbuck 1972), and chlorophyll (Adedipe et al. 1973a, Runeckles and Resh 1975a), and increases in anthocyanins and polyphenols (Koukol and Dugger 1967, Howell 1970, 1974, Kremer and Howell 1974). There is also a decrease in RNA content current with an increase in RNase activity (Craker and Starbuck 1972), increases in peroxidases (Dass and Weaver 1972, Flueckiger et al. 1979, Endress et al. 1980), respiration (Pell and Brennan 1973), and stress ethylene production (Abeles and Abeles 1972, Tingey et al. 1976b, Flueckiger et al. 1979, Stan et al. 1981). Premature leaf abscission (Jensen 1973, P. R. Miller et al. 1969, 1977, Flueckiger et al. 1979, Williams 1980) and fruit drop (Thompson and Taylor 1969, Thompson et al. 1972) also occur.

These reactions and others discussed in the injury section can affect the magnitude of oxidant impact on plant growth and yield (see Sects. 2.2.5 and 2.3.3) if specific thresholds are exceeded. In addition, for useful plants these reactions can, in part, influence plant quality. Pollutant-stressed plants may be more susceptible to biotic and abiotic stresses (Sect. 2.2.5).

2.2.5 Effects on Plant Communities and Ecosystems

In the preceding sections, the effects of photochemical oxidants focused on impacts on individual plants and homotypic populations. The responses of higher plant communities have received only limited study. The most important results of this research will be presented. With the increasing significance of photochemical oxidants in all industrial areas and their continuing widespread distribution, indications are that ecosystems even in remote areas will be impacted.

2.2.5.1 Reactions of Plant Communities Related to Air Pollutant Concentrations

As with individual plants, the responses of plant communities and ecosystems depend on pollutant dose. Based on the presentation of Smith (1974), Guderian and Küppers (1980b) analyzed air pollution effects on plant communities; in general, their scheme (Fig. 2.4) is followed in this section.

2.2.5.1.1 High Pollution Dosage

A characteristic of the association between high dosage and plant community response is a breakdown of community structure, which is more or less obvious depending on ecosystem complexity. The degradation of the system is characterized by a rapid change in structure accompanied and finally replaced by secondary succession processes that under continuing exposure lead to a new equilibrium.

In forest ecosystems, the direct effects (acute and chronic) are observed first on the most sensitive tree species. In high pollution-impacted areas of southern California, 24% of the trees died within a 4-year period (Cobb and Stark 1970, P.R. Miller 1973a, O.C. Taylor 1973b). The forested area around Crestline in the San Bernardino National Forest, California, is dominated by the oxidant-sensitive ponderosa pine (*Pinus ponderosa*) and Jeffrey pine (*Pinus jeffreyi*). In 1969, in a study area of 64,380 ha, intense, moderate, and no visible effects were found on 18,492, 21,568, and 24,320 ha respectively. Approximately 1.3 million individual trees were affected: 82% moderate injury, 15% intense injury, and 3% dead (Wert et al. 1970).

In oxidant-injured forests, the amount of dead wood is increased through enhanced morbidity and mortality. This dead material is a significant source of fuel for forest fires. Thus, while fires in injured forests may not occur more frequently, they are more intense, producing higher temperatures. These higher temperatures cause more injury to soil organisms and to sensitive plant organs in the soil (O. C. Taylor 1973 a).

When pollutant impacts open the crowns of sensitive trees, air pollutants can penetrate deeper into the stand and successively destroy the shrub, herb, and moss layers. Such complete vegetation destruction has been observed in the vicinity of individual sources emitting high concentrations of phytotoxic compounds (Niklfeld 1967, Guderian and Stratmann 1968, Hajdúk and Ruzička) 1969, Brandt and Rhoads 1972). There are no reports of such intense damage resulting from photochemical oxidant exposure. However, under high pollution exposure, the elimination of sensitive species may have far-reaching consequences for the total ecosystem: previously the dense stands would protect the shrubs and herb layers, but as they are impacted the sensitive species will disappear. At least ten shrub species in the shrub layer of the coniferous forests of the San Bernardino Forest are now endangered by this means (Cobb and Stark 1970). It is no longer possible to determine how many species have disappeared during the last 20 years of pollution impacts.

Concurrent with the degradation of the original vegetation, secondary succession processes are initiated: pollution-resistant species whose ecological amplitude corresponds with the biotic and abiotic conditions immigrate and, with the remaining pollutant-stressed species, develop a newly structured ecosystem of less complexity. As is typical of ecotones, few species with high abundance occur in agreement with the second biocoenotypic principal. Wolak (1971) has described such a secondary plant community (industriogenous) resulting from continuing exposure to sulfur dioxide, zinc, and lead in the Upper Silesian district of Poland and named it the Industrial Climax Community. These ecosystems are characterized by reduced diversity, reduced productivity, and possibly reduced stability.

2.2.5.1.2 Intermediate Pollution Dosage

Intermediate air pollution dosages are ecologically and economically significant because their either chronic or subtle, direct or indirect effects on individual species can set the stage for changes in community structure and function. Immediate, direct damage (i.e., primary injury) is the most common under high dosages, while under intermediate and low dosages the secondary effects such as reduced vigor including increased susceptibility to abiotic and biotic stresses (Wentzel 1965, Donaubauer 1966, Heagle 1973, Huttunen 1980, Laurence 1981) become more significant. At moderate doses, primary and secondary injury lead to disruptions in growth, reproduction, and vigor of the more sensitive plant species within the community. Resistant species may display a higher productivity based on more favorable competitive conditions (P. R. Miller and Kickert 1980) and attain a higher dominance within the plant community (Harward and Treshow 1971).

In productive ecosystems, decreased yield resulting from reduced plant productivity can be reliably identified and evaluated. However, the conditions necessary to determine pollution-induced reductions in the ecological performance of plants and plant communities are generally insufficient. Therefore, in the highly structured forest communities with their delicate equilibrium of plant strata, small decreases in growth and changes in crown structure for individual members of the stand can induce a series of secondary effects (J.H. Bennett and Hill 1975, Murphy et al. 1977). Such changes proceed slowly, and only the summation of annual effects gradually leads to higher morbidity. A slight change in the horizontal canopy structure will influence habitat factors such as incident radiation, precipitation, and pollution exposure to the lower-lying vegetation, causing extensive structural changes to the shrub and herb strata (Harward and Treshow 1975).

In the aspen-fir forests of the Wasatch Mountains of Utah, aspen (*Populus tremuloides*) is the dominant species and the most sensitive to photochemical oxidants (Treshow and Stewart 1973). Shade from mature aspen is needed for optimal development of white fir (*Abies concolor*) seedlings. The increased light intensity from a thinning of the aspen stands as a result of photochemical oxidants endangers the natural reproduction of white fir. The opening of the stands also alters the precipitation relationships, displacing other species.

An eastern deciduous forest (Ohio River Valley, USA) experiencing intermediate dosages from a pollutant mixture showed a decline in species richness, evenness, and a decrease in the Shannon diversity index within all strata of the community (McClenahen 1978). For example, the relative importance of sugar maple (*Acer saccharum*) declined in all strata, while the importance of yellow buckeye (*Aesculus octandra*) increased. Opposing trends in density were observed in some strata. Decline in the tree and herb strata was accompanied by increase in the subcanopy and shrub strata. These changes were attributed to more favorable light conditions in the lower strata and increases in shade-intolerant trees. Westman (1979) observed similar changes in the woody vegetation of the southern California coastal area and correlated them with changes in photochemical oxidant dosages. According to Harkov and Brennan (1979), the species most susceptible to ozone frequently belong to early successional stages and may therefore be especially endangered by ozone.

2.2.5.1.3 Low Pollution Dosage

The impacts of low dosages on vegetation lie on the borderline between the normal, i.e., unaffected, and significantly impacted vegetation. Depending on the concentration and exposure duration, effects can range from stimulations (see Chap. 2.4.1) to reductions in growth, reproductive capability, or susceptibility to abiotic and biotic stresses (see Chap. 2.3.3). Changes in competitive relationships may also occur, causing alterations in species composition (Treshow and Stewart 1973). Under typical conditions, such effects can be detected only above a specific detection limit, i.e., response threshold. The detection limit has been lowered through the development of exposure systems using either filtered air with or without pollutant addition and field exposure systems (see Chap. 2.3.4). However, these systems can only be used on small to moderate-sized plants. Before detectable reductions appear in, e. g., growth, various changes occur in biochemical, physiological, or substructural levels. The difficulties of establishing relationships between such processes and plant or community function are discussed in Chap. 2.3.1.2.

Vegetation may remove air pollution from the atmosphere without detectable effects on plant performance. The possible long-term effect on plants and plant communities is determined by the behavior of the specific substances in the ecosystem. Pollutants such as ozone and PAN which decompose rapidly cause effects through the summation of individual reactions over one or several years. With other components, e. g., heavy metals or N-containing components, they may enter into the nutrient cycles of the ecosystem, causing cumulative effects (Krämer 1976, Siccama and Smith 1978) and disturbing the balance of especially sensitive ecosystems.

In summary: under the influence of air pollution, two counter-processes are initiated in plant communities. From direct and indirect effects, changes in the structure and function of the community occur. Parallel to this degradation, spontaneous or man-supported processes occur in which either original, i.e., adaptively resistant members of the remaining community, and also immigrants undergo secondary succession (Guderian and Küppers 1980 b).

2.2.5.2 Causes for Observed Responses in Plant Communities

The effects of a given air pollutant on plant communities, as described in previous sections, range from insignificant changes to total destruction and are determined through:

- genetically controlled resistance of individual community members;
- moderating influence of environmental conditions on relative resistant relationships; and
- pollutant-induced changes in inter- and intraspecific relationships.

The following presentation (Fig. 2.4) attempts to clarify these relationships at the level of the individual, the homotypic population, considering intraspecific competition and the level of a heterotypic plant community.

The relationships among genetic, environmental, and resistance relations presented in Chapters 2.4.2 and 2.4.3 for individual and also homotypic populations also hold for plant communities; therefore, a discussion of them will be omitted. This presentation will focus on the community viewpoint.

The significance of the relationships shown in Fig. 2.4 will be highlighted by the results from limited experiments on the relations between populations of different species. Studies on pure and mixed stands of ryegrass (*Lolium multiflorum*), hairy vetch (*Vicia villosa*), and crimson clover (*Trifolium incarnatum*) showed that pollution-dependent changes in the composition of the plant communities were not totally explained by the direct pollutant effects on the differentially susceptible species (Guderian 1967); under the influence of SO_2, changes in the interspecies competitive conditions occurred. Thereby, the primary impact on the more susceptible members of the stand was so increased that they could no longer compete effectively for essential growth-determining factors such as mineral nu-

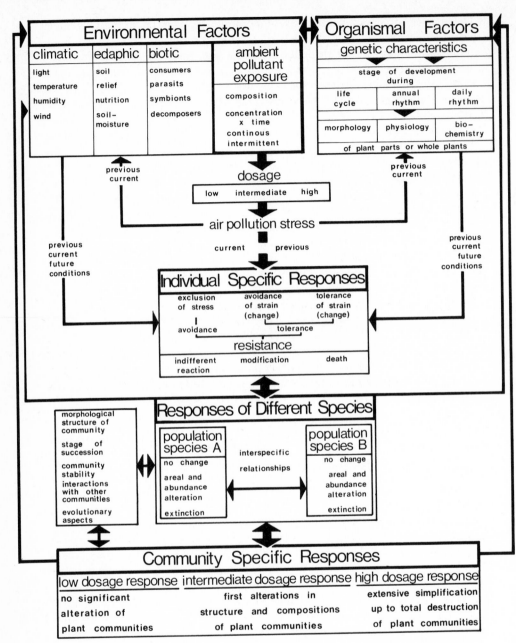

Fig. 2.4. Effect-determining factors for various responses of plants at individual, species, and community levels. (Guderian and Küppers 1980b)

trients, water, light, and growing space; consequently, they lost their importance in the community structure. As a result of changed competitive conditions in the community, the decline of the more sensitive members permitted the enhanced growth of the more resistant species under low dosages. The total yield decreased less than would have been expected from the loss of the more susceptible species. Similar results were reported for ozone (J. P. Bennett and Runeckles 1977), ultraviolet (UV-B) radiation (Fox and Caldwell 1978), and ionizing radiation (McCormick 1963).

The extent of changes in plant communities in response to a given pollutant burden depends on the condition of the community itself in addition to the previously mentioned genetic, environmental, and resistance factors. The importance of morphological structure, community strata, relief, and vegetative cover are discussed in Sections 2.2.5.1.1 to 2.2.5.1.3 as are the interrelations between successionary stages and system responses. The stability of a community influences the reactions of individual members as well as of the whole community to a given pollutant exposure. Based on their complex feedback systems, highly productive ecosystems usually rapidly regain their equilibrium following minor disturbances. Under higher exposures, one must deal with obvious visual changes when individual key species are particularly sensitive to a given air pollutant, e. g., the dominant ponderosa pine in the San Bernardino Mountains (O. C. Taylor 1973 a, b). But even low concentrations can cause considerable changes in seminatural plant communities, especially when the aut- and synecological amplitudes of the indigenous community members are very divergent. Additional stresses from air pollutants, even at low doses, can induce drastic reductions in the vigor of plants already living outside their ecological optimum. The relatively high susceptibility of Norway spruce (*Picea abies*) in the Erzgebirge of Czechoslovakia (Materna 1972) and in the boreal coniferous forest in Finland (Huttunen 1979) may result from their relatively unsuitable habitat. From this perspective, it seems problematic to transfer dose-response relationships derived from agricultural ecosystems in which plants encounter favorable growing conditions to natural ecosystems.

The relationships discussed here illustrate how pollution can degrade plant communities. However, it must be noted that spontaneous adaptations to air pollution stress have been observed which insure survival of the species not only through morphological changes such as "Kümmerwuchs," in which trees form a dense-layered mat (Halbwachs and Kisser 1967), but apparently through genetic changes also. Exposure of *Marchantia polymorpha* to lead (Briggs 1972) or various grass species to copper and zinc (Bradshaw 1971, 1972, 1976) induced, in a short time through directed selection, the development of more tolerant populations. In a region experiencing variable SO_2 stress, tolerance to SO_2 developed in *Geranium carolinianum* over a 30-year period. Populations selected from around the source were more tolerant to SO_2 than those that had not evolved under the sulfur pollution (G. E. Taylor and Murdy 1975). The consistent differences in SO_2 response among individuals was genetically controlled and quantitively inherited (G. E. Taylor 1978 b). Similar results have been reported for *Lolium perenne* and *Rumex obtusifolius* under chronic decade-long exposure (Bell and Clough 1973, Bell and Mudd 1976, Horsman and Wellburn 1977).

Dunn (1959) found that subspecies of *Lupinus bicolor* from the Los Angeles Basin which had developed in the presence of photochemical oxidant air pollutant were more resistant to smog than their counterparts from nonpolluted locations. He hypothesized that the greater resistance to photochemical oxidant arose through the process of natural selection. The results of multi-year chronic SO_2 exposures to native grasslands (Preston and Gullett 1978) also indicate that, under anything less than acute concentrations, spontaneous adaptations may occur in the course of forming a new secondary equilibrium which may also interrupt possible long-term injury (Preston 1979). The evolution of plant resistance to air pollution has its costs, for example, in the reduced growth of resistant plants and altered adaptations in affected plant communities (Roose et al. 1982). Accordingly, the pollutant concentration and its frequency of occurrence should be controlled over the long term so that plant communities maintain the capacity for using the above-mentioned evolutionary processes to fulfill their function to the fullest in natural and agrarian ecosystems (Guderian and Küppers 1980a). The genetic-controlled variation in populations as expressed in spontaneous adaptations also provides the basis for breeding population resistance through selection and propagation of relatively tolerant individuals (see Chap. 2.4.4).

2.2.5.3 Effects on Semi-Natural and Agro-Ecosystems

Pollution-induced transformations in plant communities and impacts on individual plants can lead to permanent impacts on economic, ecological, and environmental performance of vegetation. In agriculture and forestry, the damaging effects of reduced growth, impaired quality, and additional labor requirements are well documented. For individual sources of primary air pollutants, reliable pollution-induced performance reductions can, in general, be established. In contrast, for widely distributed pollutants such as photochemical oxidants, it is substantially more difficult to establish the economic loss. Only recently has man been concerned with the impacts on the ecological and environmental performance of vegetation in semi-natural ecosystems. Moreover, it should be determined how air pollutants impact vegetation functions such as air filtering, climate stabilization, regulation of water and nutrient cycles, prevention of soil erosion, and the preservation of healthy habitats for zoo- and phytocoenoses.

In the agrarian ecosystem, the economic performance of the crop is the main interest. Photochemical oxidant-induced reductions in yield and quality can have direct economic consequences. In perennial crops such as fruits and forests, the reductions in growth, vigor, and/or reproduction can limit the biological basis of production. As a result of such documented impacts, researchers and agricultural constituents have attempted to assess the extent of pollution-induced injury as a basis for establishing possible mitigative actions. In addition, such data are useful from the economic and political perspectives for the decision-making process concerning alternative policies and the establishment of standards.

A reliable determination of photochemical oxidant-induced yield reductions in agrarian ecosystems is difficult to document. Even in agricultural systems devoted to annual crops, which are more easily cultivated and managed than perennials, the reports of photochemical oxidant-induced injury are divergent. Nevertheless, there are concrete proposals for obtaining reliable data. In forestry, the impacts of reduced height and radial growth on total wood production can be determined. Although there have been documented changes in secondary succession processes in some terrestrial ecosystems, the economic and ecological consequences of photochemical oxidant attacks on primary producers, consumers, and decomposers – including host-pathogen relations – are not obvious.

2.2.5.3.1 Damage Responses in Agriculture and Horticulture

The first report of photochemical oxidant air pollution impacts on agricultural and horticultural crops in the Los Angeles basin cited yield losses of US $480,000 (Middleton et al. 1950). Since then, there have been repeated attempts to quantify the air pollution-induced agricultural losses. Two estimating methods have been used:

– Surveys of crops in the pollutant-impacted areas with assessments made for the locale and region.
– Use of dose-response relationships to make assessments for regional and multi-regional levels.

Both techniques provide only approximations. With local surveys, there are the inherent risks of false diagnosis; the damage intensity cannot be determined precisely because the necessary controls for the widely dispersed pollutants are lacking. Also, the most common response measure, foliar injury, is not closely related to reduced growth and, in actuality, growth reductions may occur in the absence of visible symptoms or when the symptoms are not recognizable. No less serious are the potential sources of error for the second method: basically, dose-response relationships can only provide an approximation; assessment of pollutant exposures are frequently imprecise and their characterization may not indicate reliably the magnitude of risk to vegetation; the temporal and spatial extrapolation does not coincide exactly with actual exposure, and the climatic and edaphic factors deviate from the assumed conditions. Finally, the environmentally and genetically controlled resistances of the model plant do not correspond to the actual resistance of the impacted cultivars. The following discussion is presented keeping these constraints in mind.

The field survey method has been extensively used in California to estimate the magnitude of yield reduction. For example, Middleton (1961) calculated the loss to agricultural plant species at 8 million US dollars. Based on reports of 450 trained agricultural experts, the California Department of Food and Agriculture estimated yield losses to agriculture in 1970 at 26 million US dollars (Millecan 1971). Losses to citrus plants alone accounted for about 19 million US dollars. Reductions not associated with visible injury were not considered except for grapes and citrus, nor were monetary losses of native vegetation, including forests and landscape plantings, estimated. Approximately 50% of the damage resulted from ozone; PAN contributed an additional 20%. In 1977, the California Department of Food and Agriculture (1977) estimated pollution-induced crop losses at

31.6 million US dollars for the five counties in the South Coast Air Basin (SCAB). In San Bernardino County, which received the highest ozone exposures, the mean yields of several crops were reduced about 20%.

The dose-response method has been used to estimate impacts at the regional and national levels in the USA. Based on the dose-response relationships from Larsen and Heck (1976), and using subjective conversion factors that related foliar injury to yield reductions (Millecan 1971), Adams et al. (1982) estimated the primary losses to 12 vegetable crops, cotton, and sugar beets in four southern California regions at approximately 32 million dollars or about 5% of producers output. An additional 20 million US dollars in losses to consumers of these crops was calculated.

Estimated losses to crop production in the USA are shown in the following examples. Moskowitz et al. (1982) used the dose-response function of Oshima et al. (1976) and estimated yield losses for alfalfa at least 4%. Adams and Crocker (1982) developed an ex post estimate of the policy worth of the current National Ambient Air Quality Standards in terms of air-pollution crop losses. The analysis focused on corn, soybean, and cotton, which account for about 60% of US crop production. In each case, they used the species-specific dose-response relations from the National Crop Loss Assessment Network (NCLAN) and Heck et al. (1982) and four levels of ozone exposure to estimate the Marshallian economic surplus value (an economic measure approximating direct and indirect effects). The authors assumed that, without an ambient air quality standard, the ambient ozone concentration would increase to a level comparable to a standard of 0.18 ppm. A reduction from this level to the current USA ambient air quality standard of 0.12 ppm (1-h averaging time not to be exceeded more than 1 day per year) produced an economic surplus for the selected crop species of about 3.4 billion dollars. Hence, the current standard was assumed to be yielding benefits of this magnitude for agriculture. The improvement of air quality associated with standards of either 0.10 or 0.08 ppm would produce an additional economic surplus value of approximately 1.1 billion dollars for each of the incremental improvements in air quality, yielding an additional surplus of 2.2 billion dollars with a standard of 0.08 ppm. The calculations assumed that these standards were achieved in the major agricultural regions of the United States.

Heck et al. (1982) estimated the annual USA crop loss at 1 to 2 billion US dollars if the current National Ambient Air Quality Standard were obtained in all areas of the country; this corresponds to 2 to 4% of the total crop production. Ozone either alone or in combination with SO_2 or NO_2 accounted for 90% of the damage.

The National Crop Loss Assessment Network (NCLAN) was established to provide more accurate data on air pollution-induced losses to agriculture. "The NCLAN consists of a group of government and nongovernment organizations cooperating in field work, crop production modeling, and economic studies to assess the immediate and long-term economic consequences of the effects of air pollution on crop production. The program will define the relationships between yields of major agricultural crops and doses of O_3, SO_2, NO_2, and their mixtures. These relationships will be used to assess the primary economic consequences of the exposure of agricultural crops to these pollutants. The program is also designed to advance the understanding of cause-effect relationships with the intent of developing simulation models" (Heck et al. 1982). The NCLAN experimental procedures are intended to provide reliable data concerning the magnitude of crop production losses in the USA. A brief description of the NCLAN program and its initial results are presented. The data should be developed under conditions as nearly natural as possible to insure that the results are representative. Based on a detailed comparison of currently available methodologies, Heagle et al. (1979 d) concluded that "open-top" field exposure chambers would produce the fewest artifacts. In the first year, 1980, of the NCLAN program, experiments were conducted in "open-top" field exposure chambers at four locations in the USA having different climatic conditions: Ithaca, New York – Northeast (NE); Raleigh, North Carolina – Southeast (SE); Argonne, Illinois – Central States (CS); Riverside, California – Southwest (SW). Different species were exposed at each site:

NE – red kidney bean (*Phaseolus vulgaris*, cv. California Light Red)
SE – peanut (*Arachis hypogaea*, cv. NC-6) and turnip (*Brassica rapa*, four cultivars)
CS – soybean (*Glycine max*, cv. Corsoy)
SW – head lettuce (*Lactuca sativa*, cv. Empire)

At each location, four replications of six ozone exposures were used:

1. AA – ambient air plot without a chamber;
2. CF (Control) – charcoal-filtered air; this resulted in O_3 concentrations of 20 to 50% of ambient O_3 due to influx through the open top (in some parts of the country, this represents an approximate "clean" air background concentration);
3. NF-1 – nonfiltered air with sufficient O_3 added for 7 h day^{-1} to compensate for system losses (about 0.015 ppm);
4. NF-2 – as NF-1 plus an additional 0.03 ppm of O_3 for 7 h day^{-1};
5. NF-3 – as NF-1 plus an additional 0.06 ppm of O_3 for 7 h day^{-1};
6. NF-4 – as NF-1 plus an additional 0.09 ppm of O_3 for 7 h day^{-1}.

Initial results showed that the ambient air exposures (AA), when compared to the controls (CF), reduced the yields of the following crops: soybean 10%, peanut 14–17%, for one turnip cultivar 7%; head lettuce 53–56%; red kidney bean 2%. When a background ozone concentration of 0.025 ppm (seasonal 7 h day^{-1} mean) was considered, yield reductions were observed in all crops at seasonal 7 h day^{-1} (0900–1600 h standard time) mean ozone concentrations of 0.06 to 0.07 ppm when compared to the control concentrations. Results of subsequent research (Table 2.3) showed that, when yields between the control (CF) and ambient air (AA) exposures were compared, the yields of cotton and soybean were reduced by 14 and 5 to 19%, respectively. Yield losses were similar between the ambient air plots (AA) and the chambers (NF-1) which received a similar seasonal mean ozone concentration. Yields of 14 corn cultivars were not significantly reduced by the ambient air exposures.

Using results from (1) NCLAN, (2) county-levels estimates of the seasonal 7 h day^{-1} ambient ozone concentrations, and (3) county-level statistics for crop dis-

Table 2.3. Reduced yields of soybean and cotton in relation to ozone concentration. (Results 1981 NCLAN program)

Treatment	Crop/location					
	Soybean/Northeast[a]		Soybean/Southeast[a]		Cotton/Southwest[b]	
	Ozone conc. (ppm)	Reduced yield (%)	Ozone conc. (ppm)	Reduced yield (%)	Ozone conc. (ppm)	Reduced yield (%)
CF	0.017	0	0.025	0	0.021	0
AA	0.035	5	0.053	19	0.077	14
NF-1	0.035	8	0.055	19	0.071	20
NF-2	0.060	20	0.069	28	0.111	40
NF-3	0.084	31	0.086	40	0.143	41
NF-4	0.112	41	0.106	46	0.185	58

[a] Mean ozone concentration (7 h day^{-1}) for 70 days
[b] Mean ozone concentration (7 h day^{-1}) for 111 days

tribution and crop production, estimates were made for the yield losses to the four major agricultural crops in the USA (Shriner et al. 1982). The county-level yield reductions were determined for peanut (sensitive), soybean (sensitive-intermediate), wheat (moderately sensitive), and corn (insensitive). These crops represent 62% of the acres harvested and account for 63.5% of the dollar value. Shriner et al. (1982) expressed summary data as yield increases expected if the ambient ozone concentrations were reduced to an assumed background of 0.025 ppm. Crop loss estimates developed by this assessment indicated that the combined annual losses from ozone impacts may be as high as 3.1($\pm$1.2) billion dollars for these four crops. Of that dollar loss, soybean accounted for approximately 64%, corn 17%, wheat 12%, and peanut 7%. The major agricultural regions of the United States generally coincide with the regions experiencing elevated ozone concentrations. These economic crop losses represent only the primary impacts on producers; additional losses to consumers through higher prices and to agricultural processors would probably exceed those for producers. Beyond the crop impacts, the losses to forestry, ornamentals, landscape plantings, and the nonmonetary losses to the ecological and environmental functions of vegetation have not been considered.

2.2.5.3.2 Effects on Semi-Natural and Forest Ecosystems

Agricultural and horticultural systems are composed primarily of annual plants, while in semi-natural ecosystems the vegetative cover contains a significant portion of perennial plants. Therefore, semi-natural ecosystems are particularly endangered from chronic oxidant exposures because small alterations accumulate over a period of years, producing a greater long-term risk than those same changes in an annual agricultural crop.

a) Primary Producers

The long-term studies on photochemical oxidant impacts on vegetation in the San Bernardino Mountains have centered on ponderosa pine populations, a key

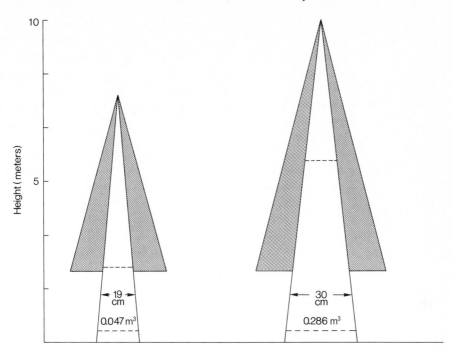

Fig. 2.5. Calculated average growth of 30-year-old ponderosa pines in polluted (*left*) and nonpolluted air (*right*) based on data from 1910 to 1940 and 1941 to 1971. Growth was determined as height, diameter at breast height, and total wood volume. (P. R. Miller et al. 1977)

species in the stands (Parmeter et al. 1962, O. C. Taylor 1973a, b, 1974a, P. R. Miller 1973a, Kickert et al. 1977, P. R. Miller and White 1977). Parmeter et al. (1962) found a decreased height growth in trees both with and without visual injury symptoms. Analysis of tree ring widths from period 1941 to 1971 showed that photochemical oxidants reduced the average annual ring about 2 mm compared to a previous corresponding period (P. R. Miller and White 1977). Basal area, as a criterion, is better suited to measure effects on radial growth than annual ring width. P. R. Miller et al. (1977) estimated the reduction in total wood production by 30-year-old ponderosa pines in the San Bernardino National Forest (Fig. 2.5). Total height, diameter at breast height, and wood volume for logs with 15-cm tops were determined. A simulation model of an eastern deciduous forest has been used to describe the changes in species composition and reduction in total biomass production for selected tree species (West et al. 1980).

Air pollutants can influence the reproductive capacity of plants through impacts on the number, size, nutritive content, and viability of seeds, as well as the mode of seed distribution, as has been shown in studies on SO_2 effects (T. C. Scheffer and Hedgecock 1955, Pelz 1963). Restrictions in the reproductive capacity of sensitive plants in the ecosystem is an additional competitive pressure.

The seeds from various pine and oak species provide a significant portion of the diet for numerous small mammals, such as western gray squirrel (*Sciurus griseus anthonyi*) in the San Bernardino Mountains. The squirrels store a major portion of the seeds they collect, contributing to their distribution. When seed production from Jeffrey and sugar pines (*Pinus lambertiana*) is decreased by photochemical oxidant, the squirrels consume a significant portion of the less desirable ponderosa pine seeds before they mature, decreasing natural reproduction (EPA 1978 b).

The briefly sketched results can be summarized as follows: if the present pollution exposures continue for several decades, then the pessimistic prognosis of Cobb and Stark (1970) will be verified: there will be a conversion from well-stocked forest dominated by ponderosa pine to poorly stocked forests of less susceptible tree species in the San Bernardino Mountains.

b) Consumers

The effects of photochemical oxidants on consumers are both direct and indirect. The effects on indigenous animals and insects will be briefly described. If and to what extent photochemical oxidants directly influence the health, biomass production, and reproduction of indigenous animal populations cannot at the present time be assessed; however, certain direct effects on selection are nevertheless possible. Richkind and Hacker (1980) established that deer mice (*Peromyscus californicus*) from an area in Los Angeles experiencing high oxidant exposures were less sensitive to acute ozone exposures in laboratory studies than similar populations from areas receiving low exposures.

The reduced number of species in the San Bernardino Mountains in comparison to similar but unpolluted habitats (P. R. Miller et al. 1977) rests completely or principally on the indirect effects or changes in food webs and habitats. The loss of specific plant species produces major impacts on species heavily dependent on it for food. From a habitat perspective, numerous animal species have only a limited ecological amplitude; consequently, alterations in the vegetation are an additional stress on the system that alter the variable relationships among animal species.

As observed in SO_2-impacted areas (Templin 1962), bark beetles (*Dendroctonus brevicomis* and *D. ponderosae*) increased on ponderosa pine in the oxidant-impacted forest stands in the San Bernardino Mountains (Stark et al. 1968, Stark and Cobb 1969, O. C. Taylor 1973 a, b, 1974 a, Dahlsten 1978). The infection, i. e., mortality, in the forest stands clearly increased with photochemical oxidant exposure (Stark and Cobb 1969). Long-term studies (1952 to 1972) in this area showed a significant increase in infections by these secondary pests, consequently increasing numbers of trees injured by bark beetles had to be logged (P. R. Miller et al. 1977). The higher tree mortality in forests over wide areas of the USA (US Department of Agriculture 1973) was caused by air pollution-induced increase in these infections by various weak parasites (Loucks and Williams 1980). On theoretical grounds, Woodwell (1970) suggested that the increased occurrence of insects would cause a brief increase in specific vertebrate animals such as insectivorous birds.

The pollution-dependent alterations in the interrelationships in ecosystems can be very complex. For example, the gradual thinning of forest stands in areas heavily impacted by photochemical oxidants in the San Bernardino National Forest has caused a greater drying of soils; consequently, specific soil fungi will not form fruiting bodies. These fruiting bodies are significant food sources for squirrels and at some times of the year contribute more than a third of their diet. The squirrel's increased reliance on seeds from coniferous tree species will have a major impact on natural reproduction of stands (P. R. Miller and White 1977).

c) Decomposers

It is not expected that photochemical oxidants will directly affect soil-dwelling decomposers. In fact, soils readily absorb ozone (Aldaz 1969, Garland 1976) where it reacts with the organic material or moist surfaces limiting its penetration (Blum and Tingey 1977). In the San Bernardino Mountains, microarthropod populations in moderately injured ponderosa pine stands were reduced by 38% compared to uninjured stands (EPA 1978 b), but the cause was probably indirect. Initially, organic substrates were elevated by increased leaf and fruit drop (P. R. Miller et al. 1977); however, over the long term, photochemical oxidants reduced the biomass production of higher plants, providing less substrate for the decomposers. Under high oxidant impacts, there is a thinning of the plant canopy, promoting drying of the soils and soil erosion, which creates negative consequences for the decomposers. An increased loss of organic and inorganic materials through wind and water erosion from the ecosystem and decreased mineralization resulting from reduced and disturbed litter layers leads to a loss of mineral nutrients for primary producers, whereby biomass production is further decreased. In addition to the foregoing results that were primarily based on data from the San Bernardino Mountains, it is likely that additional effects will also occur, leading to the gradual destruction of photochemical oxidant-impacted ecosystems (Skelly et al. 1979).

2.2.5.3.3 Symbionts and Plant Pathogens

The symbiotic association between two organisms is based on a balanced exchange of metabolites. The equilibrium between the organisms is very labile and to a large extent susceptible to perturbations. Such perturbations can include air pollutants. This association also explains the high sensitivity of lichens, which are symbiotic organisms, to air pollutants (see Sect. 2.2.5.4). The symbiotic equilibrium can also be perturbed when only one of the symbiotic partners is affected by air pollution. An altered metabolite transport from the assimilatory organs to the roots influences the response of symbionts – in the form of rhizobium-induced nodules on legumes or mycorrhizae on various plant species to photochemical oxidants. Ozone effects on the aboveground portion of legumes reduced the number, size, and weight of rhizobium nodules (Manning et al. 1971, Tingey and Blum 1973, Blum and Tingey 1977, Letchworth and Blum 1977). The reduced nitrogen assimilation can have negative consequences for both the impacted plants and the ecosystem. Chronic ozone exposures of alfalfa reduced symbiotic nitrogen fixation (Tingey 1977). Ambient levels of ozone in Ontario, Canada, were sufficient to reduce nitrogen fixation in red clover (Ensing and Hofstra 1982). In addition, leaf leachates from fescue plants exposed to ozone have been shown to inhibit nodulation in white clover (Kochhar et al. 1980).

Disruptions of mycorrhizal symbiosis have been observed on both ponderosa and eastern white pine: the number of mycorrhizal roots were reduced and, in lieu of mycorrhizal fungi, saprophytic fungi increased, destroying the feeder roots (Parmeter et al. 1962, P. R. Miller 1973 a). Long-term ozone exposure reduced the infection of citrus seedling (*Poncirus trifoliata* × *Citrus sinensis*) roots by mycorrhizal fungi which resulted in reduced seedling growth (McCool et al. 1979). A short-term ozone exposure reduced spore production by the mycorrhizal fungus *Glomus fasciculatus* by 50% (McCool et al. 1977).

As with symbionts, air pollutants impact the interrelations between the host plant and pathogen. Both qualitative and quantitive changes in host-pathogen relations can occur as influenced by the air pollutant, its dose, specific sensitivity of the host and parasite, and their developmental stages at the time of exposure. Air pollution is a widely dispersed allogenic factor which contributes significantly to the spread of multiple-pathogen diseases on plants (Heagle 1973). It is difficult to establish the cause and extent of these complex diseases.

The system relationships between air pollution, pathogens, and their hosts can be analyzed from the following perspectives:

– the direct effect of air pollution on the pathogen;
– the indirect effect of air pollution on the pathogen;
– the influence of the pathogen infection on the reaction of the host plant to air pollution.

Selected examples from this more specialized field of air pollution effects will be used to describe this problem in its basic themes, compilation, interpretation, and evaluation of detailed experimental results are contained in several summaries (Heagle 1973, Saunders 1973, Treshow 1975, Manning 1975, Treshow 1980, Laurence 1981).

Air pollution can both enhance and suppress infection, invasion, and activity of plant parasites; pathogens may also modify the sensitivity of host plants. Ozone displays bactericidal bacteriastatic, and also fungicidal and fungistatic activity. Therefore, under specific conditions, effects on the pathogenic organism are expected. Spores of the barley powdery mildew fungus produced fewer lesions when the plant received several subacute ozone exposures following inoculation (Heagle and Strickland 1972). However, there were no differences in the spore germination rate, in contrast to *Uromyces phaseoli*, the cause of bean rust following relatively low sulfur dioxide doses (Weinstein et al. 1975). Ozone had a major impact on the development of a functional host-parasite relationship in *Erysiphe graminis* f. sp. *hordei* by inhibiting the development of appressoria and penetration pegs (Schuette 1971). Ozonation of *Botrytis cinerea* in vitro and in vivo decreased conidial germination, germ-tube length, and pathogenicity (C. R. Krause and Weidensaul 1978 a). Exposure of *Fomes annosus,* on an artificial medium, to low ozone doses inhibited colony growth, the formation of air hyphae, and sporulation (Hibben and Stotzky 1969). However, such results are not significant for practical environmental protection, as shown in numerous publications (Heagle 1973, Treshow 1980, Laurence 1981), and as discussed in the following presentation. An example of numerous studies in which ozone induced no detectable effects on phytopathogenic organisms is the study of Heagle and Key (1973a) with wheat stem rust. Ozone exposures following inoculation during the penetration and infection phases, which were expected to be sensitive stages, had no influence on infection or the specialized infection structures. Stimulatory effects of ozone were found by Hibben and Stotzky (1969) during their studies on isolated spores of 14 fungal species at concentrations less than 0.25 ppm.

Air pollution can cause various alterations in the phyllo- and rhizosphere of host plants. Ozone injury to potato leaves provided an entry port for *Botrytis cinerea,* increasing the infection compared to uninjured control plants (Manning et al. 1969). Pollution-induced cellular leakage may also provide a growth substrate for pathogens (Pell et al. 1977). On the other hand, extensive foliar injury can reduce the infection-susceptible leaf area (Laurence 1981). In *Geranium,* ozone induced the formation of a "flocculent material" that by chemical or physical means hindered *Botrytis cinerea* infection (C. R. Krause and Weidensaul 1978 b). In the host plants, ozone can induce the formation of various compounds such as phenols (Sect. 2.2.4.3), isoflavonoids (Keen and O. C. Taylor 1975), and increase peroxidase activity that may be injurious to microorganisms. Laurence and Wood (1978) hypothesized that the inhibited infection of soybean by *Pseudomonas glycinea* resulted from bacteriastatic effects of an isoflavonoid. It is not possible to determine if the lack of infection, in this case, resulted solely from the biocidal characteristic of the isoflavonoid. Tanaka (1976) hypothesized that the disease severity on poplars in polluted urban areas was the result of the simultaneous infection with *Marssonia brunnea* and ozone increasing the rate of plant-produced ethylene.

The significance of indirect effects on the interrelationship among air pollution, pathogens, and plants, according to our point of view, is the pollution-induced alterations in the predisposition of plants to biotic and abiotic factors. There are known secondary effects of air contaminants such as SO_2 (Wentzel 1965, Grzywacz and Wazny 1973, Huttunen 1980) and, given the nature and extent of photochemical oxidant-induced change in higher plants, they support the conclusion of secondary effects from photochemical oxidants also, even though the data are limited. As described in Section 2.2.4, ozone and PAN induce various changes in metabolism that are frequently associated with reduced plant growth and vigor. Such weakened plants are especially susceptible to weak parasites such as bark beetles and facultative parasitic fungi. According to Manning (1975), the colonization of ozone-injured white pine needles was increased for the saprophytic *Pullularia pullulans* and reduced for the obligate parasite, *Lophodermium pinastri.* In contrast, Costonis (1968) found in inoculation experiments that both fungal species were more prevalent on ozone-injured needles than on control plants. In the Ruhr area, which is polluted primarily with SO_2, *Lophodermium pinastri* is widely distributed on both pole and mature age stands of Scotch pine (*Pinus sylvestris*); however, in clean air, this weak parasite occurs extensively only in young plantation stands with unfavorable water relations. *Fomes annosus,* a root parasite of conifers, is widely distributed in the temperate zones of both North America and Europe. In Central Europe, this basidiomycete has attained a wide distribution in the last few years (Schütt 1981 b), and the increased incidence may follow the weakening of host plants by air pollutants such as SO_2 and photochemical oxidants, both singly and in combination. In this connection, the studies of P. R. Miller et al. (1977) are particularly important and differ from the laboratory studies of Hibben and Stotzky (1969). Ponderosa and Jeffrey pine seedlings exhibited a higher infection rate than unexposed control plants. Stump inoculations of differentially injured trees suggested that ozone injury increased the susceptibility of pine stumps to *Fomes annosus* (James et al. 1980).

Air pollution changes not only the predisposition to biotic, but also abiotic, stress factors. There occurs an interaction between SO_2 and winter stress in the boreal forests of Northern Europe, producing an injury response that was not observed previously (Huttunen 1980). This response cannot be explained on the basis of dose-response relations determined under the less extreme cultural conditions of Central Europe. Additional risks to vegetation result from the co-occurrence of air pollution and water stress.

Numerous studies have determined the influences of pathogenic viruses on their host plant's reactions to ozone. A series of studies showed that virus infections reduced plant sensitivity to ozone for tobacco mosaic virus on tobacco (Bisessar and Temple 1977) and tobacco ringspot on soybean (Vargo et al. 1978). The cause of the reduced sensitivity was attributed to reduced pollutant uptake through the stomata and also to changes in plant metabolism (Laurence 1981). In contrast to the above trends, tobacco streak virus infection increased the sensitivity of tobacco plants (Reinert and Gooding 1978).

For practical environmental protection, three questions result from these briefly sketched themes:

– To what extent will the interactions among air pollutant, host plant, and pathogen alter the relationship between the air pollutant and the plant?
– Under what conditions should such changes be considered?
– How can such changes be recognized and quantified under practical conditions?

The last question relates to a diagnosis and the use of bioindicators, which are discussed in Chapters 2.3.1 and 2.3.2. The first question is important for the assessment of the long-term risk to vegetation and is directly relevant for dose-response relationships used to establish standards. In agreement with Smith (1980), one should consider air pollution, above a specific concentration and for longer durations, as an allogenic factor which in natural and semi-natural ecosystems dominates the autogenic processes of succession. This includes changes in the mineral content of soils and plants and the sparingly studied host-pathogen relations and their importance to plant-community composition and their succession.

The conditions under which biotic and abiotic factors influence the relationships between air pollutants and plants depends on numerous factors, as is explained in Chapter 2.4. An example of these factors – ambient concentration – will be given from the research of Grzywacz and Wazny (1973) conducted in the vicinity of an SO_2 source. This study clearly showed that severity of infection depended on the pollutant concentration (Fig. 2.6). As described by T. C. Scheffer and Hedgecock (1955), the conditions for the possible existence of fungi are lacking under high pollutant impacts. With decreasing exposure, the incidence of infection increased and then decreased to the level of the unexposed plants.

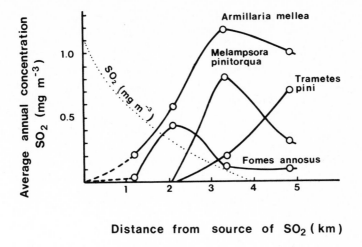

Fig. 2.6. Average percent of trees infected with phytopathogenic fungi in relation to distance from an SO$_2$ source. (Grzywacz and Wazny 1973)

2.2.5.4 Community and Economic Consequences of Functional Disruptions an Ecosystems

The value of nature's free "goods" such as pure air, clean water, a harmonious climate, and productive stands of vegetation containing numerous plant and animal species, has a significance for man that can only partially be evaluated. Westman (1977) attempted to determine the pollution-induced reductions in the "non-monetary" performance of vegetation in areas such as climate regulation, stabilization of water relations, sorption of air pollution, and reduction in soil erosion for the San Bernardino National Forest. He estimated a 27 million US dollar annual cost just for the removal of sediment from water impoundments resulting from increased soil erosion from air pollution-degraded plant stands.

A closed vegetation canopy increases the water storage capacity of the soil and reduces surface runoff, maximizing increases in groundwater. This benefits man not only through groundwater reserves, but with beneficial consequences for the climate. A healthy vegetative cover as part of an attractive landscape provides important opportunities for recreation.

At the present time, photochemical oxidants have been implicated as the cause of visible symptoms on various tree species in several national parks in the USA: Sequoia, Yosemite, Lassen, and Kings Canyon National Park (Williams et al. 1977, Williams 1979).

The preservation of species abundancy in nature is indispensible when evaluating man's present and future use, without even considering the ethical and esthetic motives. The gene pool preserved in large species diversity provides a security that in the future biogenic and anthropogenic environmental changes can be responded to by flexible changes in the animal and plant world.

2.2.6 Effects on Lower Plants and Microorganisms

Few studies have investigated the effects of photochemical oxidants on lower plants and microorganisms. Higher plants have been the center of interest, primarily because of economic considerations. The fungicidal and bactericidal effects of ozone are well understood and, in practice, it is used as a disinfectant and preservative. Ozone effects on parasitic leaf fungi are also well documented. Single-celled green algae have been used to prove the biochemical and physiological modes of ozone action.

At the cellular level, it is assumed that the mode of action is similar for both higher and lower plants. There are differences in pollutant uptake as a result of morphological and anatomical differences in plant architecture. The thallophytes, in particular, display a high uptake rate when their thalli are in a hydrated condition. They lack cuticles and most other protective tissues. Based on the limited data, it is not possible to conclude that, in lower plants, these conditions result in a generally high sensitivity to photochemical oxidants. Nevertheless, it is surmised that, within an ecosystem context, the effects on lower plants are more extensive and intensive than previously noted.

2.2.6.1 Ferns

The effects of photochemical oxidants on ferns have been observed only in the United States (Bobrov-Glater 1956) and Japan (Kadota and Ohta 1972), even though ferns generally are relatively sensitive. This sensitivity is a result of the anatomical and histological structure of the fern leaves, which facilitates the distribution of absorbed pollutants throughout the spongy mesophyll tissue of the leaf. The absorption into the cells is expedited by the limited suberization of the cell walls. As a rule, the palisade mesophyll is not extensively developed. Therefore, the injury symptoms are substantially different than for higher plants. Bobrov-Glater (1956) described the characteristic symptoms on fern leaves following periods of elevated photochemical smog in the Los Angeles basin as "tan spotting of leaflets, followed by local or general dehydration of the affected areas, ending in necrosis of the entire leaf". The complete necrosis of numerous leaves was striking, and leaf maturity had no influence on leaf sensitivity. However, the growing tip, roots, and vascular tissue were not affected unless the injury was extensive.

2.2.6.2 Mosses and Lichens

Mosses, based on their morphological, anatomical, and physiological organization as thallophytes, have a high capacity for pollutant uptake when they are hydrated. Several moss species can accumulate relatively high concentrations of heavy metals without injury, permitting their use as accumulator-bioindicators (Maschke 1981); other species are very sensitive (Rao et al. 1977). Acidic gases also cause substantial injury to mosses (LeBlanc and Rao 1975a, b). Experimental studies on the effects of photochemical oxidants on mosses have received very little attention. Comeau and LeBlanc (1971) found that a 4-h exposure to 0.25 to 1.0 ppm ozone stimulated the regenerative capacity of moss leaves, while a 6- to 8-h exposure was inhibitory.

A depletion in moss flora, particularly bark-dwelling (corticolous) mosses, has been observed repeatedly in the vicinity of large pollution sources and congested industrial areas. The extent of loss of sensitive moss species from photochemical oxidants is unknown.

The reduction of lichens in industrial and highly developed areas has received more study than mosses. It has been established, without a doubt, that lichens are good indicators of air-pollution exposure. In numerous locations the distribution of epiphytic lichens has been used to establish zones of pollution exposure. Studies have shown that lichens are very sensitive to acidic gases (see, e.g., Ferry et al. 1973, Hawksworth and Rose 1976).

Little is known of lichen sensitivity to photochemical oxidants or their modes of action. Sigal and Nash (1983) reported that the number of foliose and fruticose lichen species in southern California had decreased from 91 (Hasse 1913) to 34. In the San Bernardino Mountains, which are ecologically well studied, the number of species decreased by 50%. In a similar natural habitat, Cuyamaca Rancho State Park, which has experienced no significant pollutant exposure, there was essentially no decrease from the original number of lichen species (Sigal and Nash 1983).

In the San Bernardino Mountains, the foliose lichen *Hypogymnia enteromorpha* occurred relatively frequently. However, compared to the Cuyamaca population the mean thallus size and fertility was reduced. At ozone doses of up to 150 ppm h^{-1}, the cover of *Hypogymnia enteromorpha* and *Letharia vulpina* in the San Bernardino Mountains was essentially normal (Sigal and Nash 1983). At increased doses up to 285 ppm h^{-1} there was a gradual decline in their cover until they were almost eliminated. Studies of epiphytic lichen species inhabiting black oak (*Quercus kelloggii*) revealed a 25% reduction in the richness of lichen species in the San Bernardino Mountains compared to similar habitats in the Cuyamaca area (Sigal and Nash 1983). For other species, there was a significant negative correlation between ozone dose and lichen cover. In contrast, several resistant species were apparently unaffected by the photochemical oxidants.

Although the first observable ultrastructural changes occurred in the chloroplasts of the algal cells (Eversman 1982), it was assumed that, as in higher plants, the primary site of effect on lichens was the membranes of the algal and fungal cells (L. C. Pearson and Henriksson 1981). Because metabolite exchange between the symbionts occurs on membranes within the lichen thallus, membrane injury will disrupt the equilibrium between fungi and algae, leading to a rapid disintegration (Jahns and Neumann 1981).

There are no reports on the effects of photochemical oxidants on the exchange of metabolites between symbionts. However, effects on photosynthesis have been studied. *Parmelia sulcata* and *Hypogymnia enteromorpha* were exposed for 12 h with 0.5, or 0.8 ppm ozone. Both species demonstrated a significant reduction in gross photosynthesis at these ozone concentrations (Nash and Sigal 1979). However, *Parmelia sulcata* was more sensitive, which may explain the observed reduction in the cover of this species in the San Bernardino Mountains (Nash and Sigal 1980). Similar results were obtained following a 1-week exposure with a low concentration of PAN (Sigal and O.C. Taylor 1979). In *Pseudoparmelia caperata*, both short-term (0.1 to 0.5 ppm for 12 h) and long-term (0.10 ppm for 6 h day^{-1} for 5 days), ozone exposures reduced gross photosynthesis, but the same exposures had no significant effect on *Ramalina menziesii* (L.J. Ross and Nash 1983). Field observations in the Santa Monica Mountains in southern California, a region where ozone concentration may exceed 0.10 ppm between 50 and 100 days year^{-1}, suggested that *P. caperata* was undergoing stress (L.J. Ross and Nash 1983).

The observation of Rosentreter and Ahmadjian (1977) is confusing, as they reported that a 1-week exposure of the lichen or its algal component to 0.1 ppm ozone increased the chlorophyll content. Higher ozone concentrations had no effect. However, Brown (1980) suggested that this observation could be the result of inadequate experimental procedures.

Whether or not lichens can function as indicators to establish the cause of the presently occurring forest injury in Central Europe is unclear. On dead fir and spruce trees in the Bavarian and Black Forests of West Germany, Prinz et al. (1982) observed a more abundant lichen growth. Because epiphytic foliose and fruticose lichens are generally more sensitive to acidic gases than are coniferous trees, they concluded that the forest injury must be the result of other factors, such as photochemical oxidants, or the lichens would not have outlived the trees. Brown and Smirnoff (1978) suggested that lichens were relatively tolerant of acute ozone exposures. They found that photosynthesis of *Cladonia rangiformis* was not affected by a 2-h exposure to 2.0 ppm ozone. In contrast, laboratory studies of three lichen species in Tennessee showed that they were more sensitive to ozone than acidic gases or acidic precipitation (Sigal 1982). The sensitivity to photochemical oxidants of lichen species indigenous to Central Europe is not known. Other factors, such as improved light relations as a result of the thinning of the crowns or the leaching of nutrients from injured needles, could contribute to the phenomenon of the lichens outliving the coniferous trees. This problem provides additional scope for studies of lichens and pollution ecology.

2.2.6.3 Algae

The relatively simple single-celled algae have been used to study the biochemical and physiological effects of photochemical oxidants on plants. These studies considered effects on membrane permeability, lipid budgets, and photosynthesis, and are discussed more fully in Section 2.2.4. In this section only a few significant aspects will be repeated.

To investigate changes in membrane permeability, extensive laboratory studies have used the ozone-enhanced loss of K^+ from *Chlorella sorokiana* (Chimiklis and Heath 1972, 1975, Heath and Frederick 1977, 1979). With short-term exposures the membrane injury was reversible, but with higher concentrations or longer exposures it was not. There was an increased lipid oxidation, formation of malondialdehyde, and a decline in cell viability (Frederick 1973, Frederick and Heath 1975).

In addition to membrane injury and related disturbances, ozone reduced respiration (Heath 1978) and photosynthesis (De Koning and Jegier 1968 a, b, Verkroost 1974). Heath et al. (1982) suggested that ozone effects on the photosynthetic processes were attributable to ionic imbalances brought about by ozone interacting with the plasmalemma rather than a direct effect on the chloroplast. In *Chlamydomonas reinhardtii*, PAN reduced photosynthesis, increased the loss of carotenoides and reduced chlorophyll a and b (Gross and Dugger 1969). A 1-h exposure of *Euglena gracilis* to a combination of ozone and sulfur dioxide showed that they acted in an additive manner to reduced photosynthesis (De Koning and Jegier 1970).

Results of these studies have provided data to help clarify the modes of action of photochemical oxidants on plants. Algal cultures were used because they can serve as an experimental model for higher plants and they are easily standardized (Heath et al. 1974). It is doubtful that the air pollutants, ozone and PAN, affect aquatic algae. It is not known if photochemical oxidants impact terrestrial algae growing on stones, soil, or tree bark.

2.2.6.4 Fungi

In natural conditions, fungi that live within the soil or in living or dead organic material are probably not affected by photochemical oxidants. The fungal fruiting bodies that project above the substrate, fungi dwelling on surfaces, and spores may be impacted (Watson 1942). Studies concerning the effects of photochemical oxidants on fungi are mainly restricted to parasitic fungi because of their economic significance. The discussion in this section will be limited to the effects of photochemical oxidants, particularly ozone, on the fungi themselves and not the host-parasite relationship (see Sect. 2.2.5.3.3).

At relatively low concentrations, ozone is fungistatic or fungicidal and has been used in fruit storage (Harding 1968). On an artificial cultural medium mycelial growth, the development of aerial hyphae, and spore production were restricted (Kormelink 1967, Kuss 1950). Rich and Tomlinson (1966, 1968) found that 0.1 ppm ozone for 4 h or 1.0 ppm for 2 h stopped the apical cell division of conidiophores and resulted in the collapse of the apical cell walls of *Alternaria solani,* a foliar parasitic fungus. The apical and subapical cells swelled and elongation of the conidiophores was discontinued (Rich and Tomlinson 1966). At a similar concentration range, ozone decreased the mycelial growth and spore production of the foliar parasitic fungus, *Colletotrichum lindemuthianum* (Treshow et al. 1969).

In several fungal species, ozone stimulated growth and spore formation. On agar culture media, increased spore production was observed in *Alternaria oleracea* (Kormelink 1967, M.C. Richards 1949, Treshow et al. 1969), *Mycosphaerella citrullina* (M.C. Richards 1949), and *Colletotrichum orbiculare* (Hibben and Stotzky 1969). Kuss (1950) found an increased spore production in 8 of 30 species grown on agar.

Spores generally appear to be relatively sensitive to ozone; their germination is easily restricted. Differences in the sensitivity of spores vary among species and are influenced by spore morphology, moisture, and the nature of the substrate (Hibben and Stotzky 1969). Germination can be reduced over an ozone concentration of 0.1 to 1.0 ppm as shown by Hibben (1966) in laboratory exposures of ten parasitic fungal species grown on agar plates. In addition to reduced germination, the germ tube length and pathogenicity and/or virulence of *Botrytis cinerea* spores exposed to ozone (0.15 to 0.30 ppm for 6 h) in vivo and in vitro was reduced (C. R. Krause and Weidensaul 1978 a).

The types of culture medium can influence the ozone response. Fungal spores suspended in a liquid culture medium were less injured by ozone than spores on agar plates (Magdycz 1972, Watson 1942, Hibben and Stotzky 1969). A portion of the ozone may have reacted with oxidizable substrates in the liquid culture medium, causing this difference. The type and hydration of the surface on which the spores are placed also influence the ozone response. Moist spores were substantially more sensitive to ozone than dry ones, and spore germination was reduced to a greater extent on yeast extract-rose bengal agar than on water agar (Hibben and Stotzky 1969). The substrate-dependent differences in reduced germination observed in laboratory studies are expected to be more significant in natural conditions. In the field, the substrate (plant organ) on which the fungi are growing may be modified by photochemical oxidants, influencing the relationship between the parasitic fungi and its host. This association is discussed in Section 2.2.5.3.3.

2.2.6.5 Bacteria

Ambient air pollutant concentrations may also injure bacteria. De Mik (1973) subjected a microbial aerosol that was adhering to a gauze netting to various concentrations of air pollution. The rate at which the microorganisms died increased with the air pollution concentration. Corresponding studies with *Escherichia coli* showed that the death rate of this species was most correlated with the ozone concentration of the measured air pollutants in the ambient air (De Mik and De Groot 1973). In other exposure studies (Pan et al. 1961), it was established that ambient photochemical oxidants reduced the growth of *E. coli* colonies. The magnitude of the reduction was primarily associated with the PAN concentration.

In humans, elevated ozone concentrations irritate eyes and mucous membranes. Similar concentrations are bacteriostatic. Goetz and Tsuneishi (1959) developed the "Biological Irritation Analogue" (BIA) test in which bacteria were exposed to ambient air. The degree of growth reduction was analogous to the potential eye irritant from photochemical oxidants.

The bactericidal and fungicidal effects of ozone have extensive industrial applications. Germicidal ozone concentrations are used to disinfect drinking and waste water (Torricelli 1959, Rice and Netzer 1982, McCarthy and Smith 1974, Burleson et al. 1975, Kinman 1975, Schalekamp 1982) and in the storage of fruits and vegetables (Smock and Watson 1941, Schomer and McColloch 1948, Nagy 1959, Ridley and Sims 1966). Even the storage life of eggs is increased with ozone (Krivopishin 1973). Attempts to use germicidal ozone concentrations in climate-controlled rooms have not realized full potential because of human health risks (Witheridge and Yaglou 1939).

The toxicity of ozone to bacteria, as with fungal spores, is dependent to a large extent on the relative humidity and the moisture content of the substrate. When moist bacterial aerosols or suspensions were deposited on various surfaces, an ozone exposure (0.25 ppm for 0.5 h) at 70% relative humidity eliminated almost all of the bacteria, but at a lower humidity the bactericidal activity was low, even at 1.0 ppm ozone (Elford and van den Ende 1942).

As established in studies with animals, plants, microorganisms, and cell culture, ozone is a mutagen (Fetner 1958, 1962, I. Davis 1959, Feder and Sullivan 1969, Zelac et al. 1971). Prior to mortality, mutations were induced in *Escherichia coli* with short-term exposures to 0.1 to 1.0 ppm ozone (Hamelin and Chung 1974a, b, 1975). The mutation frequency increased with ozone concentration and exposure duration. Some of the mutants were more sensitive to ozone, while others were more resistant than the "wild-type strain". The ozone resistance was apparently associated with DNA repair mechanisms (Hamelin and Chung 1976). The capacity for DNA-polymerase to repair ozone-induced defects in DNA appears to be important for survival, because there was found to be a correlation between the occurrence of DNA breaks and the mortality rate (Hamelin et al. 1977). PAN reduced the transforming activity of isolated bacterial DNA, and in solution it reacted with the bases of the nucleic acids (Peak and Belser 1969).

In conclusion, alterations in the permeability of the bacterial cell wall played an important role in injury and cell death (Scott and Lesher 1963). Ozone altered the quantitative relationship between the phospholipids in the structure of the cell wall and interior membranes (Vokk 1977).

Particular mention should be made of the reduction in the luminescence of photobacteria. This reaction was developed into a bioassay for air pollution (Serat et al. 1967). Reliable responses occurred when the ambient air contained 0.05 ppm ozone in short-term tests (Serat et al. 1969).

Chapter 2.3
Diagnosis, Surveillance, and Estimation of Effects

In areas impacted by photochemical oxidants, it is particularly difficult to trace the pollutant-induced effects and risks to the source and determine the intensity and area affected. The occurrence of multiple air pollutants makes the identification of a specific phytotoxic component difficult and the visual injury symptoms are frequently of limited value. The extent of injury, i.e., reduced growth, quality, diversity, and the degree of risk, is difficult to assess over large areas, where air pollution-induced effects are frequently small in comparison to the variation in plant growth caused by climatic and edaphic factors.

Methods of diagnosis are presented in Section 2.3.1; the feasibilities and limits of plants as indicators are described in Section 2.3.2. The various plant reactions to ozone and PAN (Chap. 2.2) are differentially suited to estimate yield reductions of impacted plant species. The perspective from which response criteria are chosen and the ways in which these choices affect the corresponding evaluation of effects are the subject of Section 2.3.3.

2.3.1 Methods of Diagnosis

The positive diagnosis of air pollution impacts on plants with regard to its origin, expression, and area affected provides an essential basis for remedial measures. In environmental protection, pollutant-induced injuries play a role in establishing standards. Additionally they are frequently the basis for mitigative measures at sources and sites of impact and may serve as the basis for injury compensation in agricultural systems (Guderian and Vogels 1982).

The general importance of diagnostic methods for photochemical oxidants is limited because it is not possible to establish a relationship between individual injury events and specific sources. In contrast to primary air pollutants, diagnosis of photochemical oxidant injury does not lead to reduction in emissions from individual sources or compensation for yield reduction and impaired quality in the agro-economic system by an individual emitter. The remaining important tasks for diagnostic methods nevertheless require differential instrumentation for reliable identification, as with any other air pollutant.

The diagnosis of pollutant-induced plant injury, when sufficiently characteristic visible symptoms are lacking, generally proceeds in two phases: (1) the exclusion of other possible abiotic or biotic causes, and (2) the recognition of the pollutant effect (Guderian and van Haut 1970, Malhotra and Blauel 1980). Using various physiological and biological methodologies, it is possible to establish that the observed symptoms are not the result of other abiotic causes, such as unfavorable climate or soil relationships, or of biotic origin, such as animal injury, patho-genetic viruses, fungi, or bacteria or a result of a disturbance in intra- or inter-specific associations.

Middleton et al. (1950) were the first to establish the relationship between air pollution and injury on leafy vegetables, ornamental plants, and field crops that had been observed in the Los Angeles basin since 1944. Plant injury was thus the first indicator of the biological effects of photochemical smog (Haagen-Smit 1952, Hull and Went 1952, Thomas et al. 1952, Middleton et al. 1953, Stanford Research Institute 1954). B. L. Richards et al. (1958) identified ozone as a phyto-toxic component during investigations of leaf injury in California vineyards. In the following years, ozone was recognized as the cause of injury to numerous ag-ricultural and horticultural plant species and deciduous trees in many of the states in the USA (Heggestad and Middleton 1959, Daines et al. 1960, O. C. Taylor et al. 1960, Hill et al. 1961). Also an injury of unexplained etiology, known as "ozone injury," or "chlorotic dwarf" on white pine in the east, or "chlorotic de-cline" on ponderosa pine in the western USA, was shown to result from exposure to photochemical oxidants, mainly ozone (Berry and Ripperton 1963, Dochinger 1968, Dochinger and Seliskar 1970, Dochinger et al. 1970, Miller et al. 1963).

PAN-induced injury on spinach, beet (*Beta vulgaris*), lettuce, and mangel (*Beta vulgaris* var. cicla) was first established as an injurious component of photo-chemical smog by Middleton et al. (1950). The publications of Haagen-Smit et al. (1952), Bobrov (1952, 1955a), and Noble (1955) showed that the characteristic syndrome was injury to the lower leaf surface. Stephens et al. (1961) established that organic peroxides in photochemical smog were the cause of the injury, con-firming a previous hypothesis of Haagen-Smit et al. (1952).

The direct diagnosis of air-pollutant-induced plant injury follows the exclu-sion of other abiotic or biotic causes, and is supported by the results of the fol-lowing investigations (Guderian and Vogels 1982).

- Analysis of visible symptoms.
- Analysis of subtle injuries.
- Investigations of relative species resistance.
- Plant analyses, chemical and physiological.
- Soil analyses.
- Air monitoring as well as simulation of the pollutant exposure.
- Exposure of indicator plants.

As a diagnostic method, the chemical analysis of plant and soil samples as in-dicators of pollutant accumulation is not useful as neither ozone nor PAN are bioaccumulated. The feasibility and limitations of indicator plants for diagnostic purposes are discussed in Section 2.3.2. Of the remaining approaches, pheno-menological observations are doubtless of the greatest value, since ozone and PAN-induced injury symptoms are reasonably characteristic.

2.3.1.1 Analysis of Visible Injury Symptoms

Morphological and anatomical changes in plants vary not only with the air pollutant, but also with its concentration and duration of exposure. The symptoms are influenced by factors such as leaf age and developmental stage, which significantly impact the symptom distribution on the leaf and throughout the whole plant (cf. Chap. 2.4.3.1).

2.3.1.1.1 Symptoms of Ozone Injury

Four types of ozone injury can be distinguished on leaves: pigmented lesions (stipple), bleaching, chlorosis, and bifacial necrosis (Hill et al. 1970). The first three types of injury are generally characteristic reactions for chronic doses; bifacial necrosis results mainly from short-term acute exposures to high concentrations. Ozone preferentially attacks the palisade cells in the leaf mesophyll, a contributing factor to the ozone syndrome (Fig. 2.7). A histological study of soybean leaves showed that although more palisade parenchyma cells were injured than spongy parenchyma cells, the percent injured cells was similar for both cell types (Pell and Weissberger 1976). In this study, epidermal cells were not injured.

Pigmentation (stipple), the most common injury form on deciduous trees, shrubs, and herbaceous plants, occurs mainly on the upper leaf surface. The sharply defined small dot-like lesions or flecks result from the pigmentation or death of groups of palisade cells. According to the plant species, the lesion may be either dark brown, black, red, or purple; these colors result from the ozone-induced biosynthesis of specific anthocyanins and polyphenols (Koukol and

Fig. 2.7. Ozone injury. Flecking (necrotic and chlorotic) and stipple effects on the leaf. On sectioning the injury is localized primarily in the palisade tissue. (Brandt 1962)

Dugger 1967, Howell 1970, 1974). The pigments that accumulated in the injured areas of some plant species occasionally diffuse into the adjacent tissue or veins. The primary injury occurs in the tissue bounded by the smallest veins so that angular or square lesions with rounded corners predominate. The epidermis above the injured palisade parenchyma usually remains uninjured. The symptoms are best observed at low magnification with transmitted light. The injury symptom described for grape leaves (B. L. Richards et al. 1958) is particularly characteristic.

As with stipple, bleaching occurs mainly on the upper leaf surface of many herbaceous and woody species. Usually only the palisade cells are injured unless the injury is severe, and then the upper epidermis may also be affected. Numerous small necrotic, unpigmented lesions, or angular clearings of the leaf lamina impart a mottled appearance to the leaf. As the cells collapse, they usually remain connected to the healthy cells immediately above and below, and the resulting air-filled spaces impart a light gray, milky white, or tan-colored appearance to the leaf. On numerous dicotyledonous plant species, e.g., tobacco, pinto bean, soybean, and plane tree, the symptom occurs in "classic" form. On species such as small grains and grasses that lack a differentiated mesophyll tissue, injury may occur on either leaf surface.

Chlorosis, as with the previously described symptoms, occurs almost exclusively on the upper surface of leaves with palisade parenchyma. The primary lesions are usually limited to small groups of palisade cells. The overlying epidermal cells and the underlying spongy mesophyll cells usually remain uninjured. The lesion size ranges from only a few cells to flecks approximately 1 mm in diameter. With chronic or repeated exposures, the injured areas may coalesce, giving the leaf a mottled appearance. Microscopic examination showed that most of the chlorotic cells were still alive, but the chloroplasts were disrupted and the chlorophyll content was reduced (Hill et al. 1961). Chlorotic mottling or chlorotic flecks are common symptoms on pines, but on other plant species they occur less frequently than stipple or bleaching. On grasses or plants with an undifferentiated mesophyll, a fine chlorotic fleck usually develops between the veins on both leaf surfaces. The outer spongy mesophyll cells adjacent to the vascular bundle sheath and cells overlying small vascular strands appear to be most readily affected but chlorosis may extend through the leaf (Hill et al. 1961). Under long-term, persistent exposures or several days after higher concentrations, the leaves will prematurely senesce and abscise (Ledbetter et al. 1959).

Bifacial necrosis occurs when the mesophyll tissue between upper and lower epidermis is destroyed. The upper and lower epidermis are drawn close together to create a thin papery lesion that, depending on the species, ranges in color from white to red-orange. Small veins are usually injured, while the large veins survive unless higher ozone concentrations occur. Bifacial necrosis and other injury symptoms, appearing only on the upper leaf surface, frequently occur on the same leaf.

The course of injury development from high-concentration, short-term acute exposures can provide diagnostic information. The first symptom of injury is a swelling of the palisade cells and raising of the epidermal cells. The injured leaf portions appear water-soaked or oily as a consequence of increased membrane permeability and the movement of water into the intracellular spaces (Smith 1970). At this stage, injury development is reversible; if the pollutant concentration decreases or a pollutant-free period occurs the symptoms will disappear. With continuing exposures, the palisade and spongy mesophyll cells shrink and desiccate, producing bleaching and bifacial necrosis.

The value of visible symptoms as diagnostic evidence is significantly increased when the injury distribution on the leaves and the whole plant are considered. The influence of leaf age and development stage on the type, range, and distribution of visible symptoms is considered further in Chapter 2.4.3.1.

The previously described ozone symptoms do not adequately depict the types that occur on conifers. Three classic syndromes of ozone injury on conifers are: "ozone injury" and "chlorotic dwarf" of eastern white pine and "chlorotic decline" of ponderosa pine.

The name ozone injury originated with Sinclair and Costonis (1967), and Costonis (1968) for the ozone-induced injury that was previously identified as "emergence tipburn", "white pine blight," or "white pine needle die-back" of eastern white pine (Berry and Ripperton 1963). Histologically, injury to conifer needles begins with the collapse of the palisade cells adjacent to the stomatal pores, forming silver flecks which radiate from the stomata. Initially, the endodermis and stele are unaffected, but when shrinkage and chlorosis occur they are disrupted (Costonis and Sinclair 1969a, b). The syndrome of this widespread disease in the eastern USA is expressed through tip die-back on the young – 1–6 weeks old – developing needles. The partially mature tissue is the most severely injured; nevertheless, immature and fully developed tissue are also affected. The disease should not be confused with "semi-mature-needle blight" (SNB) which is not ozone-caused (Linzon 1967). In general, the ozone injury develops from chlorotic flecks into pink lesions and bands followed by a spreading orange-red necrosis which may take 1 to 2 weeks to reach the needle tip on sensitive trees, while the resistant trees may display only chlorotic flecking or mottling. The injury reduces needle retention from approximately 3 years on healthy trees to approximately 1 year; on sensitive trees the needles senesce prematurely, leaving only current-year needles by mid-summer with consequences for shoot, radial, and root growth.

"Chlorotic dwarf" of eastern white pine, which has been identified for more than 70 years (Spaulding 1909), reflects a physiogenetic condition characterized by stunted root and shoot growth, mottled needles, and premature needle drop (Dochinger and Seliskar 1965, Dochinger 1968, Dochinger et al. 1970). The needles on genetically susceptible trees start to develop normally, but chlorotic flecks soon develop. The older needles turn prematurely yellow and are shed before the young needles have fully developed. In the final stages of the disease, particularly following a drought, the current-year needles develop a tip-burn. The disease occurs most commonly on young trees in plantations. Severely injured trees tend to succumb within the first 15 years. In a population, individual trees exhibit considerable differences in susceptibility. While the less susceptible individuals display only mild mottling and nearly normal needle length, the more susceptible individuals display yellow mottling, possibly curled needles and stunting of all plant parts. Following etiological studies, Dochinger et al. (1965, 1970) demonstrated that "chlorotic dwarf" was caused by air pollutants, including ozone or a combination of ozone and sulfur dioxide.

"Chlorotic decline," "X-disease," and "ozone needle mottle" are three synonyms for an important air-pollutant-induced disease on ponderosa pine in California. The name X-disease stems from 1953 (Asher 1956) when an injury of unknown etiology was first observed, while the terms ozone needle mottle (B. L. Richards et al. 1968) and chlorotic decline (Parmeter et al. 1962) were coined later. Ozone transported into the forests from urban areas on the Pacific Coast was identified as the causative agent (P. R. Miller et al. 1963, 1969). Chronic ozone exposure induces chlorotic mottle symptoms which develop from needle tip to base, eventually followed by necrotic tip die-back. The older needles senesce prematurely, with the chlorotic areas coalescing and turning a uniform tan color. Because of the premature needle drop, severely affected trees retain only 1-year-old needles, while on uninjured trees 3- to 5-year old needles occur. Extreme variation in susceptibility to chlorotic decline is found in the forests with dead and dying trees growing alongside visibly unaffected ones.

Ozone injury occurs less frequently on fruits and blossoms than on assimilatory organs. Ozone injury on apples appeared as either a raising of the lenticels or a collapse of the tissue around them (P. M. Miller and Rich 1968). The necrotic flecks were gray to dark brown, 1 to 2 mm in diameter, and separated from the healthy tissue by a corky layer approximately 1 mm thick. On bean pods, rows of necrotic lesions developed during growth (Weaver and Jackson 1968).

Flower formation was reduced by long-term chronic ozone exposures of petunia, geranium, and carnation (Feder and Campbell 1968, Feder et al. 1969a, Feder 1970a). Flower formation was particularly impacted if the exposure occurred during flower induction (Feder and Campbell 1968). Ozone-induced yield reductions for citrus and grape crops in California are thought to result, in part, from reduced blossom formation (Thompson 1970, Thompson and Kats 1970).

Pollen is particularly sensitive to ozone. Tobacco pollen germination and pollen-tube elongation were reduced 40–50% and 50%, respectively, following exposure to 0.1 ppm ozone for 5.5 h (Feder 1968). Similar results were obtained from exposure of corn pollen (Mumford et al. 1972).

In addition to the described symptoms, photochemical oxidants can cause premature fruit drop (Weaver and Jackson 1968, Thompson et al. 1970), which has caused reduced yields in citrus groves in California, USA (B. L. Richards and L. C. Taylor 1961, Thompson 1968), and field beans in Ontario, Canada (Weaver and Jackson 1968).

2.3.1.1.2 Symptoms of PAN Injury

The first symptoms of what is now called PAN injury were described by Middleton et al. (1950) in their report, *Injury to Herbaceous Plants by Smog or Air Pollution*. The higher homologs of PAN (peroxyacetyl nitrate) such as peroxypropionyl nitrate (PPN), peroxybutyryl nitrate (PBN), and peroxyisobutyryl nitrate ($P_{iso}BN$) produce symptoms indistinguishable from PAN. Characteristics of the syndrome include glazing, silvering, and bronzing of the lower leaf surface of dicotyledonous plants such as spinach, beet, Swiss chard (*Beta vulgaris cicla*), or

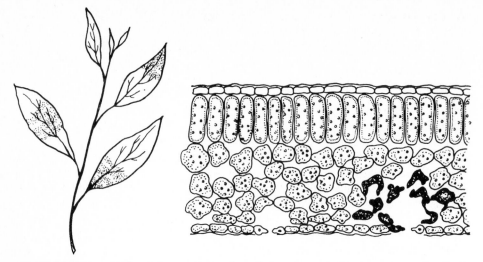

Fig. 2.8. PAN injury. Leaf age influences the location of injury; the susceptible tissue progress from leaf tip to base as the leaf matures. On sectioning, the injury is located primarily in the spongy parenchyma tissue. (Brandt 1962)

mangel (*Beta vulgaris var.* flavescens) (Middleton et al. 1958, O.C. Taylor 1969). It is possible that the foliar injury symptoms on tomato (R. G. Pearson et al. 1974) and petunia (Lewis and Brennan 1978) in the northeastern United States attributed to PAN may also be induced by exposure to the combinations of ozone and sulfur dioxide (Chap. 2.5.2.3). Under some conditions PAN and ozone may produce similar symptoms on petunia and annual bluegrass (Tonneijck 1983 a). Microscopic analyses indicate that protoplasts of the spongy mesophyll are collapsed, especially in the region of the stomata. The resultant air spaces contribute to the typical injury pattern (Fig. 2.8). The epidermis is not directly injured. When the PAN concentration is increased, the palisade mesophyll cells are affected, producing brown necrotic spots with glassy margins. Bifacial necrosis occurs more frequently in monocotyledonous than in dicotyledonous plants (Bobrov 1955 a, b) and on plants with undifferentiated mesophyll tissue. Also, PAN causes distinct transverse bands across the monocot leaf from a few mm to perhaps 2 cm wide with the bands occurring in the same location on leaves of the same age. On several grass species, the injury occurs as longitudinal streaks that are associated with the stomatal distribution. Symptoms of chronic PAN injury have not been previously described, but only the symptoms occurring in the field under the impact of the photochemical smog complex.

The PAN symptoms on ferns deviate from those observed on mono- and dicotyledonous species. Ferns have an undifferentiated mesophyll tissue with large intercellular spaces, permitting the rapid distribution of PAN throughout the leaf. Injury initially appears as bright yellow necrotic spots, progressing to complete leaf necrosis with continuing exposure (Bobrov-Glater 1956). Differential sensitivity of various portions of the leaf was not observed with ferns.

A diagnostic clue, in addition to the described symptoms, is the development of the injury. The first visible sign of injury is the development of a slightly oily or waxy appearance to leaves followed by glazing. The final symptom, i. e., bronzing, is an advanced stage and occurs 2 or 3 days after exposure.

2.3.1.2 Subtle Injury

When classifying pollutant effects on plants, the distinction is frequently made between "visible" symptoms, i. e., necrosis, chlorosis, or pigmentation, and "invisible" symptoms which cannot be established by external changes. This terminology, introduced by Sorauer and Ramann (1899) has been extensively debated, and with time enlarged upon and modified (Haselhoff and Lindau 1903, Stoklasa 1923, Reckendorfer 1952, Kisser 1966) and discussed under various names, e. g., hidden, latent, physiological, or subtle injury, but it is still basically pollutant-induced reductions in plant performance or vigor without visible, external symptoms.

Numerous attempts have been made, without success, to verify this logically based postulate; in this respect "invisible" injury, for example a reduction in growth, should be viewed as a result of alterations in plant metabolism, or cellular ultrastructure. During the last three decades, there has been a substantial improvement in the available methods to determine the occurrence of such injury. Since the early reports of "invisible" injury (Hull and Went 1952, Koritz and Went 1953) by photochemical smog, additional observations have been made. Chronic ozone exposures reduced alfalfa growth without visible injury (Tingey and Reinert 1975) and the height and specific gravity of poplar stems (Patton 1981). Also, growth reductions have been observed in the absence of visible injury for other pollutants: ethylene (Abeles and Heggestad 1973, Heck and Pires 1962), PAN (Temple 1982), sulfur dioxide (Tingey and Reinert 1975, Guderian 1977), and HCl (Guderian 1977).

The increasing importance of subtle changes in plants for practical environmental protection measures results from the reduced performance of plants as economic objects in agriculture and forestry at the present chronic exposures to low concentrations. Beyond that, newer results have increasingly shown how subtle effects can change the predisposition of plants to biotic (Laurence 1981) and abiotic (Huttunen 1980) stress factors. Naturally, the criteria for the early indication and detection of such effects has achieved a central importance.

The difficult-to-recognize effects include not only reductions in growth and changes in predisposition in the absence of external, visible symptoms, but also reduced performance when only chlorosis, pigmentation, or premature senescence occurs. This is especially true for photochemical oxidants that cover large areas. With frequent chronic exposures, not only can visible injury symptoms be lacking, but also given relatively homogeneous concentrations over large areas, the clearly defined injury gradients associated with sources emitting primary air pollutants are not present. At the same time, photochemical oxidants pose a particular risk to vegetation from latent reductions in performance: for example, in areas receiving typical photochemical smog exposures, there is frequently an insufficient period between exposures for plant recovery; also chronic exposures are influenced by a wider range of ecological conditions than acute exposures.

For these reasons, it is necessary to develop and use methods for early detection (Keller 1981). For this purpose, biochemical and physiological criteria from the cell or organelle levels of organization will be used to show their suitability as indicators (Guderian and Reidl 1982). Changes in enzyme activities, alterations in pigment contents, increased ethylene production in plant tissue (Tingey 1977), reduced ATP biosynthesis (Flueckiger et al. 1978), reduced photosynthetic rate or disturbances in plant-water relations can be applied as early indicators of air-pollution stress. The ozone-induced enrichment of zinc, copper, and cadmium in grain may also provide an early indicator (Blessin et al. 1979). An example of a sensitive indicator of air-pollution stress is peroxidase activity. Numerous plant species under the influence of various air pollutants displayed an increased peroxidase activity as an indicator of premature senescence (Keller 1974). In such studies it is possible, generally before the appearance of visible injury symptoms, to detect air-pollution-induced stress. This is particularly important for perennial plants in which small changes in plant metabolism can, during the course of time, cause significant reductions in performance (Huttunen 1980). The value of biochemical, physiological, and ultrastructural changes in cell organelles is limited because the changes are generally not specific and are frequently reversible, reflecting only the current conditions. It is also difficult to relate biochemical and physiological alterations to effects noted in individual plants, populations, or communities.

2.3.1.3 Investigations of Species Specific Resistance Relationships

Individual plant species, cultivars, plant provenances, individuals and their respective populations react with differential sensitivity to various air pollutants (see Chap. 2.4.3.2). This differential sensitivity can be of diagnostic value when the degree of injury on plant species in the study area is compared to established relative plant resistance listings for the suspected air contaminant. Therefore, it is important to examine the criteria used to establish the relative resistance which are different than the criteria used for detection and evaluation of an effect. Most relative resistant groupings for individual species have been based on visible foliar injury (Jacobson and Hill 1970, D. D. Davis and Wilhour 1976) rather than plant yield. The pollutant-induced changes in the vigor and survival of individual differentially susceptible plant species through, in part, altered interspecific competitive conditions should be considered when identifying impacts on plant communities (Guderian 1977, Wolak 1977). Alterations in plant communities, according to our present knowledge concerning their responses to photochemical oxidants, are less suited for etiology than for determining the intensity and spatial distribution of pollutant impacts. The uses of synecological changes as a method of bioindication will be discussed later.

2.3.1.4 Air Monitoring

The procedures presently available permit the reliable determination of the atmospheric content of ozone (see Chap. 1.5.1.2.1) and PAN (see Chap. 1.5.1.2.2). The ascertainment of a correlation between the ambient concentration and plant effects is an additional prerequisite for using monitoring equipment in the diagnosis and evaluation of oxidant impacts. However, it should be remembered that air monitoring only provides information concerning the probable risk, not the nature of effect. Whether an effect occurs depends not only on the concentration and duration of exposure to a particular air pollutant, but also on the relative sensitivity of the exposed plant species as influenced by the climatic and edaphic conditions. Air-monitoring methods only describe the instantaneous ambient concentrations; they do not provide reliable information concerning the pollutant exposure at the time of injury. This is especially true for secondary pollutants that are formed in the atmosphere and display large time-dependent concentration changes, frequently to a greater extent than primary air pollutants. Also, a high level of technical expertise is required to measure ozone and PAN in the ambient air, which diminishes the value of monitoring in diagnosis. Nevertheless, measurements of ambient concentrations have their value, because they can provide information on the probability that an effect may result. With primary pollutants it is possible to simulate the probability of pollutant exposure using emission data and diffusion modeling, but this has been less successful for secondary air pollutants.

2.3.2 Bioindicator Plants

For photochemical air pollutants, the development of preventative measures for environmental protection in either the prophylactic or retrospective sense is very difficult. Photochemical oxidants occur over large areas and form complex mixtures, composed of numerous ecotoxicological components which interact with each other and primary air pollutants, posing a particular risk to vegetation. Improved chemical and physical methods of air monitoring make it possible to identify and measure the most important phytotoxic components; however, because of technical limitations and expense, it is not possible to measure all potentially phytotoxic components or to accurately assess the potential risk of such a pollutant complex. Above all, measurements of ambient concentrations can be used only for risk prognosis, not to predict actual responses. The extent to which a specific episode may be injurious depends not only on the pollutant exposure, but also on the genetic and environmental resistances of the exposed plants. Consequently, chemical and physical air monitoring as the only methods are insufficient to determine the degree and origin of pollution-dependent risks. A reliable expression of the actual vegetation response is only possible by determining the effect. Organisms are particularly well suited for the detection, determination, and evaluation of air pollutants because they can demonstrate the effect of the pollutant. For this very reason, the use of indicator organisms is increasing in environmental protection.

Plants, in addition to other organisms, respond to external stimuli such as air pollutants in numerous ways. Their responses to this stimulation provide criteria suitable for indicator purposes. The differences in the activity of air pollutants are expressed in various quantitative plant reactions. Plants can serve as indicators of the biological activity of air pollutants because they are not only sensitive, but frequently provide characteristic responses to the prevalent air pollutants. Moreover, comparative studies are easier with plants than with animals or humans, because plants are usually found in large numbers in specific habitats and are exposed to environmental conditions that are easily determined.

A bioindicator must meet various requirements as expressed in various definitions of this concept. For example, Ellenberg (cited in Kohler 1978) defined a bioindicator as, "a group or community of organisms whose presence or other easily measured behavior is so closely correlated with the specific attribute that one can use it as an indicator or a quantitative test". This broadly based definition would also include indicator plants which reflect the characteristics of a site, for example the soil or climate. However, in this section, bioindicators will be considered as only those organisms whose reactions reflect the anthropogenically influenced environmental factors.

2.3.2.1 Bioindicator Methods

To establish pollution-induced effects, one can choose between effects on cultivated and native plants grown in their natural habitats, passive monitoring, or from standardized methods of plant culture and exposure in the areas under evaluation, active monitoring. In the first case, except for the use of the recently available exposure systems with and without air filtration, all parameters, beginning with the indicator and including the various internal and external growth factors that influence the response, must be accepted, while in the second case, they are partially regulated. The laboratory-based biotest will not be considered because it is not well suited for use with photochemical oxidants.

All three procedures can be used at various biological organization levels such as plant community, population, organism, organ, and organelle. Passive monitoring includes all reactions that have an indicator function, i.e., changes in the community composition or phenomenalistically confirmed injury symptoms on populations or individuals provide a preliminary indication of impacts on an ecosystem. Active monitoring, particularly the biotest with its controlled and reproducible conditions, provides a better basis, aided by physiological, biochemical, and electronic-optical methods, to determine alterations at the cellular or subcellular levels. Different expressions and application possibilities arise from these procedures: passive monitoring is well suited for diagnosis (Sect. 2.3.1), while active monitoring is better suited for detecting the gradation of impacts related to surveillance.

2.3.2.1.1 Passive Monitoring

a) Autecological Procedures

For the recognition of air-pollution-induced injury, the identification of its cause, and the determination of the injury intensity on cultivated and native plants in their natural habitats, the autoecological factors are more important than the synecological ones. Since the beginning of systematic studies of air pollution injury on vegetation, about 130 years ago, injury responses have repeatedly provided the first indication of pollutant impacts. The results of studies on individual plants or populations of individual species have aided in the identification of, e.g., sulfur dioxide (Stöckhardt 1871), thallium (MAGS and MELF 1980), and photochemical smog (Middleton et al. 1950) as phytotoxic air pollutants. The available diagnostic methods are described in Section 2.3.1; therefore, only a few additions are necessary.

Since the successful studies on the etiology of a widely occurring new type of injury on cultivated plants in California during the 1940's and 1950's, sporadic and systematic attempts have been made to determine the intensity and spatial distribution of the injury. Acute injury on oxidant-sensitive plants such as tobacco, grape, lettuce, bean, or potato is comparatively simple and reliable to monitor by establishing the degree of necrosis. By 1955, a monitoring program was established covering all of California using agricultural crops, fruit trees, flowers, and native weed species (Middleton and Paulus 1956) (cf. Chap. 2.2.5.3). Computer analysis was used to determine areas of differential exposure. However, difficulties exist in the determination of impacts from chronic exposure, such as growth losses, reduction in fruit and flower formation, and reduced yield and quality in agriculture. When large areas are impacted, it is only possible to approximate the suspected injurious agent because of differences in external growth factors.

In forest ecosystems, monitoring has concentrated on determining the injury on the dominant tree species, for example, ponderosa pine in California and eastern white pine in the eastern United States. Suitable criteria to establish the health of conifers include the amount of needle retention and their condition; studies on the height and radial growth of trees allow the reliable determination of growth losses to forestry (Pollanschütz 1975). Dendrochronological investigations permit conclusions about the pollutant exposures during the past years and decades.

Large-scale monitoring of air-pollutant-impacted forest stands has been accomplished using false-color photography (Wert et al. 1970). A scale of 1 : 8,000 is suitable for groups of trees, and at 1 : 584 injury can be recognized on individual trees. For example, photochemical oxidant injury in the Los Angeles and San Bernardino National Forests covered 40,000 ha and affected approximately 1.3 million trees. The evaluation of risks to forest trees is also possible by determining the condition of visibly injured native, nontree species. The common milkweed (*Asclepias* spp.) that is widely distributed in North America has been proposed for this use (Duchelle et al. 1980) because it has been shown to be very susceptible to ozone in both open-top chambers and in field studies. The presentation here has been brief because this same theme has been extensively discussed in other publications (e. g., see Heck and Brandt 1977).

b) Synecological Procedures

Because of the limited studies concerning the effects of air pollutants on ecosystems (see Chap. 2.2.5), the interactions of plant associations has not attained any significant importance as a bioindicator.

For plant communities to be suitable indicators of air pollutants, two questions are decisive: what requirements are needed to reliably establish pollution-induced changes in plant communities; and how sensitive are plant communities in comparison to their individual species? At present, it is not possible to answer either question with practical experience or experimental results from comparative studies. The sensitivity of plant communities in comparison to their individual member species can be deduced from the types of alteration and the manner in which these occur in plant communities. The mechanisms of air-pollution-induced alterations are based on:

- the variable magnitude of effects on differentially susceptible community members;
- pollution-dependent changes in the interspecific competitive conditions; and
- the secondary succession processes induced by air pollution.

Pollution-induced shifts in the interspecific competitive relationships can impair the more susceptible plant species of a community and thus increase the direct damage to those community members. Moreover, alterations in interspecific competitive conditions may occur at such low concentrations that the individual species of the community are not injured but produce stimulatory effects that can give a competitive advantage to specific plant species (Guderian 1978). Accordingly, one can see that specific associations of higher plants may be more sensitive indicators of phytotoxic air pollutants than the individual species comprising them, as has been observed in the initial studies on this topic.

The greater sensitivity of plant communities compared to the corresponding homotypic populations does not necessarily mean that they are better bioindicators. For this characteristic, it is more important to know how large the change is, and what efforts must be used to reliably detect pollution-induced shifts under typical conditions. Finally, the significance of plant communities as indicators of air pollution will depend on the results obtained from exposing specific model plant communities (Küppers 1984) and from determining chronic effects under natural conditions using permanent sample plots (BMI 1978).

Alterations in structure and composition are useful indicators of pollutant effects on plant communities. Using plant sociological mapping techniques, zones of differential impact can be established. Depending on the type and concentration of exposure, some species are aggregated, others uniformly distributed or completely absent (Niklfeld 1967, Trautmann et al. 1971, Wedeck 1980). In this context, Hajdúk (1961) used the terms positive and negative phytoindicators, while F. K. Anderson (1966) used the terms increaser and decreaser. These results were obtained around individual sources emitting primary air pollutants with sharp gradients in ambient concentrations and damage responses. Given the large areas impacted with photochemical oxidants which display similar amounts of injury, the conditions are naturally less favorable for the establishment of pollution-induced changes in plant communities and the use of these changes as bioindicators. During the last several years, efforts to determine air-pollution-induced changes in plant communities (O. C. Taylor 1974a, Williams et al. 1977, Westman 1979, Smith 1980, Blum et al. 1980, Montes et al. 1982) have increased the possibility that such changes can be used to recognize oxidant impacts, particularly to establish the extent of damage according to the intensity and spatial distribution.

2.3.2.1.2 Active Monitoring

Responses of native and cultivated plants growing in their habitats serve, as described in Section 2.3.2.1.1, primarily to establish the causes of pollutant-dependent injury and determine their spatial distribution. These characteristics are less suited for the quantification of air pollution impacts or for monitoring of pollutant effects. The plant material itself is too heterogeneous and the climatic and soil conditions are too variable to permit valid comparisons. Two approaches have been used to overcome these difficulties: standardized plant culture and the exposure of standardized plants using special methods.

a) Standardized Plant Culture

The quality of biological measurements, as with physicochemical methods, depends heavily on the reproducibility of the test results. Consequently, one should try to minimize the variability caused by external and internal factors associated with a pollutant exposure.

The standardization of plant culture produces a narrowing of the spectrum of effective factors, and thereby a reduction in experimental variation. Concurrently this approach can be used to significantly control and reproduce the conditions that influence pollutant uptake and pollutant effect. Methodological approaches for plant culture under similar conditions include the use of high-value seeds, the standardization of seeding, plant culture, and growth conditions such as soil, water, and nutrient supply.

A classic example of a standardized procedure for plant culture and exposure is the protocol for the ozone-sensitive tobacco cultivar Bel W3 developed by Heck et al. (1969). Since then, this method has been used successfully in the USA and other countries as a bioindicator for photochemical oxidants (cf. Sect. 2.3.2.2). To determine the effects of primary air pollutants, especially compounds that are accumulated in the plant, the grass culture procedure developed by Scholl (1971) has been useful, particularly as an "accumulation-indicator". The culture and exposure of Italian ryegrass (*Lolium multiflorum*) as an indicator plant including the sampling and analysis are detailed in VDI-Richtlinien 3972, Bl. 1 (VDI 1978a) and 3792, Bl. 2 (VDI 1982).

b) Exposure Systems

When using plants as air-pollution indicators to establish the nature and level of the current exposures, optimal results are obtained only when the plant culture and exposure methods are standardized. One can select between exposures in containers or field plots in the ambient air or exposure systems with and without air filtration. The culture of indicator plants in plots receiving ambient air such as the Test Plant Procedure (Schönbeck 1963) or Test Plot (Spierings 1967) has practical significance for the study of primary air pollutants. Comparable studies with photochemical oxidants are not known. In contrast, the container-culturing of bioindicators has achieved a greater significance, as illustrated by examples in Section 2.3.2.2, particularly the protocol of Heck et al. (1969). Above all, with the typically large areas impacted by photochemical oxidants, it is very difficult to determine reliably the intensity and spatial distribution of the pollutant effects, because air pollution-induced changes are frequently small compared to the variation in plant growth induced by soil and climatic conditions. Comparative studies at the same site with indicator plants in exposure systems with and without air filtration make it possible to reliably detect the extent of plant impacts from moderate exposures. Currently available methods to determine air pollution effects on plants and their particular applications are described in greater detail in Section 2.3.4.

2.3.2.2 Bioindicator Results from Various Countries

Plants have been used as bioindicators for photochemical oxidants in both in North America and other parts of the world using various active monitoring methods (Sect. 2.3.2.1.2). A few examples of the data obtained will illustrate the feasibility and limitations of the methods.

The bioindicator methods developed by Heck et al. (1969) were used to monitor the occurrence of ozone injury on Bel W3 tobacco in the vicinity around Cincinnati, Ohio, USA (Heck and Heagle 1970). The intensity of injury varied over the years – 1966 to 1968 – but was correlated with the weekly mean ozone concentrations as influenced by various external factors. Craker et al. (1974) also used the methods developed by Heck et al. (1969) when they exposed Bel W3 tobacco in the vicinity of Sudbury, Massachusetts. Based on the amounts of foliar necrosis on tobacco leaves, lines of equal injury intensity, "Isopollutans", were developed. The results from sites covering the state of Massachusetts showed that this method was suitable for distinguishing zones of differential exposure (Jacobson and Feder 1974). Kelleher and Feder (1978), with the help of Bel W3, were able to show the long-distance transport of photochemical oxidants. On Nantucket Island in the Atlantic Ocean, 340 km east-northeast from New York, photochemical oxidant injury occurred on tobacco plants primarily when the winds were west-southwest, from densely populated areas. Measured ambient ozone concentrations were in good agreement with the bioindicator results.

Outside of North America, Knabe et al. (1973) exposed various tobacco cultivars during 1967 and 1968, using the procedures of Heck et al. (1969) at four pollutant-impacted sites in the German Federal Republic. In 1963, E. F. Darley (unpublished data) established that both ozone and PAN symptoms occurred on native plants growing in the Ruhr area. The extent of foliar necrosis on experimental tobacco plants at the various sites showed that photochemical oxidants were a widely distributed type of air pollution in the Federal Republic of Germany. Foliar injury also occurred at a "clean air station" in southern Westphalia, but the extent of injury was less than at stations in the industrial areas. Since then, results from both local and regional studies, for example Cologne and the western Ruhr (Scholl and van Haut 1977), Munich (Rudolph 1977a, b), the lower Main region (Steubing 1977, 1980), and Stuttgart (Rabe 1978, Arndt and Lindner 1981) have confirmed the widespread occurrence of phytotoxic concentrations of photochemical oxidants.

Bioindicator plants have been used with good results in Great Britain. In 1972, Bell and Cox (1975) exposed three tobacco cultivars (Bel W3, Bel B, Bel C) at sites to the west of London. They found that Bel W3 was the most sensitive; injury to

Bel W3 decreased less with distance from London than for the other cultivars. The degree of foliar necrosis on Bel W3 correlated closely with the number of hours that the ozone concentration equaled or exceeded 0.04 ppm. Ashmore et al. (1978) exposed tobacco plants throughout all of Great Britain and confirmed the wide-scale distribution of photochemical oxidants. The assumption of Bell and Cox (1975) that the current oxidant exposures were injuring not only highly sensitive indicator plants but also agricultural plants has been confirmed (Ashmore et al. 1980 b). They used open-top chambers (Mandl et al. 1973) and exposed various radish and pea cultivars at the same locations used in the study of Bell and Cox (1975). Considerable foliar necrosis occurred on ozone-sensitive pea cultivars, while radish cotyledons were chlorotic and abscised prematurely. Measured ozone concentrations with peak values to 0.14 ppm were phytotoxic to sensitive plant species. Injury on corn and tomatoes in the vicinity of the exposure site could not definitely be ascribed to photochemical oxidants because of the lack of standardization.

Photochemical oxidant-induced plant injury has been observed in The Netherlands since 1965 (ten Houten 1966). The symptoms observed on native plants, annual bluegrass, little leaf nettle (*Urtica urens*), and nettleleaf goosefoot (*Chenopodium murale*), in the vicinity of an oil refinery center were attributed to PAN although the ambient concentration was not determined. The concurrent exposure of indicator plants, tobacco (*Nicotiana rustica*), spinach, lettuce, and several chrysanthemum cultivars under standard conditions suggested the occurrence of PAN. Exposures in controlled conditions have established that *Urtica urens* is a sensitive indicator of visible PAN injury (Tonneijck 1983 a). In The Netherlands, foliar injury is observed on *Urtica urens* plants in the field but most of the injury appears to be induced by O_3. Based on the lack of PAN symptoms on sensitive indicator plants in the field, and low ambient concentrations of PAN, Tonneijck (1983 a) concluded that, at present, PAN is not an important phytotoxic air pollutant in The Netherlands. A country-wide monitoring network has been established in The Netherlands since 1969 and, with the aid of physicochemical methods, determined the primary and secondary substances of photochemical smog (Guicherit 1975). In parallel, exposures of oxidant-sensitive plants have occurred since 1973 (Posthumus 1976). The bioindicator plants and the observed symptoms are listed in Table 2.4. The plants were exposed for weekly periods and then evaluated. The greatest photochemical oxidant exposures occurred around Rotterdam (Floor and Posthumus 1977).

Also in other European countries, the first indication of photochemical oxidants resulted from the use of the bioindicator, Bel W3 tobacco. Ozone effects were demonstrated not only in Switzerland, France, and Belgium, but also in the northern European countries of Denmark and Sweden (Eastmond and Skärby 1979, Ro-Paulsen et al. 1981). The Swedish investigations produced the hypothesis that photochemical oxidants resulted from both local sources and long-distance transport to the monitoring sites (Skärby 1979).

In the subtropical climate of Israel, the bioindicator Bel W3 can be used in either the field or in chamber exposure systems wih or without air filtration during the winter months (Goren and Donagi 1980). However, during the summer months, the closed exposure chambers could not be used because of overheating

Table 2.4. Indicator plants for monitoring photochemical oxidants in The Netherlands. (Posthumus 1976)

Plant species and cultivar	Symptoms	Pollutant
Little leaf nettle	Leaf undersurface bronzing (band form)	PAN
Annual bluegrass	Bifacial leaf necrosis (band form)	PAN
Tobacco cv. Bel W3	Necrotic flecks (spots) on the upper leaf surface	O_3
Spinach cv. Subito cv. Dynamo	Necrosis on the upper leaf surface	O_3

(Naveh et al. 1978). They found a good correlation between the extent of foliar injury and the ambient oxidant concentration.

Active and passive monitoring programs using bioindicators have been in operation in Japan since 1973 (Matsunaka 1973, Shiratori 1973). Buckwheat and oak were grown under standardized conditions in containers and exposed at the study area. Concurrently, injury symptoms on pines (*Pinus thunbergii* and *Pinus densiflora*) and cultivated vegetable species, onion, radish, morning glory, black radish, and taro, growing in the study area are evaluated. Foliar injury serves as the response criteria. Since then, the response of morning glory to photochemical oxidants in both controlled exposures and field studies has been calibrated so that the intensity of the plant reaction can be used to infer the mean ozone concentration during the exposure period (Nouchi and Aoki 1979).

2.3.2.3 Comparison of Biological and Physicochemical Methods

When sensitivity, specificity, and reproducibility are used as criteria to evaluate measurement methods, comparison generally favors the physicochemical ones; typically they are more sensitive, specific, and reproducible than biological methods. Also, physicochemical methods permit the exact determination of the concentration-time profile. With this assumption and in view of the significant progress recently made in physical and chemical methods for air analysis, it seems difficult to understand the growing use of biological methods in environmental protection (Jacobson 1977b, Floor and Posthumus 1977, Prinz and Scholl 1978, Arndt et al. 1982a).

An explanation of this apparent contradiction results from a comparison of the expected results from physicochemical and biological methods (Guderian and Reidl 1982) and emanates from the specific data needs for individual environmental protection measures (Table 2.5). In this respect, it should be remembered that the quantitative responses of biological objectives are not to be understood in the same sense as classical physical measurements (Schönbeck and van Haut 1971). An absolute causality, where the same cause always has the same effect, is not detectable. In its place are statistical occurrences regulated by the laws of probabil-

Table 2.5. Uses and limitations of physicochemical and biological methods for the determination and evaluation of pollution impacts. (Guderian and Reidl 1982)

Physicochemical methods	Biological methods
Quantitative measurement of individual atmospheric components	No measurement in the physicochemical sense; evaluation of air pollution exposure through numerous nonspecific reactions
Monitoring of emission and ambient air quality standards	Response thresholds possess a high relevance because air pollution legislation has the goal of reducing specific effects
No determination of effect or risk evaluation	Determination of actually occurring effects in the form of injury and pollutant enrichment at various organizational levels, an indicator of disruption in the ecosystem
Risk prognoses are possibly only for individual components	Evaluation of the total exposure, possibility of determining joint effects
The active portion of a mixture of air pollutants can frequently be only approximated	The effects reflect the active portion of the pollutant mixture
Determined emissions and ambient concentrations at a specific time (characterization of concentration-time profile)	Expression of effects during a specific time period; determination of effects on perennial plants

ity which make possible the successful use of plants as indicators of biologically active air contaminants.

The primary goal of environmental protection is not to reduce emissions or ambient concentrations but to protect humans, animals, plants, and materials from injurious environmental effects. However, that does not mean that the determination of effects alone is a sufficient or optimal information base. If specific air pollution effects are to be prevented, it follows that emissions must be restricted to a given quantity through the establishment of emission standards that are monitored by physicochemical methods. The same is true for ambient air standards. Ambient air standards provide a possibility for prophylactic environmental protection (Prinz and Scholl 1978), an essential condition particularly for the protection of humans. With biological indicators, however, only those exposures that are already in the toxic range can be determined.

The measurements of emissions and ambient concentrations cannot replace the determination of biological effects and the biological effects cannot replace the physicochemical measurements. The comparison (Table 2.5) of the specific use of physicochemical and biological methods is applicable not only to the effects of photochemical oxidants but to gaseous air pollutants in general. Moreover, it should be remembered that with such different materials, gross simplification is required.

The primary value of biological indicators lies in their expression of effects and indication of the phytotoxic components currently in the ecosystem. Bioindicators permit the evaluation of the total exposure; they react with only the biologically active components of the atmosphere and permit the reliable prognosis of risks for other components in the ecosystem (Guderian and Reidl 1982). When one considers that information from biological indicators is frequently obtained with comparatively small expenditures (Claussen 1980), their increased use is understandable. According to the nature of the question, physicochemical or biological methods may be used. As a whole, a meaningful combination of both response systems provides an optimal information basis for the prompt recognition and control of air-pollutant-induced risks.

2.3.2.4 Conclusions for the Use of Bioindicators

Bioindicators, with their ability to evaluate the extent of effects from current exposures and to predict future risks, can be used with the following objectives:

- recognition of causes of plant injury;
- delimitation of exposed areas and gradation of exposure levels with temporal relationships;
- risk prognosis for agriculture, horticulture, and forestry land uses;
- evaluation of risks to semi-natural ecosystems;
- optimal spatial planning based on environmental protection requirements;
- determination of air quality criteria;
- surveillance of permissible ambient concentrations; and
- demonstration of the effects of photochemical oxidants.

If and to what extent higher plants maintain or increase their role as air-pollution indicators for environmental protection depends when and with what success the following points are considered in conducting the corresponding research:

1. Greater consideration should be given to the selection of indicator organisms, based on the ecological performance of vegetation.
2. The selection of air-pollution indicators should specifically consider the chronic effects of low concentrations of complex pollution types.
3. The increased use of homo- and heterotypical populations for active monitoring, particularly exposure systems with and without air filtration is encouraged, to determine the type, intensity, and spatial distribution of chronic, low-level exposures.
4. Comparative studies of homo- and heterotypic populations should clarify if, and to what extent, plant communities are more sensitive than the individual species composing them. Studies to determine air-pollution-dependent risk to plant communities should be conducted through:

    - development of experimental conditions and of intensive epidemiological studies to determine the effects on highly structured systems;
    - establishment of permanent plots for successional controls; and

– studies of the characteristics of model plant communities or portions of ecosystems as indicators for ecotoxicological tests.

5. Work to standardize the culture and exposure of plants suitable as bioindicators should be intensified.
6. Comparative studies should clarify the extent that specific reactions of air pollution indicators can be used to reliably determine the consequences of risk to other plants or other components in the ecosystem.

2.3.3 Classification and Evaluation of Effects

In determining and evaluating air-pollutant impact on plants, its effects on various ecosystem functions should also be considered (Fig. 2.1). Above a specific threshold, air pollution will impair or eliminate the performance of vegetation, i. e., as a primary producer, as an economic objective, as a component of a healthy ecosystem and environment, and as a gene reservoir. To use this reduced vegetation performance as a basis for remedial or mitigative action, criteria must be selected that evaluate the usefulness of individual plants and vegetation as a whole from ecological and economical viewpoints. For the various effects of air pollutants at the individual organizational levels of the ecosystem, Vogl et al. (1965) and Weinstein and McCune (1971) developed a classification system, without consideration of the relationship of the response to the useful value of the plant. This concept was used to categorize (Table 2.6) the effects of photochemical oxidants described in Chapter 2.2.

2.3.3.1 Classification of Effects

At the cell, the lowest organizational level of the ecosystem, effects are initially expressed as altered permeability of biomembranes causing changes in cellular compartmentalization, water and mineral relations. These impacts, as well as changes in enzyme activities, altered metabolism, changes in cellular structure and organization, including fine structure changes, cause cellular perturbations and/or cell death.

Cellular changes above a specified magnitude influence the next-highest organizational level, for example, through reduced photosynthesis rates, elevated respiration, or disruption of plant water relations. A reduced CO_2 assimilation rate affects leaves and frequently processes in other plant organs, especially root growth (Tingey 1977, Oshima et al. 1979), rhizobium-induced nodulation of legumes (Manning et al. 1971, Tingey and Blum 1973, Tingey 1977), and the development of mycorrhizae on specific tree species (Parmeter et al. 1962, P. R. Miller 1973a) as a consequence of altered metabolite partitioning. Foliar injury symptoms include pigmentation, chlorosis, and necrosis, and they are frequently coupled with increased chlorophyll decomposition. It should be remembered that foliar injury does not always result in reduced growth or yield and also that

Table 2.6. Classification of photochemical oxidants effects on plants

Organizational level

Cell	Tissue	Organism	Community
Increased membrane permeability	Changes in photo-synthesis, respiration, and transpiration Alterations in the partitioning of metabolites	Changes in plant growth	Reduced plant growth and productivity Fluctuations in composition and reduction in species abundance
Altered enzyme activities	Alterations in growth and development of individual organs	Increased suscepti-bility to biotic and abiotic stresses	Changes in stand structure
Increased stress ethylene production	Pigmentation, bleaching, and chlorosis	Disturbances in fruit production	Disruption of food chain Changes in plant succession Possible changes in nutrient cycling Risks to consumers and decomposers
Ultrastructural changes in organelles Changes in cellular metabolism	Necrosis	Reduced yield and quality	Impairment of ecosystem productivity, including its stability and capacity for self-regulation
Altered cellular structure	Reduction in rhizobium induced nodulation	Altered plant competi-tive ability	
Disrupted cellular functions	Disturbance of mycor-rhizal development	Death of plants	
Cell death	Death or loss of plant organs		

growth and yield can be reduced without the accompanying visible symptoms (see Sect. 2.3.1.2). Reduced pollen production and altered pollen-tube growth (Feder 1968) can lead to reduced fruit production. At this level of organization the most severe effects include the premature senescence, death, and casting of plant organs.

Effects on cells and organs affect organismal growth, which includes not only "system synthesis" but "product synthesis", i. e., the development of such organs which are harvested will be affected, and this will in turn affect yield. In addition to "primary" effects, "secondary" ones (Donaubauer 1966, Heagle 1973, Laurence 1981), can result from reduced plant resistance to biotic and abiotic factors with consequences that can approach complete yield loss and reduced quality.

The composition and species abundance in plant communities may change as a result of various direct effects on individual, differentially susceptible plant species. Pollutant-induced changes in interspecific competitive conditions and secondary succession may also occur (Guderian and Küppers 1980 b). These effects and changes in stand structure may also endanger consumers and decomposers (P. R. Miller and White 1977). In the final result, the productivity of the ecosystem including its stability and capacity for self-regulation are impaired (see Chap. 2.2.5).

2.3.3.2 Evaluation of Effects

The briefly described effects on plants at the individual organizational levels of the ecosystem have different values for evaluating the effects of air pollutants on the useful value of plants (Guderian et al. 1960). The segregation of air-pollutant-induced reactions into "injury" and "damage" is now widely used in studies to evaluate air-pollution effects (Brandt 1962, Heggestad and Heck 1971, NATO/CCMS 1974, Tingey et al. 1979). Such a classification permits comparison of effects on different plant species.

Injury encompasses all plant reactions such as reversible changes in metabolism, reduced photosynthesis, leaf necrosis, leaf drop, altered quality or reduced growth that do not influence yield. Damage includes all effects that reduce the intended use or value of the plant (Table 2.7). These useful values, characterized by economic production, ecological functions, ideal values, and vegetation as a gene pool, can be reduced through the direct impact of air pollutants. Numerous studies describe this "primary damage". In contrast, there is little information concerning "secondary damage" as a consequence of photochemical oxidant reduced resistance to biotic and abiotic stresses (O. C. Taylor 1973a, Heck et al. 1977).

Table 2.7. Effects of exposure to photochemical oxidants on the performance on plants in ecosystems

	Injury All reactions of air pollutants Damage Effects on useful value			
Primary damage			Secondary damage	
Economic performance		Ecological functions	Ideal value	Genetic resources
Direct	Indirect			
Growth and yield reductions Reduced quality of foodstuffs, fodder, raw materials, and ornamental plants Impaired seed production and plant quality	Reduced selection of plant species, Changes in cultivar selections and cultural practices Increased expenditures for cultural and maintenance needs	Reduction in the areal functions of of vegetation Stabilization of water relations and climatic regulation Nutrient cycling Reduction of soil erosion requirements Reduction of noise and air pollution Welfare effects	Reduction of scientific, ethical and aesthetic value	Gene erosion through reduced abundance distribution or loss of plant species and ecotypes Reduction of genetic resources for breeding purposes

The economic damage to agriculture, forestry, and horticulture results from direct impacts on plants, including primary and secondary damage, and indirectly through changes in farming and cultural practices. The estimation of direct economic damage includes effects on growth, yield, and quality – in part a subjective value. Products used as food, fodder, or ornamental plants may show quantitative and qualitative losses ranging from changes in the composition of nutritional metabolites (Feder 1977, Neely et al. 1977, Pell et al. 1979) that influence production value to impaired appearance. In addition, impacts on seed quality and alteration in pollination in forests will have lasting effects on natural reproduction. Additional information on these facets of damage is contained in Chapter 2.2.5.

Economic losses may also occur because plant species especially adapted for cultivation in specific locations can no longer be grown because of their pollutant sensitivity. The necessity of cultivating relatively resistant species may cause changes in management or cultural practices that may impart considerable disadvantages to crop production. Frequently, additional expenditures for cultural measures can cause additional financial costs. These primary and secondary damages to agricultural, horticultural, and forestry plants frequently have economic consequences for the individual producer and at the national level (Jacobson 1982).

The ecological value of vegetation results from its function as a component of a healthy habitat, both agricultural and recreational, and to its contributions to the maintenance of water and climatic conditions and its role in reducing erosion (water and wind), air pollution and noise. Photochemical oxidants may disrupt these ecological functions over large areas, for example, various regions of the USA are already impacted, as shown by Westman (1977) for example in his evaluation of "non-monetary costs" of pollution to ecosystems (see Chap. 2.2.5.4 for additional details).

Photochemical oxidants occurring over large areas can cause injury symptoms on ornamental plants in parks and gardens (Feder 1970). The reduced ideal and aesthetic value of this vegetation frequently results in complaints based primarily on visible appearance rather than on effects on growth. Air-pollution effects on ecological and ideal values of vegetation must be considered in relation to the World Health Organization definition of human health, which includes not only freedom from disease but also the well-being of man.

Most investigations of oxidant effects on plants have concentrated on economically important species (Manning and Feder 1976, Linzon et al. 1975, Heck et al. 1977). For these plants, at least in the United States, criteria to evaluate the effects and response thresholds have been developed (Chaps. 2.2.5, 2.4.1 and 2.6). However, it is not clear what levels of chronic pollutant doses may occur before inducing gene erosion (Scholz 1980, 1981) through the reduced distribution, abundance, or death of individual species. In his studies of the coastal sage scrub community of the southern California coast, Westman (1979) showed that photochemical oxidants caused a reduction in number of species and cover of the native vegetation. Native species represent an important gene reservoir useful in the breeding of agricultural plants. This natural variation of genetic material is essential to assure continued food production aside from the consideration of other concerns such as scientific, ecological, ethical, and aesthetic interests.

2.3.4 Experimental Approaches
to Determine Air Pollution Effects

The methods for determining air pollution effects are as diverse as the participating disciplines. The standard approaches of air monitoring, exposure methods, statistics, as well as various biological and physical sciences, provide the tools for the diagnosis and evaluation of air-pollution effects. Air pollutants and their effects on various organizational levels of ecosystems display various particularities that have led to the development and use of additional specific experimental designs and equipment. Obviously, these methods are topics for discussion, but it is not the purpose of this section to describe the specifics of individual methods. This information can be found in the original publications as well as in recent literature reviews on the topic (e. g., Ormrod 1978, Heagle and Philbeck 1978, Heagle et al. 1979 d). Rather, this presentation will attempt to describe the criteria to use in the selection of experimental methods. Two factors should be noted: (1) the risk of selecting the wrong or nonoptimal methods to evaluate the experimental hypothesis; and (2) that by an evaluation of the data developed from these selected methods the investigator failed to recognize the limitations of a particular experiment or epidemiological survey.

2.3.4.1 Criteria for Selecting Experimental Methods

The selection of experimental methods to study the effects of air pollutants on plants depends primarily on the hypothesis and secondarily on the parameters used to determine the effect.

In the present context, the total plant response to air pollutants can be subdivided into the determination of the:

– cause of injury;
– type, intensity, and spatial distribution of the injury;
– quantitative relationships between exposure and response;
– basis of mitigative actions in the field;
– mode of action of individual components, singly and in combination, in a pollutant complex.

Each of these complex questions encompasses a variety of specific questions (see Chaps. 2.2.5, 2.3.1, 2.4.1, 2.4.3.2, 2.4.4, and 2.6). It is obvious that to find answers to these questions, different approaches, i. e., various experimental methods, are required. Soil and climate as external factors (Chap. 2.4.2) and developmental stages and leaf age as internal factors (Chap. 2.4.3) significantly influence plant response to a given pollutant exposure. Recent comparative studies with bush beans (Lewis and Brennan 1977, Beckerson et al. 1979, Meiners and Heggestad 1979) and potatoes (Reinert 1980) demonstrate that plants react differentially to ozone, depending on whether they were cultivated in closed chambers or the field. This is particularly true for the quantitative relations between exposure and effect, and seldom or never for the quality of the effect. The experimental object itself also significantly influences the experimental approach (see Sect. 2.3.4.2.1).

Table 2.8. Methods for determining effects of air pollutants. (After Guderian 1978)

Increasing difficulty in establishing cause-effect and quantifying exposure response ←

Exposure characteristics	Various environmental conditions			
	Controlled	Managed systems	Semi-natural	Natural
Complex pollutant exposure as in areas subjected to photochemical oxidants			Studies on plants in open-top chambers and plots with and without air filtration plus pollutant addition	Geobiological studies in agricultural and natural ecosystems
Pollutant exposure can be selected especially in the vicinity of a source emitting primary pollutants			Studies on plants in open-top chambers and plots with and without air filtration plus pollutant addition	Geobiological studies in agricultural and natural ecosystems
Pollutant concentration and exposure duration are controlled	Pollutant exposures in controlled-environment rooms	Pollutant exposures in greenhouses	Pollutant exposure: open-top chambers, linear-gradient chambers, Zonal Air Pollution System (ZAPS) Computer controlled exposure system	

Increasing difficulty in extrapolating experimental results to the field ←

In the final analysis, the main concern is the determination of air-pollutant-induced deviations from the norm at individual organizational levels of ecosystems. The study of pollutant effects, at least from the perspective of practical environmental protection, is not an end in itself, but provides the basis for mitigative actions. Therefore, the norm are the plants or plant communities as they occur in agrarian (agriculture, horticulture, and forestry) and natural ecosystems. The following discussion is on experimental methods which can be used to evaluate air-pollution effects on plants in their habitat. Results from studies with plants cultured in typical conditions are representative for practical exposure relationships. However, representativeness is only one basic requirement to insure quality of results. In general, the cause-effect relationship must be established as well as quantifying the effect (Chaps. 2.4.1 and 2.6).

The three basic requirements needed to reach a conclusion are: representativeness of typical conditions, assurance of cause-effect associations, and quantification of exposure-response relationships. These require the use of different experimental methods. The material in the following scheme will be used to discuss and illustrate the experimental approaches best suited for each experimental questions as well as those which are less suited.

Moving from top to bottom, Table 2.8 is divided into three sections to illustrate types of exposure relationships. Photochemical oxidant pollution is a complex of pollutants composed of ozone, peroxides, and other oxidants, in addition to the precursors of these secondary products. Because photochemical oxidants have wide spatial distribution, they are frequently associated with primary pollutants such as sulfur dioxide or gaseous fluoride. In addition to the complex composition, one must accept fluctuations in the exposure characteristics, concentration, and duration. In gaseous exposure studies, the exposure of each component including concentration, duration, and frequency of affect can be controlled at any arbitrary level. When studying the effects of primary components under actual or typical conditions, the possibility exists of evaluating the effects of a specific air pollutant component through the choice of a suitable emission source.

The various environmental conditions (Table 2.8) include edaphic, climatic, and organismal, as well as their interactions. In this context, the term natural means conditions as they occur in natural ecosystems; semi-natural implies conditions such as occur in managed forests, and other managed systems such as agriculture and horticulture. The influence of the experimental systems used to determine exposure response in semi-natural and managed systems can induce deviations from the natural conditions. Controlled conditions, as found in controlled-environment facilities, permit the regulation of the pollutant exposure as well as the growth parameters. Experimental studies on managed, semi-natural, and natural associations must consider the varying environmental conditions and in some studies a varying pollutant exposure.

2.3.4.2 Experimental Procedures

The following discussion will use the classification in Table 2.8 in an attempt to illustrate the limits of the experimental procedures used to determine air-pollution effects in natural, semi-natural, managed, and controlled systems.

2.3.4.2.1 Studies Under Natural Conditions

Geobiological studies are conducted under natural conditions and the experimental procedures should not influence the environmental conditions. These studies can be conducted in the vicinity of a pollution source as well as in large areas impacted by pollution. In studies near a pollution source, the exposure is characterized by the presence of one or only a few components, and there is typically a distinct decrease in the pollutant concentration and frequency of exposure with distance from the source, unless there is an elevated background and a high frequency of pollution-free periods during the total exposure period (Guderian and Stratmann 1962, 1968). Given the limited extent of the experimental area, one would expect only small differences in environmental conditions, so that prerequisites would be generally favorable for evaluating local impacts of the air pollutants.

In congested areas with a high density of emission sources and also in areas impacted by photochemical oxidants, numerous biologically active components can co-occur. In general, the pollutant concentrations display only a gradual decline with distance from the emission area. However, the mean photochemical oxidant concentrations may be higher away from the congested areas than near the source, where the precursors are emitted (see Chap. 1.3.2.2). The pollution-induced reductions in plant production also exhibit only small changes with distance from the emission area. Comparisons with unexposed locations (controls) are possible only at substantial distances from the experimental plots which can experience a wide range in conditions between test and control locations. These differences, depending on the research subject, may cause significant problems in drawing inferences about the variations in plant performance at various locations from the source.

Forest stands provide relatively favorable conditions for quantifying pollution-induced effects. On the basis of experimental values and yield tables, the number of injured and dead trees in a stand or the evaluation of height and radial growth can provide a reliable estimate of air-pollution impacts (P. R. Miller 1973a, O. C. Taylor 1973b, P. R. Miller and White 1977). Chronological studies of the annual growth rings can evaluate growth disturbances that occurred in past years and decades. Another example of the geobiological approach is Westman's study (1977), in which he associated changes in the composition of coastal sage scrub in the southern California coastal area to a gradient in ozone exposure. With agricultural and horticultural plant species, geobiological studies provide estimates of yield losses from photochemical oxidants through the use of pot-culture techniques. The O_3 concentration gradients in the Los Angeles basin have been used in estimating the influence of ozone dose on the yield of tomato (Oshima et al. 1975) and alfalfa (Oshima et al. 1976). Regression analyses indicated that differences in temperature and humidity at the different sites had less effect on crop yield than did the difference in ozone dose. However, in this case only the pollutant-exposure conditions can be considered natural, not the growth conditions of the experimental plants.

In summary, geobiological studies can quantify the long-term effects of pollutants on homo- and heterotypic populations under typical interplant competitive and environmental conditions. Vegetative studies, particularly on floristic systems, can provide information about changes in species numbers and species association. Finally, geobiological studies provide a representative basis for evaluating the reaction of other components (consumers and decomposer) in the ecosystem (cf. Chap. 2.2.5.3.2 and 2.2.5.3.3). Geobiological studies are not appropriate for determining photochemical oxidant-induced yield reductions in agricultural and horticultural cropping systems. Gradations in injury intensity can be established by evaluating external injury symptoms. Difficulties in recognizing the particular phytotoxic components and accurately characterizing the exposure limit the broader use of this method, as does its substantial cost.

2.3.4.2.2 Studies Under Semi-Natural Conditions

In contrast to geobiological studies under natural conditions, studies in semi-natural conditions use various experimental aids that may influence the plant environment. According to the definition, we are discussing studies in semi-natural conditions, where the experimental apparatus will have minimal influence on the plant reaction to the air pollutant. Special efforts are made to ensure that the experimental conditions approach a natural situation, where at most the plants are laterally enclosed with a transparent film which is open at the top so they are exposed to the atmosphere. In contrast to closed chambers typically used in controlled-environment chambers and greenhouses, the "greenhouse effect", producing elevated temperatures and altered light quality, does not occur.

This method is characterized by the use of a series of similar systems, at a given location, under semi-natural conditions in which the systems receive:

− air with filtration;
− air without filtration; and
− air with pollutant addition with or without air filtration.

The motivation for developing these systems was: (1) the difficulty in extrapolating experimental results for controlled-environment or greenhouse studies to typical field conditions; and (2) the difficulties associated with studies in natural conditions in establishing cause and effect and the problem of establishing the relationship between exposure and response (Heagle et al. 1979d, Heck et al. 1982). Four examples will illustrate the uses and limitations of these methods.

a) Open-Top Chambers

During the early 1970's, concurrent research at Boyce Thompson Institute for Plant Research (Mandl et al. 1973) and US Environmental Protection Agency/ US Department of Agriculture (Heagle et al. 1973) developed open-top chambers as a new experimental method which had significant advantages over closed-chamber and open-atmosphere approaches for evaluating air pollution exposures and their effects (Fig. 2.9). Both open-top chamber designs were cylindrical (diameter 2.16 to 3.05 m), about 2.4 m high, and covered with either corrugated fiberglass panels or clear polyvinyl chloride film. The two designs displayed some differences in the air filtration, air handling systems, and the frequency of air ex-

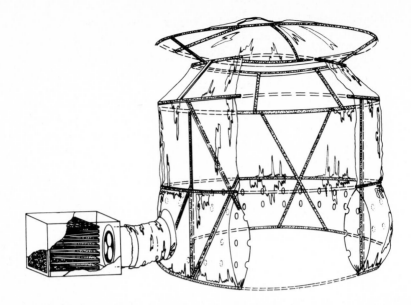

A

B

Fig. 2.9. A Construction of a modified open-top chamber. The modifications include the addition of a frustrum to reduce the ingression of ambient air, a rain cap required for particular experimental studies, and netting over the open portion of the chamber to exclude insects. **B** Experimental location using a number of modified open-top chambers (EPA, Corvallis/Oregon)

change (Heagle and Philbeck 1978), however, the performance of the chambers was similar. The chamber can use either filtered or unfiltered air which is blown into the chamber through a perforated air duct (plenum) around the chamber base. The air flows into the bottom of the plant canopy and then moves up and out of the open top of the chamber. The movement of air out of the open chamber top restricts the ingress of pollutants. Also, the open top allows rain to enter, causes only slight light attenuation, and prevents significant temperature increases within the chamber. Comparable studies revealed only small differences in microclimatic parameters and the growth and yield of plants in chambers without air filtration and in adjacent plots without chambers (Heagle et al. 1979 c, e, Heck et al. 1982, Heggestad et al. 1980, Montes et al. 1982). The small difference in microclimatic conditions may have no significant effect on plant growth during the summer when light and temperature are generally not limiting. However, with the low temperatures and reduced solar radiation that occur during the winter in Great Britain (Greenwood et al. 1982), there is no reason for the assumption that the sensitivity of cereal plants grown in chambers over winter should be similar to that of plants growing in the field. Shinn et al. (1976) suggested that open-top chambers had several disadvantages.

The chambers can be placed over plants growing in the soil or they can be filled with potted plants. In addition to the comparative studies with and without air filtration, the open-top chambers can be used in gaseous exposure studies in which measured amounts of particular pollutants are added.

In general, open-top chambers are reasonably well suited for establishing air-pollution-induced yield losses, particularly in agricultural and horticultural plant species with a reasonable expenditure. Open-top chambers were selected for use in the NCLAN (National Crop Loss Assessment Program) program (see Chap. 2.2.5.3.1), whose goal is to determine air-pollution-induced yield reductions for agricultural crops in the United States and to study the dose-response relationships for O_3, SO_2, NO_2, and their mixtures. The NCLAN program employed these chambers because of the good results obtained in studies over several years.

b) Field Exposure Systems Without Chambers

The need to study air-pollution effects on large plots and the goal of eliminating the potential chamber influence on the plant environment led to the development of field exposure systems without chambers. Examples of these systems are the linear gradient system (Shinn et al. 1976), the Zonal Air Pollution System (J. J. Lee and R. A. Lewis 1978), and the computer-controlled system of Greenwood et al. (1982). Each system will be briefly described to show the progress in developing new methods. These systems vary in the amount of control that is used in applying the air pollutant to the plot.

The linear gradient system has been used as both an exclusion and a fumigation system in the field (Shinn et al. 1977, Reich et al. 1982). This system utilizes a series of inflated polymer film tubes, under positive pressure, to distribute either carbon-filtered air or a mixture of fumigant and filtered air over the plot. The inflated tubes are positioned between rows of plants with precalibrated holes oriented toward the plant canopy. The filtered air or fumigant gas mixture is emitted from the holes and permeates into and through the canopy. The linear gradient in gas concentration along the inflated tube is obtained by: (1) a linear increase in gas discharged from the plenum (tubes); and (2) by increasing portal areas along its length combined with a constant exit velocity of the gas mixture. Because plants can be exposed to a large range of gas concentrations simultaneously, this method allows statistical regression analysis of the dose-response data. The original system was a rectangle 7.5 m long and 3 m wide encircled with clear fiberglass panels 0.6 m high. Technically, this is also a chamber as expressed in the title, *A Linear-Gradient Chamber for Exposing Field Plants to Controlled Levels of Air Pollutants* of the publication by Shinn et al. (1977). The minimal influence of the small enclosure on the growth conditions of the plot and the use of the linear gradient system without the enclosure (Reich et al. 1982) justifies the inclusion of this method under field exposure systems without chambers. As shown by experimental results, this system and the following two systems to be discussed are strongly influenced by wind speed and direction.

The Zonal Air Pollution System (ZAPS) was designed to cause minimal disturbance to the biota, prey refuges, ground-level obstructions and pathways, incident radiation, and microclimate. The original ZAPS consisted of a network of 2.54-cm-diameter aluminum pipes suspended about 75 cm above 0.5-ha plots of native grasslands (J. J. Lee and R. A. Lewis 1978). Gas release ports (0.8 mm diameter) were positioned at 3-m intervals with more than 250 ports per 0.5-ha plot and with no location more than 5.5 m from a port. The experimental design consisted of a series of plots each receiving a different SO_2 exposure. The control plot was arranged to minimize the drift of SO_2 from the test plots. Also, the control was selected to be as similar to the test plots as possible.

The ZAPS field exposure system relies on atmospheric diffusion to dilute a very high concentration of SO_2/air mixture released from the pipes. When the rate of release is constant, the resultant concentration depends largely on wind speed and can reach high levels at night when wind speeds are low. These diurnal concentration pattern follow atmospheric mixing, with the minimum concentrations occurring during mid-day and the highest concentrations occurring at night when the air is still. However, this diurnal concentration pattern is different from that observed in rural England (Greenwood et al. 1982). Also, the pollutant distribution over the crop may be very uneven. When there is large variation, there may not be enough plants available for sequential harvests. To avoid the problem of high concentrations, the release rate can be reduced during conditions of poor dilution or when the wind is from a specified direction. The ZAPS system has also been adapted for use in determining the effects of SO_2 on crops grown under normal field conditions (J. R. Miller et al. 1980).

Greenwood et al. (1982) developed a computer-controlled system for exposing a normally managed field crop (an area of about 400 m^2) to controlled concentrations of SO_2. The system allows plants to be successfully exposed to predetermined concentrations of SO_2 under essentially natural conditions. In principle, the system is also suitable for other gaseous air pollutants. The gas is released from pipes at a rate which can be varied by a microcomputer which also controls the monitoring of the SO_2 concentration. The SO_2 release rate depends on the measured concentration and wind speed. The computer control of the SO_2 release yields a much smaller variation in concentration than in the relatively uncontrolled ZAPS system. The simple, symmetrical design of the SO_2 release system enables large numbers of plants to be exposed as in a gradient system, allowing regular removal of plants for analysis without seriously modifying the crop environment.

2.3.4.2.3 Studies in Managed and Controlled Conditions

Both types of exposure system will be discussed in a single section because of their similarities, and to highlight their differences. Both types may be used in either greenhouses or controlled-environment chambers; the plants are placed in a closed area and subjected to a definite mixture of air and air pollutant. The plants may be exposed to the pollutants in separate units, i. e., several small greenhouses or controlled-environment chambers arranged close to each other or in subunits (chambers) located within either a greenhouse or controlled-environment chambers (Guderian 1977). In the first case, the greenhouse or controlled-environment chamber serves directly as the exposure chamber; in the second case it provides only the "climatic conditions" and the exposures are conducted in specially designed chambers. The basic requirements of closed chambers (Heagle and Philbeck 1978) in managed and controlled conditions are similar in principle:

- environment resembles ambient conditions;
- uniform environment throughout the chamber and among the chambers;
- transparent covering minimizing changes in light intensity and quality;
- nonreactive surfaces;
- capability of preparing, maintaining, and monitoring defined levels of known air pollutants;
- uniform pollutant concentrations throughout the chamber and among chambers.

In addition to these characteristics, the following ones are desirable: single-pass-continuous air-flow system, negative pressure within the chamber, ease of moving plants in and out of the chamber, easy to build and clean, durable, and relatively inexpensive.

Naturally, these ideal requirements are only partially attained. In general, the extent to which these goals are reached increases with the technology used, and the financial resources and inventiveness of the individual investigators. The exposure systems developed for use in closed chambers varied in complexity from relatively simple apparatus with only a limited capability of controlling the pollutant exposure and climatic conditions to increasingly sophisticated systems for control and regulation. It should be emphasized that these individual goals may be realized to varying degrees in various experimental designs in managed or controlled conditions. The key difference between the exposures in greenhouses and controlled environment rooms is the degree to which light, temperature, and relative humidity are regulated. Studies in greenhouses usually follow the natural fluctuating environmental conditions, light intensity, photoperiod, temperature, and relative humidity with some modification via the "greenhouse effect". For special experimental studies, there may be partial control of temperature and/or relative humidity. Plants may be exposed under artificial light, but usually the artificial light is only supplementary. The advantages of experimental studies in greenhouses (with or without partial climate control), compared to controlled-environment rooms, reside in less deviation from the natural conditions and also in the larger number of plants and taller plants such as fruit or forest trees that can be exposed. The disadvantages are less control of environmental conditions and possibly pollutant exposure, which may make it difficult to reproduce experimental results.

Reproducibility is the basic advantage of controlled-environment rooms. The ability to control pollutant exposure and climatic conditions to the levels desired provides the ideal conditions for investigating the influence of individual parameters on plant response to air pollutants. Controlled-environment studies are particularly well suited for determining the mode of action of individual pollutants, for studies of cause and effect, for hypothesis testing, and for determining dose-response relationships using acute short-term injurious concentrations. With long-term exposures, there is the persistent problem of extrapolating the experimental results to ambient conditions, particularly when light quality or intensity deviates substantially from natural conditions.

In the 80 years since Wislicenus (1901) conducted the first exposures in "ventilated smoke houses" with controlled air exchange, numerous and various exposure systems have been developed for use in the greenhouse and controlled-environment chambers. The details of construction and function of these systems are contained in the individual publications. An example of the newer developments is the Continuous Stirred Tank Reactor (CSTR) system (Rogers et al. 1977). In modified form, CSTR systems are recommended for use in greenhouses (Heck et al. 1978). Greenhouses may also be used as exposure systems for determining air-pollution effects as shown in several publications (e. g., Berry 1970, Guderian 1977).

2.3.4.3 Summary Evaluation

At the beginning of this section, three basic requirements for studies of air-pollution effects were postulated: (1) they must be representative of typical conditions; (2) the cause-effect relationships must be assured, that is to say, the effect must be unequivocally related to a specific cause; and (3) finally, the exposure and its effect must be quantified. These goals may not be met with a single exposure system because of the need to determine effects at various organizational levels of the ecosystem and the numerous factors that influence plant response. These were discussed in the subsections of Experimental Methods (Sect. 2.3.4.2) and in the comparisons in Table 2.9. This generalization holds: with increasing deviations from natural conditions, it becomes increasingly difficult to establish "typical" results. Experiments using controlled or managed systems are particularly well suited for studies addressing the mode of action, to clarify etiological questions, and for hypothesis testing. In contrast, under natural conditions in congested areas, the determination of the phytotoxic components creates a problem that is not readily solved. Also, the evaluation of dose-response relationships and

Table 2.9. Comparison of experimental methods under various experimental conditions

Characteristic factors	Controlled	Managed	Semi-natural	Natural
Pollutant exposure	The pollutant concentration and exposure duration are controlled as well as the frequency and sequence of exposure	The pollutant concentration and exposure duration are controlled as well as the frequency and sequence of exposure, however, the concentration may be only partially controlled	Exposure controlled or in some cases only partial control of pollutant concentration, duration, exposure frequency, and sequence	Continuous variation in pollutant exposure
Climatic factors	Individual factors can be controlled with variable deviations from the natural climatic conditions	Unless there is partial climate control, climatic conditions vary as in ambient except for the influence of the "greenhouse effect"	In open-top chambers, the conditions approach the natural; in exposure systems without chambers, there is even less deviation from the ambient conditions	Natural conditions
Plant culture and exposure	Cultured in pots	May be cultured in pots or soil	Plants usually grown in soil but pot culture may be used	Plants grown in soil
Exposure susceptibility of the plants	Depending on the climatic conditions, approximately normal to abnormal predisposition	Approximately normal with more or less deviation from normal predisposition	Approximately normal predisposition	Normal predisposition
Evaluation	Reproducible studies; particular characteristics: evaluation of the significance of individual factors on plant response, for studies of the mode of action on or in plants, for determining cause and effect and testing hypotheses, for short-term studies up to a few months, limited suitability for determining dose-response relationships at higher concentrations	Limited multifactorial, in certain experiments limited reproducibility; particular uses: determination of the significance of particular individual factors for the study of the mode of action on or in plants and model plant communities, for cause-effects analysis, limited suitability for study of dose-response experiments possible over a period of a few years	Limited multifactorial, limited reproducibility of exposure studies, when used as pollutant exclusion systems, not reproducible; particular uses: determination of growth and yield losses, dose-response relationships, effects on plant communities, experiments possible over a period of several years	Always multi-factorial, experiments not reproducible; particular uses: determination of injurious effects on various components on eco-systems such as producers, consumers, and decomposers, determination of long-term effects on forest plants and plant communities

the spatial distribution and intensity of injury are difficult to determine because of problems in determining the emissions and exposure with sufficient accuracy. The advantages of studies in natural conditions are: no problems in extrapolating to natural conditions, as well as the ability to determine the long-term effects on various components in the ecosystem.

In the area of exposure methods, the decisive progress in the last decade has been in the development of systems with and without chambers for studies that allow minimum deviation from natural conditions. Given their significant advantages as experimental systems in controlled and natural conditions, they can be adapted for studies addressing a wide range of questions.

Chapter 2.4 Factors Influencing Plant Responses

The occurrence and magnitude of vegetation effects are dependent on the concentration, frequency and duration of exposure, time between exposures, the magnitude of the concentration changes, diurnal timing of exposures, their sequence and pattern, and ozone flux into the plant as it is influenced by canopy characteristics and the leaf boundary layer (Jacobsen 1982). Most ozone response studies have considered only the concentration and duration of exposures, while the other parameters that also influence pollutant response have received only cursory study.

Plant reactions to a given air pollutant depend on the exposure parameters which may be modified by internal and external growth factors. The previous chapters described various photochemical oxidant effects and methods to evaluate them. This chapter discusses factors that influence plant response and provides information to help understand and evaluate the dose-response relationships discussed in Chapter 2.6.

2.4.1 Concentration and Duration of Exposure

Knowledge of the quantitative relationship between the ambient exposure and effect is important for air pollution studies: (1) it provides information important to understanding pollutant-induced effects, and (2) it provides the basis for protective measures based on air quality principles. Recognition of these needs led to the development of dose-response relationships. The magnitude and expense of these studies result from the difficulty of developing reliable dose-response relationships that can be used as a basis for general risk predictions to vegetation.

The relationship between ambient exposure and effect can be characterized with sigmoid curves (Fig. 2.10). Below a specific dose (D_1), the reactions of even the most susceptible plants do not deviate significantly from the normal variation of unexposed controls. Above the corresponding response threshold, responses can be observed which may be relevant for practical environmental protection. The threshold doses are different for different measures (i. e., injury, yield, quality) of response. With increasing dose, the response rapidly increases (monotonically) to a higher response level, then the curve flattens and approaches, asymptotically the maximum response level, e. g., 100% leaf necrosis. Differential resistance of individual plant species is expressed in the correspondingly higher doses required to cause injury. The variable steepness of the response curves for the individual plant species results from changes in the dependency of differential species resistance relationships on the dose levels (cf. Chap. 2.4.3.2).

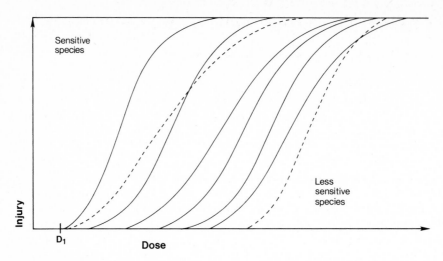

Fig. 2.10. Schematic representation of dose-response relationships for differentially susceptible plant species. Dose is the product of the ambient concentration multiplied by exposure duration

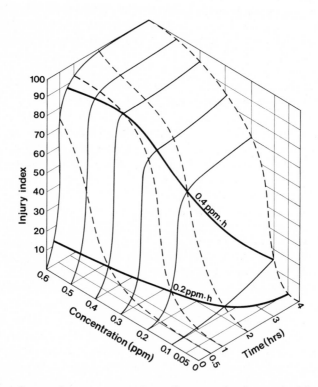

Fig. 2.11. The influence of ozone concentration and exposure duration on the foliar injury of beans (Heck et al. 1966). *Thin solid lines:* change in the injury index with increasing time at fixed ozone concentrations. *Dashed lines:* change in the injury index with increasing concentration at fixed time intervals; *heavy solid lines* show the change in the injury index at two constant ozone doses

For ozone, PAN, and other phytotoxic gases, dose alone is not a sufficient parameter to determine responses or describe effects. For a given pollutant dose, plant responses vary with concentration, exposure duration, frequency, or exposure sequence.

Injury from photochemical oxidants, PAN, and ozone increases exponentially with exposure. Heck et al. (1966) clearly showed this relationship between ozone concentration, exposure duration, and injury to bean foliage using a three-dimensional model (Fig. 2.11). The three-dimensional figure illustrates the foliar injury resulting from various combinations of ozone concentration and exposure duration. The shape of the surface depicts the variable interaction between concentration and duration in producing injury. For example, in bean plants injury occurred above a specific concentration or duration of exposure. The steepness of the concentration and time planes of the response surface illustrate how differential injury can result from only small changes in pollutant dose. In addition, it follows that the same dose in a short time will cause a stronger response than when the dose is distributed over a longer time period. The relationship between pollutant dose and effect usually does not follow the case where a specific dose always induces the same magnitude of response. The extent of foliar injury increases progressively with increasing concentration when equal products of concentration (c) and exposure duration (t) are used, as was previously observed for sulfur dioxide (van Haut 1961). Plant growth is also influenced more by concentration than exposure duration when similar products of c and t are used (Guderian 1967). Peak concentrations pose a particular risk to vegetation, especially when they follow one another rapidly.

The following examples confirm that for ozone, concentration is the important exposure parameter controlling plant response. Following a 1-h exposure at 0.50 ppm ozone (dose = 0.50 ppm h), the secondary needles of 3-year-old Virginia pine seedlings displayed 50% necrosis, while after an 8-h exposure at 0.10 ppm (dose = 80 ppm h) only 10% necrosis resulted (D. D. Davis and Wood 1973 b). During chronic exposures, the injury increased with increasing concentration. For example, the net photosynthetic rate of ponderosa pine seedlings was reduced by 25% following exposure to 0.15 ppm ozone for 9 h day^{-1} for 60 days, while doubling the concentration (0.30 ppm) and reducing the exposure duration to 30 days reduced net photosynthesis by 67%. In both studies the product of concentration and time (dose) was 81 ppm h (P. R. Miller et al. 1969). Additional examples of the greater importance of concentration compared to exposure duration have been given by numerous authors (e. g., D. N. Ross et al. 1976, Reinert and Nelson 1979, Henderson and Reinert 1979, Shertz et al. 1980 b).

The importance of concentration does not hold for all combinations of ozone concentration and time. In one study using very low ozone concentrations producing the same dose (1.2 ppm h), a 24-h exposure at 0.05 ppm reduced the growth of bean foliage more than a 12-h exposure at twice the concentration (Evans 1973). Isolated instances of growth stimulation at very low ozone (Barnes 1972 a, J. P. Bennett et al. 1974, Runeckles and Resh 1975 b, Lumis and Ormrod 1978, Leone and Brennan 1975) and PAN (Temple 1982) concentrations have been reported, as has been observed with other pollutants (Guderian 1977).

The particular importance of concentration for inducing injury responses was considered in efforts to formulate mathematical relationships between the pollutant exposure and effect. O'Gara (1922) derived the following equation from acute exposures using sulfur dioxide:

$$c = a + b/t$$

where:

c = concentration in ppm;
a = threshold concentration, that for a given exposure duration will not cause any injury;
b = constant, that reflects internal and environmental factors that influence plant sensitivity;
t = exposure duration in h.

The relationship proposed by O'Gara (Fig. 2.19) agrees well with results obtained from short-term ozone exposures. The functions (Guderian et al. 1960, Zahn 1963) established to determine the concentration-dependent risks of chronic SO_2 exposures will not be considered further because the assumption for their use with chronic ozone and PAN exposures are lacking.

Several methods have been proposed to describe chronic or seasonal ozone exposures for vegetation effect studies in the field. Heagle et al. (1979 a, b, e) used the 7 h (0930 to 1630) day^{-1} seasonal mean concentration for injury, growth, and yield studies in the field. However, this method treats low-level long-term exposures in the same way as high-concentration, short-term exposures that may produce a similar dose, and it is inconsistent with the observed dose-response relationships described above. A seasonal ozone dose has been calculated by summing the hourly ozone concentrations above a threshold concentration (Oshima 1974 b, Oshima et al. 1976). However, this method requires the calculations to be made from actual hourly data. Recently, Lefohn and Benedict (1982) proposed an exposure index that also determines the ozone exposure above a threshold, but which can be calculated from either actual data or frequency distributions. The approaches of both Oshima and Lefohn and Benedict include the concentration, duration, and frequency of exposure in the exposure index in contrast to the 7 h average of Heagle et al. (1979 a, b, e).

The establishment of a mathematical relationship between pollutant exposure and effect is made more difficult because phytotoxic concentrations of photochemical oxidants occur only under specific meteorological conditions and during specific hours of the light period. Exposure periods from anthropogenic-induced ozone are interspersed with pollution-free periods when only the natural background concentration occurs. From investigations with sulfur dioxide it is known that the pollutant-free periods serve as a recovery time if the given concentration and the exposure period are not too long and the pollutant-free period exceeds a minimum time (Zahn 1963, Guderian 1970). In any case, the results of discontinuous exposures can be different from continuous ones of the same dose.

One continuous ozone exposure of 1 h produced more foliar injury on bean and tobacco plants than two 0.5-h exposures occurring within 1 to 3 h (Heck and Dunning 1967). Tobacco plants exposed to ambient oxidant for 7 days in nonfiltered chambers developed more than twice as much injury during the period than the sum of injury to individual plants exposed to ambient oxidant for only 1 day each (Heagle and Heck 1974). In contrast, the reduction in hypocotyl growth of radish plants exposed to ozone at 7-day intervals was the sum of the individual reductions during the corresponding development stages (Tingey et al. 1973 a). In studies with sweet corn and tomatoes, Oshima (1974 b) found that the frequency of ozone exposure appeared to be as critical a factor as total dosage in long-term exposures. A greater number of relatively short-term exposures had a larger effect than fewer exposures of greater duration. Studies with kidney beans showed that exposures with constant concentrations of ozone caused the same pattern of response as variable concentrations of equivalent dose (Musselman et al. 1983). However, the variable-concentration exposures reduced plant growth to a greater extent.

These limited results do not support their use in the development of models of dose-response relationships. However, they underscore the importance of outdoor and field studies for the development of air-quality criteria.

The above presentation provides evidence that high concentration-short term exposures may produce different relationships between exposure parameters – concentration, duration, and frequency of exposure – than chronic continuous low-level or intermittent exposures. In practice, vegetation injury represents both acute and chronic concentrations that display large spatial and temporal changes (EPA 1978 a). There is general agreement (Heggestad and Heck 1971, Linzon et al. 1975, Heck and Brandt 1977) that necessary prerequisites needed to model the risks from chronic and intermittent exposures are lacking. The evaluation of short-term effects of acute injurious concentrations has allowed the development of concepts as aids for the evaluation and incorporation of the extensive experimental results into general concepts (see Chap. 2.6.2).

2.4.2 Significance of External Growth Factors on Plant Sensitivity

External growth factors include edaphic and climatic conditions. These factors influence gas phase conductance, pollutant uptake and its reactions within the plant: directly during and following an exposure and indirectly before exposure through their influences on the physiology and morphology of the plant. The stomata are the primary portals of pollutant uptake into the plant (see Chap. 2.2.1) and a significant factor controlling gas uptake. Therefore, any external factor that influences stomatal opening would also influence pollutant uptake.

In the field, these external factors can interact and modify plant responses. In some cases, the relationship between exposure and response is altered without its being known which of the environmental factors caused the change. For this reason, controlled conditions are used to study the influence of external factors on plant response.

2.4.2.1 Climatic Factors

The climatic conditions before, during, and following exposure influence the type and intensity of the plant response to a given pollutant exposure. Certain climatic conditions preceding an exposure increase the susceptibility, while others reduce it. The major significance of climatic conditions during exposure is explained by their influence on pollutant absorption and its reactions within the plant. Climatic conditions following exposure are generally of lesser importance. From the complex of possible conditions, the parameters of light, temperature, and relative humidity will be considered.

2.4.2.1.1 Light

This presentation considers the influences of the following light parameters: photoperiod, intensity, and quality on pollutant uptake and plant reactions.

a) Photoperiod

Photoperiod is the length of the light period in an alternating light-dark cycle. Pollutant absorption and injury are frequently greater in the light than in the dark, corresponding to the light-induced increase in stomatal opening. The stomata of most plants are closed in the dark; therefore exposures during this period are generally less injurious. For example, a 3-h dark treatment just prior to exposure in dim light was sufficient to decrease the stomatal aperture and to reduce ozone injury to tomato foliage (Adedipe et al. 1973 b). However, in some plants or under some environmental conditions, pollutant uptake can occur in the dark. Sulfur accumulation in plants exposed to SO_2 in the dark was one-third the accumulation in the light (Guderian 1970).

The influence of light on plant responses to photochemical oxidants goes beyond a simple influence on stomatal opening during or just prior to exposure. For the development of PAN injury in pinto beans, light was required before, during, and following exposure. The injury decreased with the duration of the light period before exposure (Dugger et al. 1963 b) and also the duration following (O. C. Taylor 1969). The injury was maximal when there was at least 3 h of light following exposure, given light both before and during exposure. A dark period as brief as 15 min preceding a 30-min PAN exposure in the light reduced injury. Given light before and during exposure, the expression of the symptom depended on the light or dark periods following exposure: 100% injury with 3 h or more of light, 0% injury with 2 h or more of darkness; shorter light or dark phases produced proportional amounts of injury at comparable PAN concentrations (O. C. Taylor 1969). The mode of action is unknown.

In contrast, for ozone there is no absolute light requirement for injury to develop. O. C. Taylor et al. (1961) found that light before, during, or following exposure was without influence, while a 24-h dark period before exposure reduced ozone sensitivity in cotton plants (Ting and Dugger 1968). A 24-h light period before exposure reduced the sensitivity of young Virginia pine plants (D. D. Davis 1970).

In addition to the direct effects of light on pollutant absorption and response, it also has a significant influence on the physiological condition of the plant. Therefore, light conditions, including photoperiod during plant culture, control the plant's predisposition to air pollutants. For example, annual bluegrass (*Poa annua*) was more sensitive when cultured under an 8-h photoperiod than under a 16-h one (Juhren et al. 1957). Cultivated plants were more sensitive when grown at an 8-h than at either a 12- or 16-h light period (MacDowall 1965a, Heck and Dunning 1967). The relatively high resistance of plants grown on long days was diminished by only a 3-day conditioning period at short days prior to exposure.

b) Light Quality

Possibly the oxidant sensitivity of plants is altered by a compound that under the influence of light is converted from an inactive to an active form (or reversed). In searching for such a substance, Dugger et al. (1963 b) determined that the amount of ozone injury depended on the wavelength of light. The most injury occurred between the wavelengths of 420 and 480 nm. In this wavelength range, the principal chromogens are carotenoids. Carotenoids react readily with atomic oxygen that is liberated by the reaction of PAN with bases (Steer et al. 1969). It is not clear if these phenomena are actually related.

Sulfhydryl compounds are among the preferential reaction sites for photochemical oxidants and, in conjunction with other compounds, influence the sensitivity of the leaf. Their synthesis was increased with increasing light intensity; the two light reactions (I and II) of photosynthesis had a differential effect on synthesis (Dugger and Ting 1968). Irradiation with 660 nm light activated photosystem II, whereby a larger amount of sulfhydryl compounds were produced than by the activation of photosystem I at 700 nm (Dugger and Ting 1968). The situation is confused if one considers that the phytochrome system absorbs in similar wavelengths with changes in its activity: illumination with 660 nm light converts P_r to P_{fr}; the back reaction follows from either dark or irradiation with 730 nm light.

c) Light Intensity

Besides the duration of light and the wavelength, light intensity before, during, and following pollutant exposure influenced plant reactions. In controlled pollutant exposures, increasing light intensity during exposure increased plant injury. With pinto bean, a positive correlation was established between the amount of ozone injury and light intensity up to 32.3 klx (Juhren et al. 1957, Heck et al. 1965). Short-term ozone exposures induced more injury in pinto bean at 43 klx than at 21.5 klx (Dunning and Heck 1973). In similar experiments, pinto beans grown at 43 klx had less injury than those grown at 21.5 and 10.7 klx, while Bel W3 tobacco reacted oppositely (Dunning and Heck 1977). Following a short-term ozone exposure, tobacco and pinto bean were more injured when grown at 21.5 klx than 32.3 klx (Heck and Dunning 1967). Plants exposed to injurious concentrations of SO_2 and O_3 in the light (16 klx) were severely injured; injury intensity decreased with decreasing light intensity and only minor injury resulted from exposures in the dark (Yamazoe and Mayumi 1977). Light or dark for 24 h before or following exposure did not influence the amount of injury.

Although these observations are not without contradictions and are based on only three plant species, a generalization can be made; low light intensity during plant growth increased the sensitivity to photochemical oxidants, while increasing light intensity during exposure increased injury. With longer exposures in ambient conditions, both light responses are superimposed so, for example, a high light intensity during pollutant exposure will momentarily increase the injury, while concurrently the plant sensitivity to a later exposure may be reduced.

In the dark, stomata are generally closed – except for plants of the Solanaceae family – and during photochemical oxidant exposure, plant injury is reduced; at approximately 1,000 lx with stomatal opening, injury is observed (Heck et al. 1965). High light intensities during exposures in the field are less likely to cause maximum foliar injury than under laboratory conditions because the plants in the field experience intense solar irradiation, temperature extremes, and possibly water stress, which leads to partial stomatal closure.

2.4.2.1.2 Temperature

The temperature during plant growth and exposure has a significant influence on plant sensitivity to photochemical oxidants. Studies with various plant species clearly showed that plant sensitivity to ozone increased with growth temperature over the range of 3 ° to 30 °C (Dunning et al. 1974, Shinohara et al. 1973, Dunning and Heck 1977, D. D. Davis 1970, D. D. Davis and Wood 1973a, Wilhour 1970b). In radish, foliar injury increased with increasing growth temperature, but the ozone effect on root growth was similar over a range of growth temperatures (Tingey et al. 1973a). Sensitivity of various vegetable plants to ozonated hexene and other alkenes increased with increasing growth temperature (Hull and Went 1952, Kendrick et al. 1953, Middleton 1956). The influence of temperature was expressed to a greater extent the longer the plants were maintained at a particular growth temperature before pollutant exposure (Hull and Went 1952, Kendrick et al. 1953, Middleton 1956). Plant sensitivity to ozone decreased when the growth temperature exceeded 30 °C (MacDowall 1965a).

With annual bluegrass (*Poa annua*), Juhren et al. (1957) showed that the temperature response during plant growth depended on the respective diurnal temperature cycles and the particular developmental phase of the leaves. During the development of the cotyledonary leaves, the sensitivity was greatest at high temperature (30 °C day; 24 °C night); however, it was significantly reduced by the time that the third or fourth leaves developed. Under cool conditions (20 °C day; 14 °C night), the sensitivity in the two-leaf stage was low, gradually increased with age until about 6 weeks, and then decreased. In cold conditions, the plant remained uninjured until flowering started. Four-week-old plants were most sensitive when grown at a temperature range of 26 °/20 ° or 23 °/17 °C, while plants in a moderately cooler (20 °/14 °C) condition were somewhat less sensitive; the response of plants in a hotter environment (30 °/24 °C) was reduced markedly. Plants grown in the mid-temperature range (26 °/20 °C) lost their sensitivity after only 3 days of growth at a higher temperature (30 °/24 °C).

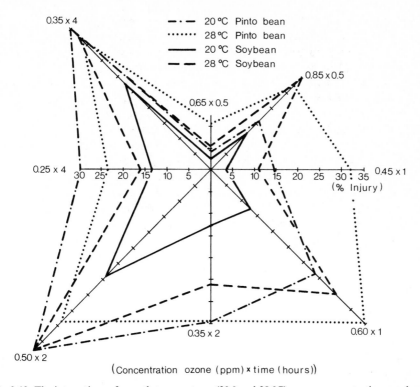

Fig. 2.12. The interactions of growth temperatures (20 ° and 28 °C), ozone concentrations, and exposure durations on the amount of foliar injury for pinto bean and soybean (Dunning et al. 1974). Increasing exposure durations are shown in a *clockwise* manner around the figure; within each exposure duration, ozone concentrations increase clockwise

MacDowall (1965a) also observed that the ozone injury level was greater on plants grown at moderate temperatures than at higher ones. High night and low day temperature also increased plant sensitivity (MacDowall 1965b).

For most plants studied, foliar sensitivity increased with increasing growth temperature up to approximately 30 °C; at higher temperatures this response was reversed. The magnitude of the temperature difference between day and night conditions may also influence the response; however, there have been too few studies to draw final conclusions. Also, the extent of time that the plant was preconditioned under a specific temperature regime influenced the amount of foliar injury (Middleton 1956).

The relationships among growth temperature, ozone dose, and their influences on the foliar responses of pinto bean and soybean (Dunning et al. 1974) have been shown diagrammatically (Fig. 2.12). While foliar injury levels for soybean increased considerably at higher growth temperature, pinto bean showed both increased and decreased sensitivity, depending on the ozone dose. Considering exposure temperature, pinto bean displayed significant differences in foliar injury depending on the ozone dose, while foliar injury on soybean was similar at both exposure temperatures (Fig. 2.13).

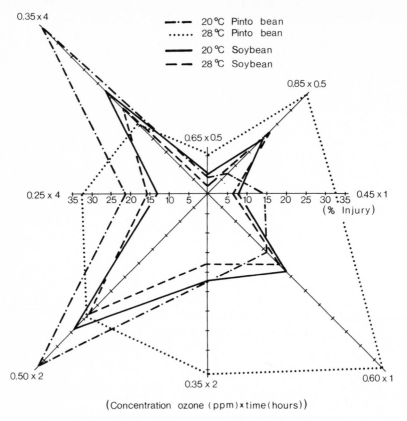

Fig. 2.13. The interactions of exposure temperature (20 ° and 28 °C), ozone concentrations, and exposure durations on the amount of foliar injury for pinto bean and soybean (Dunning et al. 1974). Increasing exposure durations are shown in a *clockwise* manner around the figure; within each exposure duration, ozone concentrations increase clockwise

Later studies (Dunning and Heck 1977) showed that 1-h exposures over the temperature range 16 °–27 °C had no significant influence on the extent of injury to pinto bean or tobacco, but at 32 °C injury on tobacco and pinto bean was significantly reduced and increased, respectively, compared to the lower temperatures. Shinohara et al. (1973) also found less ozone injury on tobacco exposed during high temperatures.

Under field conditions, ozone caused little or no foliar injury to corn when the ambient temperature was less than 32 °C. Given similar conditions except with temperatures greater than 32 °C, significant injury occurred (Cameron and Taylor 1973). It is difficult to differentiate the influence of temperature and light intensity under field conditions (Juhren et al. 1957), where they interact with relative humidity and soil water content. Even though the individual factors can influence plant sensitivity differentially, they can give a very different picture under field conditions (see Chap. 2.4.2.3.2).

Through corresponding biochemical, physiological, and anatomical features, plants have adapted to various climatic conditions which at the same time can influence plant sensitivity to photochemical oxidants. During the exposures of a large number of turfgrass species and cultivars to ozone and PAN, Youngner and Nudge (1980) observed that the cool-season types were generally more sensitive than the warm-season types.

Although carbohydrate contents have been implicated in differential plant exposures (see Sects. 2.4.3.1.2 and 2.4.3.2.1), they apparently do not account for the temperature-dependent changes in sensitivity. Dugger and Ting (1970b) concluded that sugar was not the sole factor controlling injury. Dunning et al. (1974) found no relation between the foliar carbohydrate content and the differential foliar injury response of soybean and pinto bean as it was influenced by growth and exposure temperatures. Adedipe and Ormrod (1974) found no consistent correlation between total soluble carbohydrate and ozone sensitivity as determined by growth reductions.

2.4.2.1.3 Relative Humidity

Relative humidity influences the leaf turgor and thereby controls stomatal opening and gas exchange. Pollutant uptake and plant injury increase with increasing relative humidity. In addition to the short-term effects of relative humidity on pollutant uptake, the long-term effects induce anatomical and morphological changes in leaves which may also affect pollutant uptake.

Using young Virginia pines, D. D. Davis (1970) established a correlation between the relative humidity during plant growth and the ozone-induced needle necrosis. Similar results were obtained with American ash (Wilhour 1970b). With pinto bean, the injury intensity following short-term acute ozone exposures increased with the relative humidity during growth (Dunning and Heck 1973, 1977), when the plants were grown at a light intensity of 21.5 klx, but not at 43 klx (Dunning and Heck 1973).

During plant growth, relative humidity can also influence anatomical characteristics of the leaf. According to Rentschler (1973), plants grown at a low relative humidity developed thicker wax layers than those grown at a high relative humidity. The wax crystals that partially overlay the stomata were larger, more branched, and entwined, thereby reducing transpiration and also pollutant uptake through the stomata and cuticle. Plants grown at a low humidity also developed smaller stomata.

Otto and Daines (1969) and Rich and Turner (1972) showed that stomatal opening increased with relative humidity. Subsequent studies showed that foliar ozone uptake also increased with relative humidity (McLaughlin and Taylor 1981). The extent of foliar ozone injury was correlated with relative humidity in American ash (*Fraxinus americana*) (Wilhour 1970a, b), pinto bean and tobacco Bel W3 (Otto and Daines 1969, Dunning and Heck 1973, 1977). With ozone exposures of tobacco, Menser (1963) established an increase in injury with increasing relative humidity. This principle has been shown by others (Hull and Went 1952, Thomas and Hendricks 1956, Leone and Brennan 1969a). The relative humidity following exposures does not apparently influence foliar injury (Otto and Daines 1969).

A water film on the leaf surface may elevate the relative humidity increasing the phytotoxicity of photochemical oxidants; however, the results are contradictory. Rain (MacDowall et al. 1963) or dew (Costonis and Sinclair 1969a, Leone and Brennan 1969a, b) increased ozone injury on various plant species; however, MacDowall (1966) found no effect of dew on the ozone sensitivity of tobacco.

The discussed relationship between injury and relative humidity is important not only in the laboratory and greenhouse but also under field conditions. Under the generally more humid conditions in the eastern USA, ozone concentrations of 0.14 to 0.22 ppm as daily means may cause extensive plant injury, while ozone concentrations of 0.2 to 0.4 ppm may produce only limited injury in California, where low humidity conditions can persist over extended periods (Heggestad et al. 1964, Heck et al. 1977, EPA 1978a).

Increasing relative humidity does not appear to have a practical influence on PAN effects. O. C. Taylor (1969) hypothesized that PAN would not occur in phytotoxic concentrations at elevated relative humidities because PAN is unstable at high humidity.

2.4.2.2 Soil and Nutrition

2.4.2.2.1 Soil Water Content

Soil water content, especially soil water deficiency, can have both short- and long-term effects on pollutant uptake. Short-term water stress impairs plant hydration, decreasing stomatal aperture and reducing pollutant uptake. Long-term water stress during plant growth causes physiological and morphological alterations, reducing gas exchange and thereby pollutant uptake.

Fully turgid plants exposed to ozonated hexene were more severely injured than plants with a deficient water supply (Hull and Went 1952, Koritz and Went 1953). Water-stressed tomato plants lost only a small amount of chlorophyll when exposed to ozone as compared to well-watered plants (Khatamian et al. 1973, Adedipe et al. 1973b). Similar results were found for bean and barley exposed to ozone and SO_2 (Markowski and Grzesiak 1974). Pollutant uptake was reduced, as shown by the decreased sulfate enrichment in the water-stressed plants. The stomata of tobacco under water stress were generally closed, while the stomata of well-watered plants were open even during ozone exposures (Rich and Turner 1972). Water-stress-induced protection from ozone injury can be obtained within 1 to a few days, depending on the severity of the water stress (G. S. Taylor et al. 1960, Tingey et al. 1982). The differential ozone response between two corn cultivars was associated with a greater sensitivity of one cultivar to water stress which induced stomatal closure reducing gas exchange (Harris and Heath 1981).

The influence of water stress on plant sensitivity to photochemical oxidants is important in agricultural practice. G. S. Taylor et al. (1960) observed that field-grown tobacco plants, during drought periods, were protected by some means from photochemical oxidants. Foliar injury to tobacco in Ontario, Canada (Walker and Vickery 1961), and Florida, USA (Dean and D. R. Davis 1967), increased proportionally to the amount of irrigation water applied. In particular areas of the USA where tobacco is endangered by ozone, the tobacco farmers are informed over the radio of predicted periods of high photochemical oxidants and advised to restrict their irrigation as much as possible (Dean and D. R. Davis 1972). Following a period of photochemical smog in California, well-watered grain plants exhibited extensive ozone and PAN injury, while plants grown in fields that had not been irrigated were uninjured (O. C. Taylor 1974 b). Obviously, water stress cannot be used as an avoidance mechanism for photochemical oxidant injury over the long term because water stress can significantly reduce plant productivity.

Tobacco plants that were grown under water-stressed conditions and then watered to field capacity a few hours before ozone exposure exhibited less injury than plants that were well watered during the total growth period (MacDowall 1965 a). Studies with water-stressed bean and tobacco plants showed that they became as sensitive to ozone as well-watered plants within a few days after the water stress was removed (G. S. Taylor et al. 1960, Tingey et al. 1982).

2.4.2.2.2 Elevated Salt Concentrations in the Soil Solution

Oertli (1959) grew sunflowers in sand culture using nutrient solutions of different osmotic potentials and salt contents to test their influence on plant response to photochemical smog. A high salt concentration in the solution produced a "physiological drought", reducing water uptake by the roots. Therefore, a negative correlation was found between the salt concentration in the nutrient solution and photochemical smog-induced foliar injury. High salt concentrations not only caused a physiological drought, but the elevated salt content in the plant tissue may injure cellular metabolism and thereby alter plant response to photochemical oxidants.

Increased salt in the soil decreases the osmotic potential of the soil solution, restricting water uptake by the plants. With time this effect can, in part, be compensated for as a result of the increased passive uptake of salt which decreased the osmotic potential of the cell sap. In addition, the increased salt content of the tissue disrupts the ionic equilibrium in the cytoplasm, producing changes in membrane permeability and alterations in cellular metabolism (Levitt 1980).

Regardless of the cause, if the plant growth medium contains elevated salt concentrations, the plants are less susceptible to photochemical oxidant injury. Chronic exposure to elevated soil salts coupled with increased evapotranspiration reduces plant hydration, causing a partial stomatal closure (Levitt 1980). Studies with beet (*Beta vulgaris*), alfalfa, and pinto bean showed that plant growth was decreased as the salt concentration in the solution was increased to an osmotic potential of -84 kPa by alterations in metabolism and gas exchange (Ogata and Maas 1973, Maas et al. 1973, Hoffman et al. 1973, 1975). Following an ozone exposure (0.10 to 0.35 ppm for a few h day^{-1} for 2 to 9 weeks), plant growth reductions were correlated with increasing ozone concentrations for the "salt-free" plants but with increasing salt concentrations the magnitude of the ozone effect was decreased. A similar antagonistic response was found with studies on young red maple (*Acer rubrum*) trees. The addition of 4,000 ppm NaCl reduced growth more than 2,000 ppm; the growth of plants at the high salt level was scarcely impacted by 0.25 ppm ozone (Dochinger and Townsend 1979). In many areas, salt is used on roads as a de-icer and it may influence vegetation along the edge of highways, raising the issue of possible interactions between salt and photochemical oxidants.

2.4.2.2.3 Plant Nutrients

Numerous elements are essential for plant growth; they are separated into two groups: macronutrients (nitrogen, potassium, calcium, magnesium, phosphorus, and sulfur) and micronutrients (chlorine, boron, iron, manganese, zinc, copper, molybdenum, and cobalt for nitrogen-fixing legumes). Nutrient deficiencies as well as excesses can induce disease symptoms on the plants. In any case, the physiological condition of a plant is controlled by the corresponding nutrient supply. Because the physiological condition of the plant influences its susceptibility to photochemical oxidants, the plant nutrient status can significantly influence the plant response to a given ozone dose. However, this influence is neither qualitatively nor quantitatively reproducible because of "the number of experimental investigations is still quite small and presents an incomplete and controversial picture" (Guderian 1977). General conclusions are not possible because of contradictory results. These discrepancies occur, in part, because of dissimilar experimental conditions and the use of various plant species at different developmental stages. The result is that it is not possible to give a general fertilization recommendation that will increase plant resistance to photochemical oxidants.

The effect of nitrogen nutrition has been studied by several groups. White Gold tobacco plants grown at an excess or deficiency of nitrogen were more severely injured than plants which received the optimal or slightly suboptimal levels (MacDowall 1965a). The mechanism is not clear even though a correlation between stomatal opening and nitrogen nutrition was found. Leone et al. (1966) reported different results for tobacco; injury was most severe on plants which received the optimal nitrogen level and less for plants that received either an excess or deficiency; they also observed that the plants grown on the optimal nitrogen level had the lowest carbohydrate contents. Low carbohydrate concentrations have been correlated with reduced plant susceptibility in several plant species (see Sect. 2.4.3.1.2). In their study the nutrition level was based on the foliar concentration rather than the concentration in the nutrient solution, as was done in most other studies. This different reference base is undoubtedly an additional cause for the lack of agreement among the various studies.

In field studies with tobacco, Menser and Street (1962) showed that the injury severity and incidence decreased with increasing nitrogen fertilization, even though the yield increased. The authors assumed that the nitrogen fertilizer delayed leaf maturity and shortened the ozone susceptible stage. In multi-year field studies, grapes (*Vitis labrusca* cv. Concord) that were not fertilized had more photochemical oxidant injury than plants that received 56 or 112 kg N ha^{-1} (Kender and Shaulis 1976). However, nitrogen nutrition had no influence on the response of white clover (*Trifolium repens*) and tall fescue (*Festuca arundinacea*) to long-term ozone exposures in the open-top chamber (Montes et al. 1982).

The influence of phosphorus nutrition on sensitivity of tomatoe to ozone was associated with carbohydrate content. In tomatoe grown in sand culture at three levels of phosphorus nutrition and exposed to ozone (Leone and Brennan 1970), phosphorus-deficient plants displayed only slight injury, but the injury intensity increased with the phosphorus concentration in the nutrient solution. The phosphorus-deficient plants had an elevated carbohydrate reserve prior to ozone exposure, but it decreased rapidly following exposure.

The effects of potassium nutrition on plant response are contradictory. Pinto bean and soybean grown at a low level of potassium exhibited more foliar injury following ozone exposure than plants grown at a high level (Dunning et al. 1974). The difference in sensitivity was not associated with carbohydrate content; the authors suggested that the increased injury at low potassium was associated with increased stomatal opening. Silver maple (*Acer saccharinum*) seedlings were grown in half-strength Hoagland nutrient solutions containing either 0, 2, or 117 ppm potassium and exposed to 0.3 ppm ozone for 6 h day^{-1} for 2 days (Noland and Kozlowski 1979). Seedlings grown at 0 or 2 ppm potassium showed less foliar injury than plants grown at 117 ppm potassium. The increased stomatal opening that occurred in the high potassium treatment was associated with an elevated ozone uptake and more foliar injury than in the potassium-deficient plants.

At high nitrogen levels, the addition of phosphorus and potassium decreased the injury to mangel and spinach (Brewer et al. 1961). At a low soil phosphorus level, injury increased proportionally to the added potassium, but at a higher soil phosphorus level, potassium addition decreased injury. High phosphorus and/or potassium increased the injury to duckweed when the remaining nutrients were held constant at one-third the level of Hutner's solution (Craker 1971 b). In studies using various growth media, the amount of ozone-induced foliar injury in soybean depended on the concentration of N, P, and K in the nutrient solution (Heagle 1979 b). Plants that received no supplemental nutrients exhibit only slight injury; the addition of a moderate amount of fertilizer increased the ozone sensitivity. When higher fertilizer rates were used, the plants grew better than at moderate rates, but the foliar sensitivity to ozone decreased.

Sulfur nutrition influenced plant susceptibility to ozone (Adedipe et al. 1972b). The tissue sulfate content increased proportional to the sulfate concentration of the nutrient solution. Plants grown at a high sulfate level were more resistant to ozone injury, possibly through an increase in the sulfhydryl content of the tissue.

The pH of the nutrient solution may have a significant influence on the intensity of plant injury (Yamazoe and Mayumi 1977). The most intense foliar injury from ambient photochemical oxidants occurred on plants grown on the acidic nutrient salts, ammonium sulfate and ammonium chloride; moderate foliar injury occurred when ammonium sulfate was used with a decreased superphosphate content; and only slight injury occurred when ammonium nitrate or urea, which are essentially neutral, were used. The addition of lime raised the pH of the nutrient solution and decreased plant sensitivity. It is not clear if these changes in plant sensitivity represent a direct effect of pH or if pH altered the uptake of other elements which contribute to the changes in plant response.

There is no general agreement concerning the influence of plant nutrient supply on ozone sensitivity. However, an association between forest decline (see Chap. 2.5.5) and Mg and Ca deficiencies has been observed.

2.4.3 Significance of Internal Growth Factors on Plant Sensitivity

The influence of internal growth factors which provide the basis for differential plant responses is expressed in species, cultivar, and individual differential sensitivity which can change in conjunction with external growth conditions (see Sect. 2.4.2) and plant development.

2.4.3.1 The Influence of Developmental Stage and Leaf Age

To understand and evaluate the ontogenic differences in sensitivity, it is necessary to differentiate between the development of the individual and its organs. Both developmental phases influence the type and degree of plant reaction to a given air pollutant. The existing results allow only a fragmentary evaluation of the influence of plant developmental stage on sensitivity.

2.4.3.1.1 Leaf Age

Leaves, roots, fruit, and pollen are the plant organs most sensitive, either directly or indirectly, to photochemical oxidants. The most obvious visible responses occur to the chlorophyll-containing leaves which are the principal sites for autotrophic reactions, i.e., photosynthesis. The leaves or needles of higher plants are especially good indicators of the degree of air pollution from photochemical oxidants.

Analogous to other air pollutants (Guderian 1977), the sensitivity of leaves to ozone and PAN varies substantially with leaf age. The schematic representation (Fig. 2.14) from the work of Glater et al. (1962) illustrates these changes. Following exposure of 100 young tobacco plants to photochemical smog, only the middle-aged leaves, 1 to 5, showed foliar injury, while the fully differentiated leaf 6 and the younger expanding leaves in the apical meristem were uninjured. Zones of differential sensitivity were also observed on individual leaves. For example, the basal portion of leaf 4 displayed 87% foliar injury, while the similar position on leaf 3, the next youngest, had only 25% foliar injury.

A series of studies confirmed the high sensitivity of middle-aged leaves (Hill et al. 1970). Leaves of dicotyledonous plant species were most sensitive during the later stages of leaf enlargement (Tingey et al. 1973b), after they had passed the period of maximum enlargement rate, but prior to achieving their final size (Ting and Mukerji 1971, Townsend and Dochinger 1974, Evans and Ting 1974). These data confirm the results of Ting and Dugger (1968) and Bennett (1969), in which

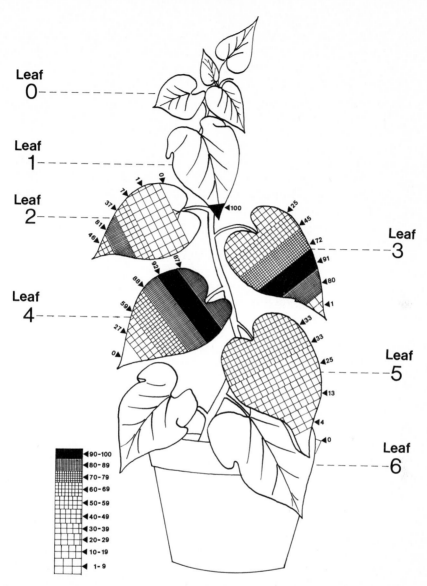

Fig. 2.14. The distribution of injury to tobacco leaves showing the influence of leaf age and leaf developmental stage on sensitivity to photochemical smog. (Glater et al. 1962)

leaves between 65 and 95% of final expansion were most sensitive to ozone. This period of maximum sensitivity corresponded with the formation of intracellular spaces and the stomata becoming functional, allowing the diffusion of gases throughout the leaf. In a few plant species, mature leaves can be injured (Shertz et al. 1980b), but young leaves are uniformly resistant to injury (Hill et al. 1970).

Every leaf passes through phases of differential sensitivity; however, the chronological leaf age is less important than its physiological one (Dugger and

Ting 1970a, b, O.C. Taylor 1973c). During the course of shoot elongation and development, for plants such as tobacco and field bean (*Vicia faba*), the most susceptible leaf position rises from the basal leaves toward the apex. Sensitive petunia cultivars acquired a marked ozone tolerance following the visible differentiation of the flower bud, and the tolerance was displayed several nodes below the flowering node (Hanson et al. 1975). Differential injury levels on leaves of individual shoots reflected the differential exposures of the preceding days, weeks, and months (Ashmore et al. 1980a), one of the values of indicator plants (see Chap. 2.3.2). Plants with a rosette growth habit, e.g., lettuce, celery, or beets, exhibit injury earlier on the outer rosette leaves than the inner ones during the early stages of plant development. As with sulfur dioxide (van Haut and Stratmann 1970), leaf susceptibility is displaced toward the center of the rosettes and corresponds to the developing leaves. To assess the total effect of a single exposure, it is necessary to consider the susceptible developmental stages of the total number of leaves on the plant.

Conifer needles display an age-dependent change in susceptibility, similar to dicotyledonous plant leaves. Ponderosa pine needles that were 5 to 7 weeks old (60 to 90 mm long) and still elongating were most sensitive (D. D. Davis and Coppolino 1974a). The susceptibility decreased as the needles reached full elongation, 100 to 120 mm. In a second similar experiment, ozone injury was the most severe on 5- to 8-week-old needles (D. D. Davis 1977). Virginia pine seedlings were exposed to 0.25 ppm ozone for 4 h; cotyledon, primary, and secondary needles were most severely injured when the needles were 3 to 5 weeks old (D. D. Davis and Wood 1973b).

The high susceptibility of young, developing needles during early summer is not necessarily the time when the most severe injury occurs in ozone-impacted areas. With chronic exposures, the effects accumulate to a maximum toward the end of the growing season. Long-term, low-concentration exposures may also injure mature tissue of certain species, i.e., pine or beans (Hill et al. 1970).

The tissue-age-dependent differences in susceptibility of various zones of the leaf (Glater et al. 1962, Evans 1973) are considerable and range from symptom-free portions to variable injury severity on others. A generalization for ozone injury to dicotyledonous plants can be made: on the youngest susceptible leaves, visible injury occurs primarily on the leaf tip; on the next oldest leaves, injury occurs in the middle, and on the oldest leaves, at the base of the leaf blade (Menser et al. 1963, Hill et al. 1970). Corresponding to the progressive differentiation of the tissue, the transverse susceptibility zones shift from leaf tip to leaf base (cf. Fig. 2.14).

Conifers display a similar age-dependent relationship. During needle expansion, primarily the tips are injured (Berry and Ripperton 1963). With increasing needle age, the zone of "recently mature" tissue migrates to the needle base (D. D. Davis and Wood 1973b). On fully differentiated needles, injury symptoms are distributed uniformly over the needle length (D. D. Davis and Coppolino 1974a).

The state of tissue development influenced the distribution of photochemical oxidant injury (Bobrov 1955a). In agreement with progressive cell differentiation from leaf tip to leaf base, bifacial necrosis on half-developed annual bluegrass

(*Poa annua*) leaves occurred on the tip of the youngest affected leaf, in the middle of the lower next oldest leaf, and at the base of the oldest leaves.

On dicotyledonous plant species such as petunia or small nettle (*Urtica urens*) and monocotyledonous species such as annual bluegrass, PAN exposure caused banding injury symptoms that corresponded to leaf development (O. C. Taylor et al. 1960, O. C. Taylor and MacLean 1970, Posthumus 1977). At moderate PAN exposures, visible symptoms were restricted to the lower leaf surface of dicotyledonous plants because injury occurred primarily in the spongy mesophyll. PAN studies on 10 bean cultivars found that the primary leaves were more susceptible than secondary leaves (Starkey et al. 1976). According to Dugger et al. (1963 b), the youngest leaves of pinto bean were most susceptible to PAN, while the more developed ones were most susceptible to ozone. However, deviations from this relationship can occur with changes in growth conditions.

After a single PAN exposure, the following sequence of injury symptoms can be observed on the leaflets of compound leaves such as tomato and potato (O. C. Taylor and MacLean 1970). Corresponding to the succession of tissue maturation, symptoms appeared at the apex of the terminal leaflet on the youngest susceptible leaf; the next oldest leaf was usually injured near the base of the terminal leaflet and at the apex of the first two lateral leaflets. The third oldest susceptible leaf was injured at the base of the first two lateral leaflets and at the apex of the second pair of lateral leaflets, while the more mature terminal leaflet remained symptomless. PAN exposures on successive days caused separate and distinct injury bands on the leaf blade. On monocotyledonous leaves, PAN symptoms appeared as distinct transverse bands from a few mm to perhaps 2 cm wide (Bobrov 1955a, Juhren et al. 1957, W. M. Noble 1965).

2.4.3.1.2 Developmental Stage

While leaf- and tissue-age-dependent sensitivity to photochemical oxidants has been extensively studied, only a few studies have been concerned with the sensitivity of the whole plant during the course of its development. Short-lived plants are especially sensitive to ozone during their early developmental stages. With few exceptions, 9- to 14-day old seedlings of numerous grass species were more injured by ozone (0.05 to 0.5 ppm) than 66- to 71-day-old plants (G. A. Richards et al. 1980). Ozone reduced radish hypocotyl growth the most if the exposure occurred during the period of rapid hypocotyl enlargement (Tingey et al. 1973a). Hypocotyl growth was reduced 37% from a single ozone exposure to 14-day-old plants, while growth was reduced 25 and 15% in 7- and 21-day-old plants, respectively. Pinto beans (Heck and Dunning 1967) and soybeans (Blum and Heck 1980) were most susceptible to ozone at 9 to 17 and 8 to 11 days old, respectively. Young cereal and spinach plants were generally more susceptible than mature stands (Hill et al. 1970). Sensitive sweet corn cultivars were essentially uninjured by ozone once they reached a height of 45 cm. Studies with white beans in areas impacted with photochemical oxidants indicated that crop maturity (stage of development) regulated the time of symptom expression, and crop vigor its severity (Haas 1970). Petunia hybrids became less sensitive to ozone when the flower buds differentiated (Hanson et al. 1975).

In general, the extent of foliar injury does not allow a reliable estimate of the magnitude of yield loss that will result. If the foliar injury occurs during the early developmental stages and good growth conditions prevail, the plant yield will approach or equal the yield of uninjured plants. It is not known if the later stages of cereal development – spike emergency to the milk stage – are "critical periods" during which ozone may have a permanent impact on yield, however, these stages are especially susceptible to SO_2- and NO_2-induced yield reductions (van Haut 1961, van Haut and Stratmann 1967). Wheat yield was reduced by 30% from 4 h exposures of 0.20 ppm ozone for 7 days during anthesis (Shannon and Mulchi 1974).

For woody plants, there are only limited data available concerning their resistance reactions, and these are generally limited to very early developmental stages. Seedlings of four conifer species were most sensitive to ozone at age 2 weeks or younger, while for sulfur dioxide they were first sensitive after 8 weeks (Berry 1974). Virginia pine seedlings that had cotyledonary, primary, or secondary needles were exposed to 0.25 ppm ozone for 4 h and foliar injury assessed on a scale of 1 to 7 (D. D. Davis and Wood 1973 b). All three needle types were maximally sensitive at 3 to 5 weeks following needle emergence. However, the cotyledonary needles were most severely injured (injury scale 6) with the primary and secondary needles exhibiting injury indices of 3.5 and 4, respectively. With annual plants, the cotyledonary leaf stage is the "critical developmental stage" for sulfur dioxide (van Haut 1961). Necrosis on cotyledons has a strong influence on the subsequent development of the plant.

The high susceptibility of the very early developmental stage was confirmed by Townsend and Dochinger (1974) on studies with red maple (*Acer rubrum*) seedlings. At the youngest growth stages (plants were 4 to 11 cm high), the plants displayed more foliar injury and larger growth reductions than seedlings in the oldest growth stages (24 to 32 cm high).

To evaluate the risk to forests, the sensitivity of individual tree and understory species during their later developmental stages needs to be understood. These are difficult studies to conduct, witness the decades of studies with sulfur dioxide. Field observations showed (Wentzel 1963) that conifers in the later pole stage to the timber stage were particularly susceptible to sulfur dioxide; similar conditions also prevailed for mature trees. During this developmental phase, large growth reductions and disintegration of stands can occur (Wentzel 1962, Knabe 1970, Materna et al. 1969).

Three factors are suggested as possible causes for the differential sensitivity of individual leaf developmental stages to ozone and PAN:

– differential pollutant uptake per unit leaf area per unit time;
– differential distribution of the absorbed pollutant influencing the amount reaching the active sites within the mesophyll cells;
– biochemically controlled differences in the susceptibility of the leaf.

As shown in Chapter 2.2.1, practically all photochemical oxidants reach the leaf interior by diffusion through the stomatal pore. The number, condition, and function of the stomata consequently influence the rate and amount of pollutant absorbed. At very early developmental stages of cotyledonary or foliage leaves, few functional stomata are present (Meidner and Mansfield 1968). The lack of functional stomata or very low gas exchange explains the high "resistance" of this developmental stage to ozone and PAN. With increasing leaf age, the stomata rapidly become functional; gas exchange per unit area reaches a maximum prior to full leaf expansion, and then it declines (Guderian 1970). Intermediate-aged tobacco leaves had a higher ozone uptake rate than young or old leaves (Craker and Starbuck 1973). Young-expanding camellia leaves absorbed three times as much ozone as mature leaves (Thorne and Hanson 1972). Correspondingly, differential injury in relation to leaf developmental stages results, in part, from variable pollutant absorption.

The importance of stomata for pollutant uptake and response was confirmed by Evans and P. R. Miller (1972); ozone-sensitive pine species had a higher stomatal frequency than did resistant species. Similarly, the stomatal frequency of susceptible tobacco cultivars was about 30% higher than that of resistant ones, but there was no relationship with stomatal size (Dean 1972). In contrast, Elkiey et al. (1979) found no association between stomatal frequency or size and the differential sensitivity of various petunia cultivars. The sensitive cultivar had the lowest stomatal frequency and the largest stomatal pore size, yielding no difference in total stomatal pore area per leaf among the cultivars. However, the resistant petunia cultivar had a rougher epidermis and denser trichomes. In peas, differential cultivar response was not associated with external leaf characteristics (Dijak and Ormrod 1982). In grapes (*Vitis labrusca*), an ozone-sensitive cultivar had a lower leaf conductance and fewer stomata than its resistant counterpart (Rosen et al. 1978). As soybean leaves expanded from 25 to 100% of final size, the ozone-resistant cultivar maintained a higher stomatal frequency than its susceptible counterpart (G. E. Taylor et al. 1982).

With leaf expansion and parallel to the development of functional stomata as entrance ports, intercellular spaces form which are important for pollutant distribution, as shown by Bobrov-Glater (1956) with her studies of ferns. Turrell (1942) associated differential sulfur dioxide responses among alfalfa (*Medicago sativa*) leaves to differences in the intercellular space volume and mesophyll cell-surface area. In contrast, differential ozone sensitivity of soybean cultivars was not related to internal leaf characteristics (G. E. Taylor et al. 1982). In soybean, the formation of intercellular spaces coincided with the period of rapid leaf expansion and the cessation of cell division in the palisade layers (Decker and Postlethwait 1960), which was also the developmental stage most susceptible to ozone (Tingey et al. 1973 b). The reduced ozone injury in fully expanded leaves was associated with the formation of cutin and suberin layers on the mesophyll cells which, in turn, would reduce the penetration of ozone to its active site within the cell (Glater et al. 1962).

Differential susceptibility associated with either leaf age or plant developmental stage cannot be related solely to differential pollutant uptake and distribution within the leaf. Specific biochemical factors must also be considered.

During leaf development, the period of maximum ozone susceptibility occurred when reducing and soluble sugars were at a minimum. Studies on beans (Heath et al. 1974, J. H. Bennett et al. 1977) showed that the period of maximum leaf expansion rate coincided with a minimum sugar content and maximum ozone susceptibility. In cotton, agreement was found between the minimum soluble sugar content and maximum ozone susceptibility, which occurred after the maximum expansion rate but prior to full leaf expansion (Ting and Mukerji 1971). Similarly, in eastern white pine needles, maximum sensitivity was related to the period of lowest sugar content (Barnes and Berry 1969).

The relationship of additional plant metabolites to plant sensitivity has been investigated. In cotton and soybean leaves, maximum susceptibility corresponded with the minimum free amino acid content (Ting and Mukerji 1971, Tingey et al. 1973 b) and in tobacco with the minimum total protein content (MacDowall 1965 b). This highly susceptible phase corresponded with minimum concentrations of sulfhydryls (Dugger and Ting 1970 a, b), nucleotides (Leopold 1964), and ascorbic acid (Hanson et al. 1970). A foliar application of abscisic acid reduced ozone injury through a decrease in stomatal aperture (Fletcher et al. 1972), presumably reducing ozone uptake. Whether these relationships are causal or only correlated are generally unresolved questions.

2.4.3.2 Species, Cultivar, and Individual Plant Resistance

Individual plant species, cultivars, provenances, and even individuals of populations react with differential sensitivity to given air-pollutant stresses. In contrast to specific plant pathogenic organisms (Gäumann 1951), there is no absolute resistance to air pollutants. Every plant species can be injured by appropriate combinations of pollutant concentration and exposure duration. The relative plant resistance to air pollutants is based on the expression of genetic traits which may change during development (Dochinger et al. 1965, Taylor 1968, Hanson et al. 1976), determining the susceptibility of individual plants. The variable tolerance of various species, cultivars, provenances, and individual plants during plant development also is determined by both previous and current environmental conditions (Fig. 2.1) (Guderian and Küppers 1980 b).

2.4.3.2.1 Criteria to Evaluate Resistance

Levitt (1980) defined two mechanisms of plant resistance to stress: stress avoidance and stress tolerance. In the former case, the stress induced by a specific ambient dose was partially or completely excluded. Numerous factors control the resistance of plant organs to the ingress of damaging compounds into the cells. In addition to morphological characteristics such as leaf shape, epidermal structure, including waxes and trichomes (Rentschler 1973, Elkiey et al. 1979), stomatal number, size, distribution (Meidner and Mansfield 1968, Evans and P. R. Miller 1972 b, Dean 1972) and stomatal resistance are of importance (Ting and Dugger 1971, Thorne and Hanson 1972, 1976, Turner et al. 1972, Miyake et al. 1974, Tingey and Taylor 1982). During the course of development, the above-mentioned morphological parameters adapt to changes in plant water relations. Morphological characteristics that reduce transpiration also reduce gas exchange and may thereby decrease pollutant uptake.

Stress-avoidance mechanisms that are effective against photochemical oxidants are known. In ozone-resistant onion cultivars, the stomata closed following injury to the guard cell membranes, while in sensitive cultivars the stomata remained open, allowing additional ozone uptake and injury (Engle and Gabelman 1966). Similar observations were reported by Butler and Tibbitts (1979a) on various bean cultivars and by Elkiey and Ormrod (1979c, 1980a) on differentially sensitive petunia cultivars. Total leaf conductance of ozone-sensitive tobacco cultivars was greater than that of tolerant cultivars, but the differences were not sufficient to completely explain the differences in injury (Turner et al. 1972). In soybean and pea cultivars differential susceptibility was not related to stomatal exclusion or differential stomatal closure (G. E. Taylor et al. 1982, Dijak and Ormrod 1982). In fact, the tolerant soybean cultivar absorbed as much or more ozone than the sensitive cultivar.

Stress tolerance assumes the absorption of the particular pollutant into the cells (G. E. Taylor 1978a). When the absorbed material is assimilated, tolerated, or buffered, no morphological or physiological change (strain) occurs and strain avoidance exists. Above a specific intracellular concentration, that is to say, when biochemical thresholds are exceeded, injury occurs which is either reversible (elastic strain), i. e., subtle changes in the photosynthetic rate, or irreversible (plastic strain), i. e., leaf necrosis. Following perturbations, living systems attempt to reestablish normal metabolic states through repair and/or compensatory processes of the perturbation and its consequences (Levitt 1980, Tingey and G. E. Taylor 1982). Repair processes remedy the cellular perturbation, removing the cellular dysfunction. However, with compensation processes, the perturbation and its physiological consequences remain, but the cell responds with mechanisms that counter the detrimental effects. Recovery from the perturbation or dysfunction may range from partial to complete, and consequently injury will vary as well.

Biochemical thresholds characterize the transition between strain avoidance and strain tolerance, as well as elastic strain and plastic strain, and determine the tolerance of individual plants. From this explanation, it follows that the individual responses can range from minor changes in the plant to the death of the affected plant part or plant, subject to the intensity of the ambient stress and the actual resistance of the plant.

Recently, Tingey and G. E. Taylor (1982) proposed a conceptual model to explain the causes of environmental, developmental, and genetically controlled differences in plant responses to ozone. The model (Chap. 2.2) proposed that plant injury resulted from ozone diffusion to susceptible sites within the cells and that three general processes controlled the magnitude and expression of the injury:

– processes that controlled the flux of gases into the leaves;
– processes that influenced pollutant distribution within the leaf, including ozone-scavenging reactions;
– processes that repaired or compensated for the perturbation.

The active interplay of these processes controlled the amount of injury; the relative importance of the processes varied among plants and was influenced by internal and external conditions.

The basis for the genetically controlled differences in resistance to ozone and PAN is, in general, not resolved. However, in addition to the role of gas absorption and distribution, various plant metabolites and metabolic scavenging or repair reactions have been suggested.

Cultivar differences in sensitivity have been associated with differences in soluble sugars, including reducing sugars, as have developmental differences (see Sect. 2.4.3.1.2), but the mechanism of the increased tolerance is not known. The older leaves of the ozone-insensitive tobacco cultivar Bel B contained a significantly lower soluble sugar content than the older leaves of the ozone-sensitive Bel W3 (Dugger et al. 1966). In contrast, the young ozone-insensitive leaves of both cultivars exhibited similar high sugar concentrations.

Differential foliar concentrations of ascorbic acid were correlated with the oxidant sensitivity of petunia cultivars (Hanson et al. 1970, 1971). The exogenous application of ascorbic acid to isolated chloroplasts decreased the effect of PAN on photosynthesis (Dugger and Ting 1970 b). In plant cells, ascorbic acid functions as a reducing compound with absorbed sulfur dioxide (Keller and Schwager 1977) and a weak oxidizing agent with PAN and ozone (Mudd 1980). However, ascorbic acid may not have a general role in plant resistance to photochemical oxidants. Differentially susceptible tobacco and petunia cultivars exhibited no differences in their foliar ascorbic acid contents (Menser 1964, Thorne and Hanson 1976). An oxidant exposure decreased the foliar concentration of ascorbic acid in petunia leaves (Hanson et al. 1970), but caused no change in ascorbic acid content of ponderosa pine seedlings following chronic ozone exposures (P.R. Miller et al. 1969).

The chemical reactivity of ozone in the liquid phase within the cell may affect its movement and distribution. Ozone may react with unsaturated fatty acids, sulfhydryl and ring-containing compounds or various scavenging systems (see Chap. 2.2.1.1.2).

The role of sulfhydryl compounds in the expression of variable resistance is unclear. Ozone and PAN react preferentially with sulfhydryl compounds, oxidizing them to disulfides and sulphonic acids (Tomlinson and Rich 1970, Mudd and Freeman 1977, Freeman et al. 1979). Sulfhydryl compounds, particularly SH enzymes, have various functions in cellular metabolism, and their reduction or inactivation will be reflected in a corresponding loss of function. Tobacco cultivars insensitive to ozone contained lower levels of sulfhydryls and ATP than sensitive cultivars (Tomlinson and Rich 1968). The significance of the total sulfhydryl content in relation to sensitivity is unclear; the deciding factor may be the function of the sulfhydryl compounds and whether their concentrations increase or decrease.

Plant species or cultivars which displayed extensive foliar necrosis following ozone exposure produced more polyphenols, which following oxidation by peroxidases and polyphenol oxidases to quinones can polymerize with amino acids and proteins (Howell 1974). These enzymes were activated sooner and to a higher level in sensitive soybean cultivars than resistant ones (Tingey et al. 1976 a). However, in differentially sensitive pea cultivars, leaf peroxidase activity increased about the time foliar injury was visible, but it was unrelated to cultivar sensitivity (Dijak and Ormrod 1982).

In addition to genetic factors, differential responses are influenced by developmental stage (see Sect. 2.4.3.1), edaphic and climatic factors (see Sect. 2.4.2), and biological interactions (see Chap. 2.2.5). The moderating or sensitizing effect of environmental factors before, during, and after pollutant exposure are, in part, responsible for the striking deviations in the relative resistance listings and groupings proposed by various authors. However, the genetic and environmentally controlled sensitivity relations of plant species, cultivars, and individuals provide an important basis for practical measures for environmental regulation. If these relationships are optimally used for diagnostic purposes, planting recommendations for pollution-impacted areas, the breeding and selection of relatively resistant cultivars, and the determination of dose-response relationships, then, in addition to the previously discussed factors, the selected response criteria and the concentrations of ozone and PAN must be considered when establishing and evaluating the degree of relative sensitivity.

The studies of Heagle (1979 a), using various ozone concentrations on soybean cultivars, confirmed the observation that plant sensitivity rankings depended on pollutant concentration (Wentzel 1963, Tingey et al. 1972). With chronic ozone exposures, soybean cultivars displayed a different susceptibility sequence than with high, acutely injurious concentrations. The importance of the response criteria used to determine relative resistance was also emphasized (Heagle 1979 a). When visible injury was used to evaluate differential susceptibility, the soybean cultivars exhibited a different sensitivity ranking than when production was used. Jensen (1973) reached a similar conclusion from his studies of the sensitivity relationships of various deciduous trees. In many species, foliar injury is not always correlated with plant yield. Sensitivity rankings based on foliar injury are not always the same as those based on yield (e. g., Oshima et al. 1977). Also the relative ozone resistance of soybean cultivars, based on growth, determined during the vegetative phase, was different from that based on seed yield (Heagle and Letchworth 1982). In only a few cases has the relative resistance determined by visible injury (F. A. Wood and Coppolino 1972) agreed with that determined by growth. It is preferable to determine relative resistance based on the reduced performance of the affected species; the function of the plant species in question should determine the criteria used to evaluate the response (Guderian 1977).

Only in recent years have there been comprehensive studies of the criteria used to assess relative plant resistance to photochemical oxidants. Following the considerations of Kulagin (1973), Harkov and Brennan (1979) proposed the consideration of an ecophysiological perspective for the evaluation and determination of the ozone sensitivity of deciduous trees in the northeastern USA. The inclusion of, for example, rate and dynamics of growth (Jensen 1973), reproduction (Harward and Treshow 1975, Smith 1980), and the position of the species in plant succession (P. R. Miller and McBride 1975) in resistance evaluations provides a reference point from which to evaluate the probable survival of trees in oxidant-impacted areas. Comparative studies of the resistance relations of the same species and cultivars in controlled pollutant exposures and in ambient-pollutant impacted areas provide additional opportunities to obtain reliable results (Karnosky 1978).

The available sensitivity groupings are based primarily on the use of visible symptoms as response criteria. In addition, the majority of the studies used short-term pollutant exposures to acute injurious concentrations of ozone and PAN, therefore caution is required before using these data to estimate the risk to a plant species or its suitability for growing in photochemical oxidant-impacted locations.

2.4.3.2.2 Relative Ozone Sensitivity

Table 2.10 contains a synopsis of the relative sensitivity of shrubs, ornamentals, coniferous, and deciduous trees to ozone. In developing the listing of sensi-

Table 2.10. Relative ozone sensitivity of woody plants grown in North America. (D. D. Davis and Wilhour 1976)

Sensitive species	Sensitive species
Ailanthus, tree of heaven *(Ailanthus altissima)*	Oak, Gambel *(Quercus gambelii)*
Ash, European mountain *(Sorbus aucuparia)*	Oak, white *(Quercus alba)*
Ash, green *(Fraxinus pennsylvanica)*	Pine, Austrian *(Pinus nigra)*
Ash, white *(Fraxinus americana)*	Pine, jack *(Pinus banksiana)*
Azalea, campfire *(Rhododendron kaempferi* Campfire*)*	Pine, Jeffrey *(Pinus jeffreyi)*
	Pine, loblolly *Pinus taeda)*
Azalea, Hinodegiri *(R. obtusum* Hinodegrii*)*	Pine, Monterey *(Pinus radiata)*
Azalea, Korean *(R. poukhanensis)*	Pine, ponderosa *(Pinus ponderosa)*
Azalea, snow *(R. kurume* snow*)*	Pine, Virginia *(Pinus virginiana)*
Bridalwreath *(Spirea vanhoutii)*	Poplar, hybrid-mixed *(Populus maximowiczii* × *trichocarpa)*
Cherry, Bing *(Prunus avium* var. Bing*)*	
Cotoneaster, rock *(Cotoneaster horizontalis)*	Poplar, tulip *(Liriodendron tulipifera)*
Cotoneaster, spreading *(C. divaricata)*	Privet, londense *(Ligustrum vulgare* var. pyramidale*)*
Grape, Concord *(Vitis vinifera var.* Concord*)*	
Honey locust (thornless) *(Gleditsia triacanthos inermis)*	Serviceberry, Saskatoon *(Amelanchier alnifolia)*
	Snowberry alba *(Symphoricarpos alba)*
Larch, European *(Larix decidua)*	Sumac fragrant *(Rhus aromatica)*
Lilac, Chinese *(Syringa chinensis)*	Sycamore, American *(Platanus occidentalis)*
	Walnut, English *(Juglans regia)*
Intermediate species	Intermediate species
Apricot, Chinese *(Prunus armeniaca* var. Chinese*)*	Pine, lodgepole *(Pinus contorta)*
	Pine, pitch *(Pinus rigida)*
Boxelder *(Acer negundo)*	Pine, scotch *(Pinus sylvestris)*
Cedar, incense *(Calocedrus decurrens)*	Pine, shortleaf *(Pinus echinata)*
Cherry, Lambert *(Prunus avium* var. Lambert*)*	Pine, slash *(Pinus elliottii)*
Current, northern black *(Ribes hudsonianum)*	Pine, sugar *(Pinus lambertiana)*
Elder, black bead *(Sambucus melanocarpa)*	Pine, Torrey *(Pinus torreyana)*
Elm, Chinese *(Ulmus parvifolia)*	Privet, common *(Ligustrum vulgare)*
Forsythia, Lynwood Gold *(Forsythia intermedia spectabilis* Lynwood Gold*)*	Redbud, eastern *(Cercis canadensis)*
	Rhododendron *(Rhododendron catawbiense album)*
Gum, sweet *(Liquidambar styraciflua)*	
Hemlock, eastern *(Tsuga canadensis)*	Rhododendron *(Rhododendron nova zembla)*
Larch, Japanese *(Larix leptolepis)*	Rhododendron *(Rhododendron roseum elegans)*
Mock Orange, sweet *(Philadelphus coronarius)*	Snowberry, vaccinioides *(Symphoricarpos vaccinioides)*
Oak, black *(Quercus velutina)*	
Oak, pin *(Quercus palustris)*	Viburnum, linden *(Viburnum dilatatum*
Oak, scarlet *(Quercus coccinea)*	Viburnum, Tea *(Viburnum setigerum)*
Pine, eastern white *(Pinus strobus)*	

Table 2.10 (continued)

Less sensitive species	Less sensitive species
Apricot *(Prunus aremeniaca)*	Oak, bur *(Quercus macrocarpa)*
Arborvitae *(Thuja occidentalis)*	Oak, English *(Quercus robur)*
Azalea, Chinese *(Rhododendron mollis)*	Oak, northern red *(Quercus rubra)*
Avocado *(Persia americana)*	Oak, shingle *(Quercus imbricaria)*
Beech, European *(Fagus sylvatica)*	Oregon Boxwood *(Paxistima myrsinites)*
Birch, European white *(Betula pendula)*	Pagoda, Japanese *(Sophora japonica)*
Box, Japanese *(Buxus sempervirens)*	Peach *(Prunus persica)*
Dogwood, gray *(Cornus racemosa)*	Pear, bartlett *(Pyrus communis* var. Bartlett*)*
Dogwood, white *(Cornus florida)*	Pieris, Japanese *(Pieris japonica)*
Euonymus, dwarf winged *(Euonymus alatus compactus)*	Pine, digger *(Pinus sabiniana)*
	Pine, singleleaf pinyon *(Pinus monophylla)*
Fir, balsam *(Abies balsamea)*	Pine, red *(Pinus resinosa)*
Fir, Douglas *(Pseudotsuga menziesii)*	Privet, Amur north *(Ligustrum amurense)*
Firethorne, Laland's *(Pyracantha coccinea Laland)*	Redwood *(Sequoia sempervirens)*
	Rhododendron, Carolina *(Rhododendron carolinianum)*
Gum, black *(Nyssa sylvatica)*	Sequoia, giant *(Sequoia gigantea)*
Holly, American [female and male] *(Ilex opaca)*	Spruce, Black Hills *(Picea glauca* var. *densata)*
Holly, Hetz Japanese *(Ilex crenata)*	Spruce, Colorado blue *(Picea pungens)*
Holly, English *(Ilex aquifolium)*	Spruce, Norway *(Picea abies)*
Juniper, western *(Juniperus occidentalis)*	Spruce, white *(Picea glauca)*
Laurel, mountain *(Kalmia latifolia)*	Viburnum, Korean spice *(Viburnum carlesi)*
Linden, American *(Tilia americana)*	Viburnum, burkwoodii *(Viburnum burkwoodii)*
Linden, little-leaf *(Tilia cordata)*	Wood's rose *(Rosa woodsii)*
Locust, black *(Robinia pseudoacacia)*	Yew, dense *(Taxus densiformis)*
Mahonia, creeping *(Mahonia repens)*	Yew, Hatfield's pyramidal *(Taxus media hatfieldi)*
Maple, bigtooth *(Acer grandidentatum)*	
Maple, norway *(Acer platanoides)*	
Maple, sugar *(Acer saccharum)*	

Note: To prevent the misunderstanding that there is an absolute ozone resistance the term "less sensitive" rather than "tolerant" is used. Within a sensitivity grouping, species are arranged alphabetically by common name

tive, intermediate, and less sensitive species, D. D. Davis and Wilhour (1976) used results from controlled ozone exposures, as well as observations from the field. They indicated that their listing was not only a compilation of earlier research, but also contained new unpublished results.

Table 2.10 illustrated that numerous economically and ecologically important woody plant species exhibited sensitive and intermediate sensitive reactions to ozone, in contrast to PAN. In addition, various species important to European agriculture, i.e., grapes, fruit tree crops: cherry, walnut, and currants; forest species: European and Japanese larch, Scotch pine and Austrian pine are ozone-sensitive. Notable is the relative ozone tolerance of European beech and Norway spruce, which are among the tree species most sensitive to sulfur dioxide and hydrogen fluoride (Wentzel 1968). There is also a wide range in relative sensitivity among individual species within a genus, i.e., *Quercus* and *Pinus* which have species in all three sensitivity groupings. It is not possible to establish a clear difference in relative ozone sensitivity between coniferous and deciduous trees as it is for the acidic gases, SO_2 and HF.

Table 2.11. Relative sensitivity of agricultural crops and weeds to ozone. (EPA 1976)

Sensitive species	Sensitive species
Alfalfa *(Medicago sativa)*	Oat *(Avena sativa)*
Barley *(Hordeum vulgare)*	Onion *(Allium cepa)*
Bean *(Phaseolus vulgaris)*	Potato *(Solanum tuberosum)*
Buckwheat *(Fagopyrum esculentum)*	Radish *(Raphanus sativus)*
Citrus *(Citrus* sp.*)*	Rye *(Secale cereale)*
Clover, red *(Trifolium pratense)*	Safflower *(Carthamus tinctorius)*
Corn, sweet *(Zea mays)*	Smartweed *(Polygonum* sp.*)*
Grape *(Vitis vinifera)*	Soybean *(Glycine max)*
Grass, bent *(Agrostis palustris)*	Spinach *(Spinacia oleracea)*
Grass, brome *(Bromus inermis)*	Tobacco *(Nicotiana tabacum)*
Grass, orchard *(Dactylis glomerata)*	Tomato *(Lycopersicon esculentum)*
Muskmelon *(Cucumis melo)*	Wheat *(Triticum aestivum)*

Intermediate species	Intermediate species
Cabbage *(Brassica oleracea)*	Parsnip *(Pastinacea sativa)*
Carrot *(Daucus carota)*	Pea *(Pisum sativum)*
Corn, field *(Zea mays)*	Pepper *(Capsicum frutescens)*
Cowpea *(Vigna sinensis)*	Peanut *(Arachis hypogaea)*
Cucumber *(Cucumis sativus)*	Sorghum *(Sorghum vulgare)*
Endive *(Cichorium endivia)*	Stevia *(Pigueria trinervia)*
Hypericum *(Hypericum* sp.*)*	Timothy *(Phleum pratense)*
Parsley *(Petroselinum crispum)*	Turnip *(Brassica rapa)*

Less sensitive species	Less sensitive species
Beet *(Beta* sp.*)*	Mint *(Mentha* sp.*)*
Cotton *(Gossypium* sp.*)*	Piggy-back plant *(Tolmiea menziesii)*
Descurainia *(Descurainia californica)*	Rice *(Oryza sativa)*
Jerusalem cherry *(Solanum pseudo-capsicum)*	Strawberry *(Fragaria* sp.*)*
Lamb's quarters *(Chenopodium album)*	Sweet potato *(Ipomoea batatas)*
Lettuce *(Lactuca sativa)*	

Note: To prevent the misunderstanding that there is an absolute ozone resistance the term "less sensitive" rather than "tolerant" is used. Within a sensitivity grouping, species are arranged alphabetically by common name

The relative ozone sensitivity of agricultural crops and weeds listed in Table 2.11 is based on the handbook, *Diagnosing Vegetation Injury Caused by Air Pollution* (EPA 1976).

Among the important agricultural crops that are endangered by ozone are the principal cereal crops; wheat, rye, barley, oats, and the forage crops; alfalfa and red clover and vegetable crops; potato and tomato. Endive and cabbage species exhibit a relatively high ozone sensitivity in contrast to their low level of sensitivity to SO_2 and HF.

2.4.3.2.3 Relative PAN Sensitivity

In comparison to ozone, the lesser importance of PAN, as a phytotoxic air pollutant, is reflected by the limited number of investigations into relative plant

Table 2.12. The relative PAN sensitivity of ornamentals and tress. (EPA 1976)

Sensitive species	Sensitive species
Aster *(Aster* sp.*)*	Petunia *(Petunia* sp.*)*
Dahlia *(Dahlia* sp.*)*	Primrose *(Primula* sp.*)*
Fuchsia *(Fuchsia* sp.*)*	Ranunculus *(Ranunculus* sp.*)*
Mimulus *(Mimulus* sp.*)*	Sweet-Basil *(Ocimum basilicum)*
Mint *(Mentha* sp.*)*	

Less sensitive species	Less sensitive species
Apple *(Malus sylvestris)*	Lilac, common *(Syringa vulgaris)*
Arborvitae *(Thuja orientalis)*	Lily *(Lilium* sp.*)*
Ash, green *(Fraxinus pennsylvanica)*	Locust, honey *(Gleditsia triacanthos)*
Ash, white *(Fraxinus americana)*	Maple, Norway *(Acer platanoides)*
Azalea, campfire *(Rhododendron* sp.*)*	Maple, silver *(Acer saccharinum)*
Basswood *(Tilia americana)*	Maple, sugar *(Acer saccharum)*
Begonia *(Begonia* sp.*)*	Mountain-ash, American *(Sorbus americana)*
Birch, European white *(Betula pendula)*	Narcissus *(Narcissus* sp.*)*
Bromiliads *(Bromeliaceal)*	Oak, English *(Quercus robur)*
Cactus (Cactaceae)	Oak, northern red *(Quercus rubra)*
Calendula *(Calendula* sp.*)*	Oak, pin *(Quercus palustris)*
Camellia *(Camellia* sp.*)*	Oak, white *(Quercus alba)*
Carnation *(Dianthus caryophyllus)*	Orchids (Orchidaceae)
Chrysanthemum *(Chrysanthemum* sp.*)*	Perwinkle *(Vinca minor)*
Coleus *(Coleus* sp.*)*	Pine, Austrian *(Pinus nigra)*
Cyclamen *(Cyclamen* sp.*)*	Pine, eastern white *(Pinus strobus)*
Dogwood *(Cornus florida)*	Pine, red *(Pinus resinosa)*
Fir, balsam *(Abies balsamea)*	Pine, Scotch *(Pinus sylvestris)*
Fir, Douglas *(Pseudotsuga menziesii)*	Poplar, hybrid *(Populus maximowiezii*
Fir, white *(Abies concolor)*	× *trichocarpa)*
Gum, sweet *(Liquidambar styraciflua)*	Poplar, tulip *(Liriodendron tulipifera)*
Hemlock, eastern *(Tsuga canadensis)*	Spruce, Black Hills *(Picea glauca densata)*
Ivy *(Hedera* sp.*)*	Spruce, blue *(Picea pungens)*
Larch, European *(Larix decidua)*	Spruce, Norway *(Picea abies)*
Larch, Japanese *(Larix leptolepis)*	Spruce, white (*Picea glauca)*

Note: To prevent the misunderstanding that there is an absolute ozone resistance the term "less sensitive" rather than "tolerant" is used. Within a sensitivity grouping, species are arranged alphabetically by common name

sensitivity. In spite of similar active sites within the leaf mesophyll for ozone and PAN, considerable differences in relative plant sensitivity to the two pollutants occur both within and among species. For example, the sensitivity rankings determined for ten bean cultivars were different for PAN and ozone (Starkey et al. 1976).

The woody plants that have been studied consistently show a high degree of "tolerance" to visible PAN injury. In the sensitivity lists of D. D. Davis and Wilhour (1976), as well as the EPA publication *Diagnosing Vegetation Injury Caused by Air Pollution* (1976), no woody plants are found under the listings of sensitive or intermediate plants (Table 2.12). PAN endangers primarily agricultural, horticultural, and many herbaceous ornamental species (Tables 2.12 and 2.13). Sen-

Table 2.13. Relative PAN sensitivity of agricultural plants and weeds. (EPA 1976)

Sensitive species	Sensitive species
Bean *(Phaseolus vulgaris)*	Lettuce *(Lactuca sativa)*
Celery *(Apium graveolens)*	Mustard *(Sinapis* sp.*)*
Chard, Swiss *(Beta vulgaris cicla)*	Nettle, littleleaf *(Urtica urens)*
Chickweed *(Stellaria media)*	Oat *(Avena sativa)*
Clover *(Trifolium* sp.*)*	Pepper *(Capsicum frutescens)*
Endive *(Cichorium endivia)*	Pigweed *(Amaranthus retroflexus)*
Grass, annual blue *(Poa annua)*	Tomato *(Lycopersicon esculentum)*
Ground cherry *(Prunus fruticosa)*	Wild-oat *(Avena sativa)*
Jimson-weed *(Datura strananium)*	

Intermediate species	Less sensitive species
Alfalfa *(Medicago sativa)*	Bean, lima *(Phaseolus limensis)*
Barley *(Hordeum vulgare)*	Broccoli *(Brassica oleracea)*
Beet, sugar *(Beta vulgaris)*	Cabbage *(Brassica oleracea)*
Beet, table *(Beta vulgaris)*	Cauliflower *(Brassica oleracea botrytis)*
Carrot *(Daucus carota)*	Corn *(Zea mays)*
Cheeseweed *(Malva parviflora)*	Cotton *(Gossypium* sp.*)*
Dock, sour *(Rumex crispus)*	Cucumber *(Cucumis sativus)*
Lamb's-quarters *(Chenopodium album)*	Onion *(Allium cepa)*
Soybean *(Glycine max)*	Radish *(Raphanus sativus)*
Spinach *(Spinacia oleracea)*	Rhubarb *(Rheum rhaponticum)*
Tobacco *(Nicotiana tabacum)*	Sorghum *(Sorghum vulgare)*
Wheat *(Triticum aestivum)*	Squash *(Cucurbita pepo)*
	Strawberry *(Fragraria* sp.*)*

Note: To prevent the misunderstanding that there is an absolute ozone resistance the term "less sensitive" rather than "tolerant" is used. Within a sensitivity grouping, species are arranged alphabetically by common name

sitive responses are exhibited by the cultivated plants, i. e. oat, bean, tomatoe, lettuce, and clover; and intermediate sensitivity is shown by wheat, barley, sugar beet, and spinach. Pronounced SO_2 and HF resistance are exhibited by carrot (Guderian and Stratmann 1968), but it displays an intermediate PAN sensitivity, again showing how the same plant may be influenced differentially by the various atmospheric components.

2.4.3.2.4 Differential Cultivar Sensitivity

Differential ozone responses exist not only among individual species but also among cultivars of a single species and thus are genetically based. An extensive literature documents differential cultivar sensitivity. Selected examples will be used to illustrate the range in differential cultivar resistance and its use as a practical control measure to reduce ozone injury. In addition, the differential cultivar response data provides a general framework to extend our understanding of plant resistance to ozone.

The relative sensitivity of tobacco cultivars has been extensively examined. Heggestad et al. (1964) clearly established the differential sensitivity of six tobacco cultivars to ozone and photooxidants in California and Washington, D. C. The investigations of MacDowall et al. (1963) found that White Gold was the most sensitive cultivar of the 32 examined. Heggestad and Menser (1962) selected, from breeding trials with cigar-wrapper tobacco cultivars, Bel W3 as a highly sensitive indicator plant for ozone. Since then, this cultivar has achieved worldwide success in identifying and monitoring ozone-impacted areas (Knabe et al. 1973, Jacobson and Feder 1974, Floor and Posthumus 1977, Ashmore et al. 1980a).

Large differences in the relative sensitivity of bean (*Phaseolus vulgaris L.*) cultivars were identified by the research of D. D. Davis and Kress (1974), among others. The most sensitive of the ten cultivars examined exhibited approximately 33% leaf area injured, while the least sensitive cultivar displayed only a trace of foliar injury. In this study and others, the white-flowered cultivars were more sensitive than those with colored flowers. Similar or larger differences in susceptibility were found in both controlled fumigations (Beckerson et al. 1979) and field trials with bean cultivars in New Brunswick, New Jersey (Brennan and Rhoads 1976). When the most sensitive cultivars were severely injured, the less sensitive cultivars showed no symptoms. Worthy of note was the additional observation that cultivars that exhibited the most susceptibility to ozone were the most resistant to bean rust (*Uromyces phaseoli*). In similar studies, Manning and Vardaro (1974a) found only partial agreement with Brennan and Rhoads (1976) in the relative sensitivities of the various bean cultivars, possibly because of differences in growth conditions or variable high oxidant exposures.

Other research established that soybean cultivars exhibited substantial differences in cultivar susceptibility (V. L. Miller et al. 1974, Howell et al. 1976). The research of Heagle (1979b) with four soybean cultivars established that the relative ozone sensitivity changed with the growth medium, fertilizer rate, hour and season of exposure.

The far-reaching, frequent occurrence of injury on potato in the eastern United States and yield reductions approaching 50% (Heggestad 1970, 1973) was the motivation for extensive research on the differential sensitivity of potato cultivars (Rich and Hawkins 1970, Brasher et al. 1973, Heggestad 1973). Despite deviations among individual experiments, the later maturing cultivars were generally less sensitive than the early ones (Mosley et al. 1978). There was no clear relationship between the extent of leaf injury and the magnitude of yield reduction. However, there is undoubtedly an upper limit of injury to potato foliage beyond which yield losses will occur. The timing of ozone exposure in relation to the stage of tuber development is a significant factor influencing final production.

Hanson et al. (1976) studied the differential sensitivity of seven inbred petunia lines and showed that the degree of ozone resistance did not coincide with the level of PAN resistance. The genes responsible for oxidant resistance act in an additive manner and there was evidence for partial dominance in the progeny. A high degree of inheritance for resistance and susceptibility to acute ozone injury in sweet corn leaves under field conditions was shown during two and three generations of breeding with three series of cultivars (Cameron 1975).

Long-term exposures of corn to ozone concentrations of 0.05 and 0.10 ppm for 7 h day^{-1}, under conditions simulating the natural environment, yielded clear differences in the susceptibility of the two cultivars investigated (Heagle et al. 1972). The less sensitive cultivar White Midget displayed no visible injury and biomass production and yield were not significantly decreased, although there was a trend toward reduced seed yield. However, the yield of the sensitive cultivar Golden Midget was significantly reduced at 0.10 ppm ozone. With many plant species, the reproductive stage of development appears to be more sensitive than the vegetative stage. In another study using a different corn cultivar, Coker 16, the threshold ozone doses for foliar injury and effects on vegetative growth were much lower than those for decreased kernel yield, demonstrating that field corn can withstand some injury and growth reduction with no yield loss (Heagle et al. 1979c). Tomatoe exposed to SO_2 (Guderian and Stratman 1968) or ozone (Henderson and Reinert 1979) displayed the same relation between injury and growth. In "open-top" field-chamber exposure studies using five open pedigree and six commercial corn hybrids, large differences in cultivar resistance were found with reductions in shoot fresh weights ranging between 2 and 63% (Heagle et al. 1979c).

In investigations with many grass species and cultivars (Richards et al. 1980), the "warm-season entries" *Zoysia japonica* and Tufcote Bermudagrass (*Cynodon dactylon*) were found to be less sensitive than the "cool-season entries" such as tall fescue (*Festuca arundinacea*), perennial rye grass (*Lolium perenne*), red fescue (*Festuca rubra*), Kentucky bluegrass (*Poa pratensis*), and annual bluegrass (*Poa annua*). Substantial differences in sensitivity were found among the seven cultivars of Kentucky bluegrass.

Additional information on the differential sensitivity of various agricultural and horticultural crops to ozone is available and includes, among others, investigations with cereal species (Reinert 1975, Heagle et al. 1979e, Nakamura and Ota 1977), forage legumes (Brennan et al. 1969, Howell et al. 1971), sugar beet (Menser 1974), spinach (Manning et al. 1972), radish (Reinert et al. 1972), tomato (Clayberg 1971, 1972, Reinert and Henderson 1980), celery (Proctor and Ormrod 1977), chrysanthemum (Brennan and Leone 1972, Klingaman and Link 1975), azalea (Gesalman and Davis 1978), petunia (Uhring 1978), and grape (Shertz et al. 1980a).

Among the numerous studies concerning the differential sensitivity of woody plants, the following work of Townsend and Dochinger (1974) is an example: red maple (*Acer rubrum*) seedlings reacted differently to ozone, depending on their provenances and the phenological stages studied. The stability of ozone tolerance during several plant and leaf ontogenic stages suggests strong genetic control which should facilitate the selection of ozone-tolerant lines.

The ozone injury threshold for five clones of quaking aspen (*Populus tremuloides*) varied between 0.05 and 0.20 ppm (Karnosky 1976), additional evidence that intraspecific variation in relative resistance can be as large or larger than the interspecific variation. The occurrence of cultivars from several species in all three relative resistance rankings (sensitive to less sensitive) documents this quite clearly.

The considerable differences in the sensitivity of individuals within a species (Skelly et al. 1977, Blanchard et al. 1979, Steiner and Davis 1979) is clearly shown by the early research on the chlorotic dwarf syndrome of eastern white pine.

Scions from mother trees displaying the chlorotic dwarf syndrome exhibited the same symptoms when grafted onto healthy root stock, while scions from trees without the disease remained healthy when grafted onto the root stock of trees showing the chlorotic dwarf syndrome (Dochinger et al. 1965, Dochinger and Seliskar 1965). Because of the similarity in morphology and anatomy, the genetic control of specific biochemical properties associated with differential resistance should be investigated. Coyne and Bingham (1981) measured the differential effects of ozone on the photosynthetic performance of individual ponderosa pine in relation to ecotypes. They found that photosynthesis decreased sooner in ecotypes sensitive to ozone. Needle retention was also decreased in the sensitive individuals.

In addition to the previously cited articles, the following publications contain extensive reviews of relative plant resistance to photooxidants (US Department of Health, Education and Welfare 1970, Linzon et al. 1975, Heck et al. 1977, EPA 1978a, Skelly and Johnson 1979).

2.4.4 Measures to Reduce Pollutant Effects in Plant Stands

The principal means of protecting vegetation from photochemical oxidants injury is by reducing the precursors of photochemical oxidants, thereby reducing the ambient concentrations of ozone and PAN (cf. Part 1). Certain precautionary measures at the site can, however, reduce but not prevent the damaging effects of pollution. The measures available to reduce injury are questionable because they are only practical in agriculture and horticulture, limited in forestry, and not practical in natural ecosystems.

The possibilities of countering injurious pollutant effects in plant stands includes the use of genotypes possessing relative tolerance to the pollutant (Sect. 2.4.3.2) and other modifying factors of plant response, such as external growth factors or various chemical protectants. The availability of differentially sensitive cultivars or individuals provides the possibility of breeding and growing relatively tolerant plant types. The influence of external environmental factors and chemicals to modify plant responses to photochemical oxidants leads to possible changes in cultivation practices, cultural requirements, and chemotherapeutic treatments to reduce photochemical oxidant injury to vegetation.

2.4.4.1 Differential Resistance as a Basis for Remedial Measures

Based on the considerable range in differential resistance, the cultivation of relatively tolerant species and cultivars provides the best available approach for reducing the impact of ambient oxidants at the receptor site. To be successful, however, detailed knowledge of the relative resistance relationships of the affected cultivars is needed. Also, the criteria used to determine the reduction in useful value (Sect. 2.4.3.2.1) and the influence of internal (Sect. 2.4.3) and external (Sect. 2.4.2) growth factors on the plant response should be determined. An additional problem to consider is that the relative resistance relationships of individual plant species and types vary with different air pollutants, as has been observed with ozone and PAN (Hanson et al. 1976).

Cultural methods were used to moderate the injurious effects of photochemical oxidants on vegetation before the origin or the cause of the injury was known. The selection and culture of tobacco cultivars resistant to weather fleck injury preserved the cigar wrapper industry in the Connecticut River Valley from almost certain economic disaster. Subsequent studies showed that the selected cultivars had a significantly higher ozone tolerance than Bel-C which was no longer grown in the Connecticut River Valley because of its susceptibility to weather fleck injury (Menser and Hodges 1972). Additional studies confirmed a general relationship between plant resistance to weather fleck and ozone injury (Huang et al. 1975).

In the interim, numerous studies have illustrated the wide range in intraspecific susceptibility in agricultural and horticultural plant species as well as forest tree species (see Sect. 2.4.3.2). As expected, these results have and will continue to find practical use, particularly for sensitive cultivars or plant types in areas heavily impacted by photochemical oxidants. Knowledge of differentially sensitive species will obviously be used for the propagation of relatively resistant types (Jensen et al. 1976).

There is general agreement (Kress 1976, Jensen et al. 1976, Bialobok et al. 1980) that the occurrence of autonomous resistance differences within a species provides the prerequisite conditions for breeding relatively tolerant plant types, particularly through positive selections and crossings. In conjunction with previous observations, the conditions are particularly favorable in some plant species for the artificial selection of differentially tolerant types. In onions (*Allium cepa*) for example, ozone resistance is controlled by a single dominant gene (Engle and Gabelman 1966). In *Petunia hybrida*, the genes responsible for oxidant tolerance act primarily in an additive manner, with some indication of partial dominance in the progeny (Hanson et al. 1976). Partial dominance for both resistance and sensitivity has been reported for tobacco (Aycock 1972, Provilaitis 1967, Menser and Hodges 1972, Huang et al. 1975). The data for corn indicate that the genes responsible for ozone sensitivity are either additive or incompletely dominant (Cameron 1975). De Vos et al. (1980) studied the inheritance of PAN resistance in *Petunia hybrida* using inbred parents of a susceptible F_1 hybrid and a resistant hybrid. In the experiment with the most severe PAN injury, genes for susceptibility exhibited almost complete dominance over those for resistance and epistatic effects were not significant. In other experiments with less PAN injury, resistance was partly dominant to susceptibility and one or more epistatic parameters were significant. The number of genes involved in photochemical oxidant tolerance in petunia, tobacco, and corn is not known, but it is certainly greater than one. In contrast to previous observations in *Phaseolus vulgaris* (Saettler 1975), crossing of two ozone-sensitive and ozone-resistant *Phaseolus vulgaris* cultivars demonstrated that ozone resistance was recessive and appeared to be regulated by a few major genes (Butler et al. 1979). Noteworthy in this connection is a study of 387 snap bean cultivars grown in the United States, in which approximately 70% were ozone-tolerant (Meiners and Heggestad 1979). Based on these observations,

Meiners and Heggestad suggested that it was not necessary to breed bean cultivars specifically for resistance to oxidant air pollutants. However, breeding lines should be evaluated for sensitivity to photochemical oxidant air pollutants prior to being released, to insure that susceptible types are not recommended or grown in areas impacted by photochemical oxidants. This recommendation is also true for other plant species.

Ozone or PAN concentrations above a specific level can exert a selection pressure, as do other environmental factors, in breeding experiments and in semi-natural and managed conditions. Accordingly, it appears reasonable to conclude that native plants will display a relative sensitivity level dependent on the natural ozone level. In the course of several decades, plants with a short generation cycle may develop resistant populations as a result of dynamic selection in areas experiencing increasing exposure from anthropogenic ozone and PAN (see the discussion in Chap. 2.2.5.2). Artificial selection studies conducted in areas experiencing elevated photochemical oxidant exposures should, in the course of selection, also attain a partial resistance to photochemical oxidant injury. Plant breeders generally attempt to select plants with the highest yield and the fewest injury symptoms, regardless of the cause.

Special breeding programs to develop ozone- or PAN-resistant cultivars are seldom described in the literature. However, examples include tobacco (Provilaitis and White 1966, Huang et al. 1975), loblolly pine (Heck and Brandt 1977), petunia (Hanson et al. 1974), and tomato (Clayberg 1971, 1972). Given the intensive documentation of intraspecific differential sensitivity, findings about the nature of this resistance (see Chap. 2.2 and 2.4.3.2) and specific results (Huang et al. 1975, Jensen et al. 1976, Campbell et al. 1977, Bialobok et al. 1980), it is expected that breeding programs to enhance relative tolerance should be successful. From the perspective of agriculture, horticulture, and forestry, the increased breeding of plant types tolerant of photochemical oxidants is encouraged. However, from the broader perspective of environmental protection, such a course would be undesirable. The previous efforts to limit ambient concentrations of pollutants occurred primarily as a result of injury to economically important plant species. If economic losses are successfully reduced through extensive breeding and cultivating of tolerant types, it will lessen the urgency to decrease ambient concentrations, subsequently posing risks to natural vegetation and humans.

2.4.4.2 Fertilization, Irrigation, and Cultural Methods as Protective Measures

The reaction of plants to a given pollutant exposure fluctuates with the external growth factors. Therefore, this variation may be used to reduce the injurious effects of a pollutant exposure. However, a high probability of success should not be expected from these approaches.

Given the contradictory results from studies using various levels of individual plant nutrients (see Sect. 2.4.2.2.3), it is difficult to propose a general fertilization recommendation. However, in areas impacted by photochemical oxidants, a balanced fertilizer regime should be used. This regime would insure sufficient plant nutrients, so that during periods of low or no pollution the plants would be able to compensate for the pollution-induced injury. Several small applications of fertilizer are preferred because a single large application may increase plant sensitivity. Based on the observation of Heck and Dunning (1967), the elevation of the ambient CO_2 concentration in greenhouses should reduce the risk from photochemical oxidants. However, more information about the influence of CO_2 on pollutant uptake and pollutant effects is needed.

The proper supply of water to the plant is a decisive factor (Sect. 2.4.2.1.4.1) in controlling pollutant uptake and therefore pollutant effects. In some areas particularly at risk from photochemical oxidants, tobacco farmers are warned by radio of predicted smog episodes amd encouraged to reduce irrigation to the extent possible (Dean and D. R. Davis 1972). However, the uncertainty of predicting smog episodes and the risk of reducing plant growth by decreasing the water supply when irrigation is decreased limits the usefulness of this preventive measure.

Experience with other air pollutants suggests that the control of the canopy structure by cultural measures could reduce photochemical oxidant effects. Isolated trees and shrubs, as well as plants on the edges of stands, experience a greater pollutant exposure. An early and extensive canopy closure can provide some protection through changes in air movement and by a filtering effect. The influx of polluted air is reduced in lightly thinned or compact canopies (Wentzel 1963). In fruit culture, various training and pruning techniques should be used in areas impacted by photochemical oxidants to develop a compact crown, even though it may have some disadvantages. As a consequence of air pollution, shoot growth rate is reduced sooner than would be expected to occur from aging alone. Pruning as a means of regenerating both the vegetative and reproductive activity of trees or shrubs should be considered, as well as the cultivation of locally adapted fruit types on vigorous root stocks (Guderian 1969).

2.4.4.3 Protectants Against Photochemical Oxidants

Numerous chemicals are active in protecting plants and particular microorganisms from photochemical oxidants (Epstein and Bishop 1977). The chemicals may be applied either as hydrates, aqueous solutions, emulsions, or powders that are either distributed on the plant surface, mixed with the soil, or injected to reduce injurious effects. Protective chemicals can be grouped as:

- antioxidants which may counteract the effects of photochemical oxidant air pollutants,
- biocides that are active against pathogenic fungi and animal pests, as well as photochemical oxidants,
- growth regulators that not only alter plant growth but also influence the sensitivity of plants to photochemical oxidants,
- various other chemical types that induce protection through diverse means.

2.4.4.3.1 Antioxidants

Antioxidants are compounds that prevent the reactions of organic compounds with molecular oxygen. These various chemical substances are used in numerous areas, for example protecting foods from discoloration and spoilage, as synergists in insecticides (Rubin et al. 1980), and in the fabrication of rubber products as a protectant against ozone damage.

Any naturally available reductant such as ascorbic acid can act as a protectant (Barnes 1972b). Ascorbic acid has been used in the Los Angeles area to reduce photochemical oxidant injury on numerous plant species such as bean, celery, romaine lettuce, petunia, and citrus (Freebairn and Taylor 1960). For example, bean plants displayed only about 40% as much foliar injury when sprayed with K-ascorbate (0.01 N) as unsprayed controls. Studies with K- and Ca-ascorbate showed that the effectiveness of ascorbic acid depended on its uptake into the leaf (Freebairn 1963). It is not known if the generally low protectant activity of ascorbic acid (Dass and Weaver 1968) is attributed to the lack of any specific effect on O_3-induced growth alterations, its autooxidation to dehydroascorbic acid (Siegel 1962), or its limited uptake through the roots (Freebairn 1963).

Of the numerous compounds developed as antioxidants, EDU {N-[2-(2-oxo-1-imidazolidinyl) ethyl]-N'-phenylurea} is especially effective in protecting plants from ozone injury. For example, EDU reduced bronzing and delayed leaf drop in navy beans (*Phaseolus vulgaris*) more effectively than the systemic fungicides, carboxin and benomyl (Hofstra et al. 1978) or the antioxidants, piperonyl butoxide and n-propyl gallate (Rubin et al. 1980). The protective effect of EDU, applied as either a foliar spray or soil drench, has been confirmed on petunia and chrysanthemum cultivars and numerous other herbaceous and woody species (Cathey and Heggestad 1982a, b, c). Potato plants treated with either EDU or the fungicide chlorthalonil (tetrachloroisophthalonitrile) exhibited significantly less ozone and *Alternaria solani* blight injury than nonsprayed control plants. Furthermore, EDU was more effective (Bisessar 1982). The combined effects of EDU and the fungicide, chlorthanlonil, were additive in reducing foliar ozone injury and increasing tuber weight.

The mechanism of antioxidant action may be to inhibit microsomal mixed function oxidases (m.f.o.) activity (Rubin et al. 1980). This hypothesis may explain the protection offered against injury by these types of compound. The m.f.o. system requires molecular oxygen for activity. Ozone is a "superactive" form of molecular oxygen; perhaps in plants treated with ozone the m.f.o. system becomes "superactive", abnormally oxygenating, peroxidating, or oxidating many different cellular constituents. Inhibiting the m.f.o. system with antioxidants could prevent this "superactivation", and thereby prevent ozone damage. An alternative hypothesis was proposed by E.H. Lee and J.H. Bennett (1982). Using EDU, they showed that it enhanced the activity of superoxidase dismutase (SOD) and catalase in bean leaves. They suggested that EDU enhanced the basic aerobic nature of the cells by inducing and regulating the oxidant-scavenging enzymes that protect the cells from oxyradicals formed during photosynthesis and where the plant experiences environmental stresses.

Gilbert et al. (1977) proposed that the protective mechanism of Santoflex 13 dust [N-(1,3-dimethylbutyl)-N'-phenyl-p-phenylene-diamine] on apple seedlings, bean, melon, and tobacco plants, and also in the rubber industry resulted from its scavenging of ozone at the surface with the resultant formation of a surface film of ozonized antioxidant. The antioxidant diphenylamine (DPA) is an additional example of a chemical that reduced ozone injury (Elfving et al. 1976). A combined spray application of DPA and the antitranspirant Wilt Pruf on apple leaves was especially effective in diminishing visible and histological injury symptoms.

2.4.4.3.2 Biocides

The fungicides have attained the most significance of the various biocides used in agriculture, horticulture, and forestry for protecting plants from photochemical oxidant injury. The antioxidant properties of insecticides and herbicides have received little attention.

Studies during the 1950's showed that spray or dust applications of thiocarbamate fungicides, zineb (zinc ethylenebisdithiocarbamate), maneb (manganese ethylenebisdithiocarbamate), thiram (tetramethylthiuram disulfide), and ferbam (ferric dimethyldithiocarbamate) reduced the foliar injury of plants exposed to ozonated gasoline vapors or 1-hexane (Kendrick et al. 1954). In laboratory and field studies, the degree of protection was directly related to the concentration of the chemicals (Kendrick et al. 1962). The protection was not systemic, but occurred only if the lower leaf surfaces were well covered, suggesting that the oxidants were deactivated at the surface through a reaction with the chemicals. Thiocarbamate did not reduce weather fleck injury on tobacco (Walker 1967). Initially, plant protection studies focused on reducing ozone entry into the plant. However, present research methods use chemicals to alter the sensitivity of the plants at biochemical and physiological levels.

Systemic fungicides have received special attention, particularly those with antisenescence properties (Klingensmith 1969). Numerous plant species, such as tobacco (G. S. Taylor 1970, Reinert and Spurr 1972), bean (Manning et al. 1973a, 1974, Pellissier et al. 1972a), grapevine (Kender et al. 1973), turf grasses (Moyer et al. 1974a, Papple and Ormrod 1977), azalea (Moyer et al. 1974b), and poinsettia (Manning et al. 1973b) are protected from ozone injury by applications of benomyl (methyl-1-butylcarbamoyl-2-benzimidazolecarbamate) as a spray, soil drench, or soil amendment. However, benomyl did not protect pinto beans from PAN injury (Pell 1974, 1976). High concentrations (60 to 120 μg g^{-1} of benomyl, incorporated into the soil increased the injurious effects of PAN on the leaves of White Cascade petunia (Pell and Gardner 1979). It is possible that the antisenescence properties of benomyl maintained the plants at the optimal stage of sensitivity.

The antisenescence and antiozone properties of benomyl and related fungicide residues in the benzimidazole moiety (Pellissier et al. 1972b, Tomlinson and Rich 1973). Ultrastructural studies of the palisade cells of pinto bean leaves found significant alterations in chloroplasts and other cellular structures, following an ozone exposure (0.25 ppm for 4 h), that did not occur in the benomyl-treated plants (Rufner et al. 1975). The chloroplast, plasmalemma, and tonoplast membranes were almost completely disrupted.

Other fungicides, including carboxin (5,6-dihydro-2-methyl-1,4-oxathiin-3-carboxanilide) and other 1,4-oxathiin derivatives (Rich et al. 1974, Curtis et al. 1975), thiophanate [diethyl 4,4'-o-phenyl-enebis (3-thioallophanate)] and its methyl and ethyl analogs (Moyer et al. 1974a, Seem et al. 1973), triarimol [α-(2,4-dichlorophenyl)-α-phenyl-5-pyrimidinemethanol] and EL 279 [α-(2-chlorophenyl)-α-cyclohexyl-5-pyrimidinemethanol] (Seem et al. 1972) also display some effectiveness against ozone injury.

Herbicides and insecticides can also influence plant response to ozone. Tobacco plants treated with chloramben (3-amino-2,5-dichlorobenzoic acid) lost their ozone tolerance (Carney et al. 1973) while isopropalin (2,6-dinitro-N,N-dipropylcumidine) protected greenhouse-grown flue-cured tobacco plants from ozone injury without significantly altering their leaf chemistry (Sung and Moore 1979). In the field, tobacco plants treated with isopropalin or diphenamid (N,N-dimethyl-2,2-diphenylacetamide) reduced oxidant injury during the first 2 to 4 weeks following transplanting but not later in the season (Reilly and Moore 1982). The finding that insecticides may interact with ozone is particularly important for integrated pest-management programs (Teso et al. 1979). Lannate 90 SP (methomyl) and ozone interacted to produce a synergistic injury response, while treatment of bean leaves with Spectracide 25 EC (diazinon) significantly reduced ozone injury. Biocides may interact with themselves and ozone to influence plant response. For example, benomyl and carboxin counter the nematicide-induced stimulation of ozone injury to plants (P. M. Miller et al. 1976).

2.4.4.3.3 Growth Regulators

A series of studies have established that the phytohormones – indole-acetic acid (IAA), gibberellic acid (GA), 6-benzylamine purine (BA), and abscisic acid (ABA) – interact with photochemical oxidants (Hull et al. 1954, Ordin and Probst 1962, Siegel 1962, Runeckles and Resh 1975b, Adedipe and Ormrod 1972, Fletcher et al. 1972). The cytokinin BA was more active in moderating the ozone-induced growth suppression and loss of chlorophyll in radish leaves than either IAA or GA (Adedipe and Ormrod 1972). In contrast, an application of 6-furfurylaminopurine, a cytokinin similar to BA, increased the ozone susceptibility of tobacco plants (T.T. Lee 1966). Foliar applications of benzimidazole, N-6-benzyladenine, and kinetin retarded the senescence of bean leaves, inhibited the loss of free sterols, and protected the plants against ozone injury (Tomlinson and Rich 1973). The protection against ozone was achieved only when the chemical concentrations were sufficient to inhibit the degradation of chlorophyll. Compounds with anti-senescent properties may retard the ozone-induced premature senescence (Walker 1967, Pellisier et al. 1972b, Seem et al. 1972). The observed protection induced by these chemicals was not, in general, great enough for these compounds to attain significant practical use as protectants against photochemical oxidants. However, individual phytohormones are well suited for investigating some of the detailed questions concerning the mode of ozone action in plants (Runeckles and Resh 1975 b, Tomlinson and Rich 1973).

Growth retardants are effective in reducing ozone injury on plants. For example, CBBP (2,4-dichlorobenzyltributyl phosphonium chloride) and SADH (succinic acid 2,2-dimethyl hydrazide) reduced ozone injury on petunia (Cathey and Heggestad 1972). However, to be effective, the SADH concentration had to be at least twice the concentration used to retard stem elongation. Adding L-ascorbic acid and a wax coating to the spray solution increased the protection afforded by SADH. Chemicals that did not retard petunia growth, for example chloromequat [(2-chloroethyl) trimethyl ammoniumchloride] and ancymidol [α-cyclopropyl-α(4-methoxyphenyl)-5-pyrimidine-methanol] failed to protect plants against ozone. Applications of ancymidol and chloromequat reduced ozone and sulfur dioxide induced foliar injury in eight poinsettia cultivars (Cathey and Heggestad 1973). Increased air pollution tolerance was believed to be partly the result of a series of morphological changes and also reduced penetration of O_3 and SO_2 into the susceptible tissues.

2.4.4.3.4 Other Chemicals

On the leaf surface, various chemicals protect plants against ozone injury. Folicote (an emulsion of paraffinic hydrocarbon waxes), an antitranspirant, provided good protection against ozone injury in four genera of the Solanaceae family (Knapp and Fieldhouse 1970) and beans (Pellisier et al. 1972 a, c). Spray coverage of both leaf surfaces was required to obtain complete protection. The Folicote treatment presumably reduced ozone uptake into the leaf and therefore foliar injury was reduced.

Dusts of various particulate substances – charcoal, diatomaceous, ferric oxide – reduced ozone injury on tobacco plants (Jones 1963). These various compounds apparently cause a destruction of ozone at the leaf surface.

2.4.4.3.5 Summary of the Practicality of Chemical Protectants

Even though examples were given to confirm that various chemicals can reduce ozone injury, this should not lead to the conclusion that they are practical for preventing ozone and PAN-induced injury. Regardless of the activity of a particular chemical, practical conditions limit their use:

- The protective action depends on the chemical (Ormrod 1978), its time of application (Curtis et al. 1975, Papple and Ormrod 1977), plant species and cultivars (Manning et al. 1974), as well as their developmental stage (Pell 1974, 1976). Also, the chemical has a limited period of effectiveness (Manning et al. 1973a, b, P. M. Miller et al. 1976).
- The compound must be applied from a few to many times during the plant growth period, because the effective period of the chemical is limited (Kender et al. 1973). In this case, the price of the particular chemical and its cost of ap-

plication limit its possible uses, especially if these chemicals do not also function as biocides in plant cultivation practices.

- Because of price, the application of protectants is frequently possible only during smog episodes. However, there is great uncertainty in predicting smog events.
- The protectants are not equally effective for all air pollutants. Benomyl, for example, can prevent ozone injury (see Sect. 2.4.4.3.2) but not injury from other oxidants (Pell and Gardner 1979).
- Finally, the use of protectants is limited because they may have other toxic effects or lead to the formation of compounds, in food and fodder plants, that are toxic to animals and humans (Manning et al. 1973a, Klingaman and Link 1975, Papple and Ormrod 1977).

Chapter 2.5 Effects of Pollutant Combinations

Photochemical oxidants are a widely distributed, complex array of compounds that may co-occur on either a local or regional scale with other pollutants emitted from various sources. Consequently, plants grown in the natural environment are seldom exposed to single pollutants, but rather to varying combinations and concentrations of several atmospheric components. Thus, when assessing the responses of plants in the field, the effects of pollutant combinations should be considered in addition to the impacts of individual pollutants.

2.5.1 Pollutant Combinations

There are several reviews of the effects of pollutant combinations on vegetation (e. g., Reinert et al. 1975, Ormrod 1982). Preceding the publication of Menser and Heggestad (1966), information on the effects of pollutant mixtures was limited. However, their report provided the initial impetus for extensive research on this topic. They showed that Bel W3 tobacco plants exposed to ozone (0.03 ppm) or sulfur dioxide (0.24–0.28 ppm) were uninjured, but that substantial foliar injury occurred when the plants were exposed to both gases simultaneously. The authors called this response a synergistic effect. Subsequent studies confirmed this finding and extended the observations to show that pollutant combinations can influence not only foliar injury responses but numerous other plant processes. These plant responses to pollutant combinations depend not only on the components of the mixtures, but also the concentrations of the individual pollutants, their temporal succession, and they may be modified by external and internal growth factors.

This chapter will focus on the effects of ozone in combination with other pollutants and the factors that modify the response. The majority of the published research has discussed ozone-sulfur dioxide combinations; studies concerning the effects of ozone with other pollutants are limited. In describing the effects of pollutant combinations, the authors will follow the same logical approach as used for discussing ozone and PAN in the other chapters – mode of action, diagnosis, and factors influencing plant response.

In describing the effects of pollutant combinations, a number of terms have attained general, though not universal, acceptance to illustrate the range of responses that may occur with the simultaneous or alternating occurrence of two or more pollutants. The following concepts were not derived from an understanding of the toxicology of the mixtures but rather from a cursory review of statistical terms:

- The plant response to the two pollutants is independent (no interaction). The response to pollutant A is not altered by a concurrent exposure to pollutant B.
- The plant response to the pollutant mixtures is additive. The response to the mixture is similar to the summed effects of the individual pollutants.
- The plant response may be less than additive (antagonistic). The response to the pollutant combination is less than the summed response to the individual pollutants.
- The plant response may be greater than additive (synergistic). The response to the pollutant combination is greater than the summed effects of the individual pollutants.

2.5.2 Ozone and Sulfur Dioxide

2.5.2.1 Mode of Action

Both ozone and sulfur dioxide influence various biochemical and physiological processes with many of their responses associated with changes in membrane permeability. Low concentrations of ozone and sulfur dioxide induce similar stress symptoms, including premature senescence. Slightly higher concentrations can affect individual metabolic processes. The primary biochemical responses of ozone are based on its high oxidation potential and its tendency to form free radicals; SO_2 can also form free radicals but it may also be a weak oxidant (Puckett et al. 1973), a weak acid, or as sulfite (or sulfate) perturb ion relations and enzyme regulation (Ziegler 1973). Because the mode of action of the individual pollutants is not well understood, it is difficult to predict the mode of action of pollutant combinations. This situation is made worse by the limited research in this area. However, plant respone to pollutant mixtures is controlled by the same general processes that control response to ozone: the amounts of pollutants that diffuse into the plant, the rate of scavenging reactions that reduce cellular concentrations protecting reactive sites, and the rate and amount of homeostatic processes (Chap. 2.2; also see Fig. 2.2).

2.5.2.1.1 Pollutant Uptake

The importance of stomatal control and pollutant uptake in multiple pollutant studies became apparent from earlier work on the uptake of single gases and the observations that pollutants that did not enter the plant were generally not phytotoxic. The implied hypothesis has been that the pollutant combinations interfered with normal stomatal mechanisms, allowing greater pollutant uptake than for the individual pollutants.

Leaf resistance data were compared for species that responded synergistically to the combination of ozone plus sulfur dioxide (radish, cucumber) and species that did not (soybean, white bean) (Beckerson and Hofstra 1979 a, b). For both groups, leaf resistance tended to decrease following exposure to sulfur dioxide and increase following exposure to ozone; the two pollutants combined caused a greater increase in leaf resistance than ozone alone. The stomata of ozone-sensitive petunia cultivars closed more rapidly in the presence of ozone and sulfur dioxide mixtures, while the stomata of ozone-insensitive cultivars closed more slowly (Elkiey and Ormrod 1979 c). However, neither the synergistic nor antagonistic response could be explained by the changes in stomatal behavior. Peas reacted synergistically to ozone-sulfur dioxide mixtures, but neither the amount of injury nor the synergistic response was associated with stomatal resistance (Olszyk and Tibbitts 1981).

The uptake of ozone and sulfur dioxide singly and in combination was studied in three petunia cultivars that differed to the pollutants in sensitivity (Elkiey and Ormrod 1980 a, 1981 a). Pollutant uptake for the single gases and the mixtures was generally greater for the more sensitive petunia cultivars. The uptake of ozone and sulfur dioxide was greater from single gas exposures than for the uptake of the same gases from the mixtures (Elkiey and Ormrod 1981 a). Uptake rates decreased throughout the day and from day to day, but the reduction was larger for the combinations than from the single gases. Similar results were found with *Poa pratensis* cultivars that differed in pollutant sensitivity (Elkiey and Ormrod 1981 b).

Studies of the sulfate content of petunia tissues following exposure to SO_2 with and without ozone confirmed the previous reports of stomatal behavior and gas uptake (Elkiey and Ormrod 1981 c). At 90% relative humidity, the plants contained less sulfate after being exposed to a mixture of SO_2 plus ozone than from SO_2 alone. Similar trends were observed in the older leaves when the plants were exposed at 50% relative humidity.

Data from studies of stomatal behavior, pollutant uptake, and tissue sulfate content all confirm that the uptake of individual gases from pollutant mixtures is less than the uptake rates of the individual gases alone. However, this trend should not be interpreted as a contradiction in relation to the possible synergistic effects of pollutant combinations. Even though the uptake of individual gases from mixtures is less, the total number of molecules of absorbed pollutants is greater from mixtures; this probably contributes to their greater phytotoxicity.

2.5.2.1.2 Perturbation

The modes of action for ozone and sulfur dioxide, individually, are not well understood or documented; the modes of action for pollutant combinations are even more obscure. The few relevant studies have focused on changes in membrane permeability because the primary site of ozone action is thought to be plant membranes. Three petunia cultivars that varied in their ozone sensitivity were used to study the effects of ozone and sulfur dioxide, singly and in combination, on potassium and total electrolyte leakage as indicators of membrane permeability (Elkiey and Ormrod 1979 d). Potassium efflux was not increased by a 5-h exposure, but total electrolytes were elevated in the intermediate and sensitive petunia cultivars exposed to ozone. Sulfur dioxide had no significant effect, while the pollutant combination $(O_3 + SO_2)$ increased the electrolyte leakage of all three cultivars. The effects of the pollutants individually and in combination on the leaf water potential of the three cultivars were also studied (Elkiey and Ormrod 1979 a). Ozone rapidly decreased the leaf water potential. The decline was largest in the cultivar intermediately sensitive to ozone; sulfur dioxide had no significant effect. In contrast, the pollutant mixture caused a smaller decrease and the rate of decline was less rapid than for ozone alone.

Beckerson and Hofstra (1980) studied the effects of pollutant combinations on membrane permeability, as measured by electrolyte leakage, in several species that showed either an "antagonistic" or "synergistic" injury response to the combinations. Ozone significantly increased electrolyte leakage from soybean and white bean (species that displayed an antagonistic foliar injury response) prior to visible injury. However, the mixture did not stimulate electrolyte leakage. Sulfur dioxide apparently moderated the ozone effects on the membrane. Ozone had no significant effect on electrolyte leakage from either radish or cucumber (species that displayed a synergistic foliar injury response), but the combination of ozone and sulfur dioxide significantly increased leakage.

Changes in membrane permeability as determined by changes in electrolyte leakage or leaf water potential do not show definite patterns. This may mean that potassium or total electrolytes were not the best indicators of membrane impacts or that the plasma membrane is not the membrane system most sensitive to pollutant combinations.

2.5.2.1.3 Injury

Physiological and biochemical changes induced by mixtures of ozone and sulfur dioxide have been only sparingly studied. The reduction of net photosynthesis in alfalfa (*Medicago sativa*) exposed to ozone and SO_2 or ozone and NO_2 was additive when compared to the reductions induced by the individual pollutants (Bennett and Hill 1974). In contrast, the reduction in net photosynthesis for sugar maple and white ash was more than additive after two days of exposure (Carlson 1979). The net photosynthesis of *Vicia faba* was reduced when the ozone concentration exceeded 0.05 ppm for 4 h but, unless visual injury developed, it returned to the control level following the fumigation (Black et al. 1982). In this study the addition of 0.04 ppm SO_2 caused a significantly greater reduction in photosynthesis than ozone alone. The SO_2 had the effect of reducing the ozone concentration that was inhibitory to photosynthesis. The addition of SO_2 had no effect on the photosynthetic recovery rate. In studies with sunflower, neither O_3 nor SO_2 inhibited photosynthesis, but mixtures of the gases at the same concentrations as used previously significantly reduced photosynthesis (Furukawa and Totsuka 1979).

In white bean, which displayed an antagonistic response to $O_3 + SO_2$ mixtures, the chlorophyll content was reduced by exposure to the mixture but the reduction was detected two days later than it was observed in exposures to ozone alone (Beckerson and Hofstra 1979 c). However, after 5 days the chlorophyll loss was greater in the pollutant combination than from the individual pollutants. The pollutant combination had no significant effects on protein or RNA levels. The carbohydrate and protein contents of *Ulmus americana* leaves were reduced within 24 h following exposure to mixtures of $O_3 + SO_2$, but within 4 weeks after exposure the concentrations were comparable to control leaves (Constantinidou and Kozlowski 1979a).

Poplar leaves exposed to O_3 and SO_2 singly and in combination showed no visible foliar symptoms (Krause and Jensen 1978). However, scanning electron micrograph studies showed crystalline-like inclusions in the bundle-sheath extension cells exposed to the combination. Globules were observed in mesophyll cells and the chloroplasts appeared distorted and damaged only from the mixture. Scanning electron micrograph studies of poplar leaf surfaces showed that pollutant combinations induced changes that were not detected following exposure to the single gases (Krause and Jensen 1979).

2.5.2.2 Effects on Plants

In addition to metabolic studies, numerous other reports have documented the effects of mixtures of O_3 and SO_2 on visible injury and plant growth and yield. As in the metabolic studies, these effects have shown responses ranging from antagonistic to synergistic.

2.5.2.2.1 Foliar Injury

The types of responses and the magnitudes of effects of pollutant exposures on visible injury are shown for selected species in Table 2.14. Brief exposures of tobacco Bel W3 to mixtures of SO_2 (0.24–0.28 ppm) and ozone (0.03 ppm) produced visible injury, while exposure to the individual pollutants did not (Menser and Heggestad 1966). In a series of experiments, Tingey et al. (1973d) studied the effects of brief exposures of ozone and SO_2 on 11 different plant species. Tobacco, radish, and alfalfa exhibited a greater than additive response; with cabbage, broccoli, and tomato, the injury was additive or less. The concentrations used in these studies and others (Reinert et al. 1970, Jacobson and Colavito 1976) were in the same range as measured ambient concentrations of combined pollutants. Begonias exposed to low concentrations of the combined gases for 5 days exhibited a synergistic response for visible injury (Gardner and Ormrod 1976). Studies with pea (Olszyk and Tibbitts 1982) showed that the concentration and duration of exposure to the mixture had to be significantly reduced to maintain a degree of injury similar to that induced by the individual pollutants. However, in soybean and white bean the injury intensity caused by the mixture was reduced and the rate of symptom development delayed compared to effects caused by ozone alone (Hofstra and Ormrod 1977).

The chlorotic dwarf syndrome of eastern white pine (see Chap. 2.3.1.1) was produced by chronic exposures to SO_2 and O_3 singly or in combination (Dochinger et al. 1970). Following an 8-week exposure to intermittent concentrations of O_3 and SO_2 (each at 0.10 ppm), a sensitive clone of eastern white pine displayed 16% needle necrosis. Exposures to the individual pollutants caused only

3 to 4% needle injury (Dochinger et al. 1970). Eastern white pine clones from visibly injured mother trees from pollution-impacted areas displayed a greater than additive amount of injury following exposure to a mixture of ozone (0.025 ppm) and sulfur dioxide (0.05 ppm) (Houston 1974). A SO_2-sensitive clone of eastern white pine exposed to O_3 (0.05 ppm for 2 h) developed no visible injury, exposure to SO_2 (0.05 ppm, 2 h) caused a moderate amount of injury, exposure to the mixture (each at 0.05 ppm) caused less injury. Sequential exposures to O_3 and SO_2 for 2 h each followed by an exposure to the combined pollutants 24 h later caused the most injury (Costonis 1973). Scots pine exposed to the combined pollutants displayed less injury than plants exposed to SO_2 alone; O_3 did not induce foliar injury (Nielsen et al. 1977).

2.5.2.2.2 Growth Effects

In addition to the mixture effects on physiological processes and foliar injury, most recent studies have focused on the mixture effects on plant growth and yield. Growth and yield studies have been conducted in both greenhouse and field exposure chambers. Most of the studies used O_3 but there have been a few reports of the interaction between ambient photochemical oxidants and SO_2. Selected examples of the effects of O_3 and SO_2 mixtures on the growth of various plant species are shown in Table 2.15.

Mixtures of O_3 (0.05 ppm) and SO_2 (0.05 ppm) for 8 h day^{-1}, 5 days week^{-1} for 5 weeks reduced the growth and yield of radish (Tingey et al. 1971). The pollutant mixtures caused a greater than additive reduction in shoot growth and a less than additive reduction in root growth. In other studies using the same pollutant concentrations, growth reductions for tobacco Bel-W3 (Reinert et al. 1969), soybean (Tingey et al. 1973 b), and alfalfa (Tingey and Reinert 1975) were generally additive. Field studies with alfalfa (Neely et al. 1977) showed that exposure to mixtures of O_3 (0.05 ppm, 7 h day^{-1} and SO_2 (0.05 ppm, 24 h day^{-1}) for 68 days reduced shoot, stubble, and root dry weights to a similar extent as those caused by either pollutant alone. Alterations were also observed in various metabolites. Growth reduction in soybean exposed to O_3 (0.10 ppm) and/or SO_2 (0.10 ppm) for 6 h day^{-1} throughout the growing season in field exposure chambers were approximately additive (Heagle et al. 1974). Additional studies with soybean (Heagle et al. 1983), in which the plants were exposed to O_3 (0.025 to 0.125 ppm, 7 h seasonal mean for 111 days) and SO_2 (0.0 to 0.367 ppm, 4 h seasonal mean for 101 days) indicated that the resultant growth reductions were the consequence of the independent action of the two gases. The shoot fresh weight of white bean exposed to the combination of O_3 (0.15 ppm) and SO_2 (0.15 ppm) for 5 or 10 days was reduced less by the combination than by O_3 acting alone (Hofstra and Ormrod 1977). Mixtures of O_3 and SO_2 had a greater impact on radish hypocotyl than foliage growth; when low concentrations of NO_2 were added into the mixture it tended to intensify the yield reduction even though NO_2 alone did not decrease growth (Reinert and Gray 1981, Reinert et al. 1982).

In contrast to herbaceous plants which tended to display additive or synergistic growth responses, the majority of the studies on woody vegetation have shown less than additive or antagonistic responses on plant growth from the pollutant

Table 2.14. Visual injury on various plants induced by the joint action of ozone and sulfur dioxide

Plant species	Concentration (ppm)		Exposure duration (h)	Effect		References
	O_3	SO_2		Intensity (%)	Type[a]	
Tobacco *(Nicotiana tabacum)*						
Bel W3	0.03	0.24	2	15	+	Menser and
	0.03	0.28	4	41	+	Heggestad (1966)
Bel-B	0.03	0.24	2	9	+	
	0.03	0.28	4	23	+	
Bel-B	0.05	0.25	4	4	+	Tingey et al. (1973d)
	0.10	1.00	4	34	+	
Bel W3	0.05	0.25	4	17	+	
	0.10	1.00	4	48	+	
	0.05	0.50	4	60	+	
	0.10	0.10	4	95	0	
	0.10	0.25	4	85	+	
	0.10	0.50	4	96	+	
Tobacco *(Nicotiana glutinosa)*	0.03	0.45	2	26	+	Grosso et al. (1971)
	0.03	0.45	4	39		
Tobacco *(Nicotiana rustica* var. brasilia*)*	0.03	0.34	2	35	+	
	0.03	0.34	4	33		
Bean *(Phaseolus vulgaris)*	0.19	1.70	0.8	24	+	Matsushima (1971)
	0.37	1.60	1.0	24	*	
Lima Bean *(Phaseolus limensis)*	0.05	0.25	4	0	*	Tingey et al. (1973d)
	0.10	1.00	4	9	*	
Soybean *(Glycine max)*, 8 varieties	0.30	1.00	6	13–38	*	Reinert et al. (1975)
Soybean *(Glycine max)*	0.15	0.15	6 h d^{-1}; 5 d	1	–	Hofstra and Ormrod
	0.15	0.30	6 h d^{-1}; 5 d	0	–	(1977)
Soybean *(Glycine max)*	0.50	0.25	4	0	*	Tingey et al. (1973d)
	0.10	1.00	4	1	*	
White Bean *(Phaseolus vulgaris)*	0.15	0.15	6 h d^{-1}; 5 d	1	–	Hofstra and Ormrod
	0.15	0.30	6 h d^{-1}; 5 d	0	–	(1977)
Pea *(Pisum sativum)*	0.11	0.11	4	6	+	Olszyk and Tibbitts (1982)
Broccoli	0.05	0.25	4	3	*	Tingey et al. (1973d)
(Brassica oleracea	0.05	0.50	4	17	+	
var. botrytis*)*	0.10	0.10	4	34	+	
	0.10	0.25	4	11	0	
	0.10	0.50	4	19	0	
	0.10	1.00	4	32	+	
Cabbage	0.05	0.25	4	0	*	Tingey et al. (1973d)
(Brassica oleracea	0.05	0.50	4	4	*	
var. capitata*)*	0.10	0.10	4	22	0	
	0.10	0.25	4	14	0	
	0.10	0.50	4	54	+	
	0.10	1.00	4	28	0	

Table 2.14 (continued)

Plant species	Concentration (ppm)		Exposure duration (h)	Effect		References
	O_3	SO_2		Intensity (%)	Type[a]	
Alfalfa	0.05	0.25	4	0	*	Tingey et al. (1973d)
(Medicago sativa)	0.05	0.50	4	2	*	
	0.10	0.10	4	24	+	
	0.10	0.25	4	21	+	
	0.10	0.50	4	60	+	
	0.10	1.00	4	3	*	
Radish	0.05	0.25	4	7	*	Tingey et al. (1973d)
(Raphanus sativus)	0.05	0.50	4	7	*	
	0.10	0.10	4	50	+	
	0.10	0.25	4	22	+	
	0.10	0.50	4	46	+	
	0.10	1.00	4	45	0	
Tomato *(Lycopericon*	0.05	0.25	4	3	*	Tingey et al. (1973d)
esculentum)	0.05	0.50	4	1	*	
	0.10	0.10	4	50	−	
	0.10	0.25	4	10	0	
	0.10	0.50	4	13	0	
	0.10	1.00	4	2	*	
Spinach	0.05	0.25	4	0	*	Tingey et al. (1973d)
(Spinacia oleracea)	0.10	1.00	4	1	*	
Onion *(Allium cepa)*	0.50	0.25	4	0	*	Tingey et al. (1973d)
	0.10	1.00	4	3	*	
Bromegrass	0.05	0.25	4	0	*	Tingey et al. (1973d)
(Bromus inermis)	0.10	1.00	4	4	*	
Begonia *(Begonia* sp.)	0.30	1.20	1 d	43	+	Gardner and Ormrod (1976)
Grape *(Vitis*	0.40	0.40	4	47	+	Shertz et al. (1980a)
labruscana)	0.80	0.80	4	84	−	
Eastern white pine *(Pinus strobus)*	0.10	0.10	8 h d^{-1}; 8 w	16	+	Dochinger et al. (1970)
Eastern white pine *(Pinus strobus)*	0.05	0.05	2	6.5 (Index 1 to 7)	−	Costonis (1973)
Eastern white pine *(Pinus strobus)*	0.025	0.05	6	16	+	Houston (1974)
Trembling aspen *(Populus tremuloides)*	0.05	0.20	3	11[b]	+	Karnosky (1976)

[a] Type of response
  + = greater than additive
  0 = additive
  − = less than additive
  * = not determined or not significant
[b] Injury expressed as percentage of leaves injured/plant

Table 2.15. The effect of the joint action of ozone and sulfur dioxide on plant growth

Plant species	Concentration (ppm)		Exposure duration	Effect			References
				Growth[a] reduction		Type[b]	
	O_3	SO_2					
Tobacco *(Nicotiana tabacum var. Bel-W3)*	0.05	0.25	8 h d^{-1}; 5 d w^{-1}; 4 w	32% 49%	SDW RDW	0 +	Tingey and Reinert (1975)
Soybean *(Glycine max)*	0.10	0.10	6 h d^{-1}; 43 d 92 d 133 d	53% 53% 73% 51% 66%	SFW SFW SFW pod number seed weight/ plant	0 0 0 0 0	Heagle et al. (1974)
Soybean *(Glycine max)*	0.25	0.25	4 h d^{-1}; 3 d w^{-1}; 11 w	64%	PDW	0	Reinert and Weber (1980)
Alfalfa *(Medicago sativa)*	0.05	0.05	8 h d^{-1}; 5 d w^{-1}; 2 w	18% 24%	SDW RDW	− −	Tingey and Reinert (1975)
Radish *(Raphanus sativus)*	0.45	0.45	4 h	16% 70%	SDW[c] RDW	0 0	Tingey and Reinert (1975)
	0.05	0.05	8 h d^{-1}; 5 d w^{-1}; 5 w	10% 55%	SDW RDW	0 0	Tingey et al. (1971)
Begonia *(Begonia sp.)* Early vegetative stage	0.15 0.15 0.30	0.60 0.60 1.20	5 d 5 d } succes- 1 d } sively	8% 40%	SDW[d] SDW	* +	Gardner and Ormrod (1976)
	0.15 0.15 0.30	0.60 0.60 1.20	5 h 5 d } succes- 1 d } sively	5% 23%	SDW[d] SDW	* +	
Prefloral stage	0.15 0.15 0.30	0.6 0.6 1.2	5 d 5 d } succes- 1 d } sively	11% 8%	SDW[d] SDW	* *	
	0.15 0.15 0.30	0.6 0.6 1.2	5 d 5 d } succes- 1 d } sively	10% 10%	SDW[e] SDW	* *	
Kentucky bluegrass *(Poa pratensis)*	0.15 (+0.15 ppm NO_2)	0.15	10 d (ozone: 6 h d^{-1}; 10 d)	1–50%	leaf area	0	Elkiey and Ormrod (1980b)
Red top *(Agrostis alba)*	0.15 (+0.15 ppm NO_2)	0.15	10 d (ozone: 6 h d^{-1}; 10 d)	28%	leaf area	0	
Creeping bentgrass *(Agrostis palustris)*	0.15 (+0.15 ppm NO_2)	0.15	10 d (ozone: 6 h d^{-1}; 10 d)	26%	leaf area	0	
Colonial bentgrass *(Agrostis sp.)*	0.15 (+0.15 ppm NO_2)	0.15	10 d (ozone: 6 h d^{-1}; 10 d)	90/70%	leaf area	+/0	
Red fescue *(Festuca rubra)*	0.15 (+0.15 ppm NO_2)	0.15	10 d (ozone: 6 h d^{-1}; 10 d)	90/70%	leaf area	+/0	
Perennial ryegrass *(Lolium perenne)*	0.15 (+0.15 ppm NO_2)	0.15	10 d (ozone: 6 h d^{-1}; 10 d)	13%	leaf area	0	

Table 2.15 (continued)

Plant species	Concentration (ppm)		Exposure duration	Effect		References
				Growth[a] reduction	Type[b]	
	O_3	SO_2				
Eastern white pine (*Pinus strobus*)	0.025	0.5	6 h			Houston (1974)
Sensitive clone				52%	needle length —	
Tolerant clone				32%	needle length —	
Scotch pine (*Pinus sylvestris*), 17 clones	2.0	1.0	$6\,h\,d^{-1}$; 4 d	4–39%	shoot height —	Bialobok et al. (1980)
Yellow poplar (*Liriodendron tulipifera*)	0.1	0.1	$12\,h\,d^{-1}$; 50 d	7%[f]	—	Jensen (1981b)
Cottonwood (*Populus deltoides*)	0.1	0.1	$12\,h\,d^{-1}$; 50 d	55%[f]	—	Jensen (1981b)
Hybrid poplar (*Populus deltoides × P. trichocarpa*)	0.25	0.50	$12\,h\,d^{-1}$; 24 d	59%	PDW —	R. D. Noble and Jensen (1980)
Poplar (*Populus × euramericana*)	0.035	0.06	SO_2, $24\,h\,d^{-1}$; O_3, $12\,h\,d^{-1}$; 28 d	12–25%	PDW 0/+	Mooi (1981)
Poplar (*Populus maximowiczii*)	0.035	0.06	SO_2, $24\,h\,d^{-1}$; O_3, $12\,h\,d^{-1}$; 28 d	15–27%	PDW 0/+	Mooi (1981)
Apple (*Malus domestica*)	0.4	0.4	4 h	45%	shoot growth +	Shertz et al. (1980b)

[a] SFW = shoot fresh weight
  SDW = shoot dry weight
  RDW = root dry weight
  PDW = plant dry weight

[b] Type of response:
  + = greater than additive
  0 = additive
  − = less than additive
  * = not determined or not significant

[c] 11 days following exposure
[d] 2 weeks following exposure
[e] 8 weeks following exposure
[f] Relative growth rate

mixtures. Although the combination of O_3 and SO_2 caused significant growth reduction in cottonwood, yellow poplar, and hybrid poplar, the reductions were less than expected from the responses of the individual gases (Jensen 1981 b, R. D. Noble and Jensen 1980). One week following a single exposure to SO_2 or O_3 singly or in combination, *Ulmus americana* leaf expansion was reduced by both the sulfur dioxide and combination treatments (Constantinidou and Kozlowski 1979 b). However, within 2 weeks the growth rates in those treatments had returned to the control rate, indicating that the plants had recovered, but the final leaf area was still reduced. In white pine where the chlorotic dwarf syndrome has been associated with exposures from pollutant mixtures, growth effects tended to be greater than additive (Houston 1974). In hybrid poplar cuttings exposed to chronic levels of O_3 and SO_2 in combination and individually, growth analysis indicated that the growth rate and net assimilation rate displayed a synergistic reduction when compared to the effects of the individual pollutants (Jensen 1981 c). In poplars, chronic exposures to pollutant mixtures stimulated leaf abscission (R. D. Noble and Jensen 1980, Mooi 1981). The enhanced leaf drop may, in part, be associated with the observed growth reduction.

Interactive effects were also observed when SO_2 was added to ambient photochemical oxidants. Oshima (1978) found that the addition of SO_2 to ambient oxidant caused a significant reduction in kidney bean yield and plant biomass even though the same oxidant dose without SO_2 caused no significant reduction. The yield of snap beans displayed a significantly greater reduction when SO_2 was added to the chambers than from ambient oxidant alone (Heggestad and R. H. Bennett 1981). However, the authors indicated that the SO_2 concentration was unusually high and would not represent ambient conditions. The yield of potatoes was similarly reduced from ambient oxidant, with or without added SO_2. However, the combination treatment significantly altered the nitrogen content of the tubers (Foster et al. 1983).

At concentrations of 0.05 or 0.10 ppm for O_3 and SO_2, the foliar injury response tended to be synergistic or at least additive. Chronic exposures to mixtures of $O_3 + SO_2$ reduced the growth of woody and herbaceous plants to varying amounts. Although the mixtures caused significant growth reductions, the effects on woody plants tended to be less than additive, while the effects on herbaceous plants ranged from less than additive to additive (i. e., the effects of the combination could be estimated from the effects of the individual gases). In practice, yield reductions have been observed in the concentration range of 0.05 ppm ozone and 0.05 ppm sulfur dioxide. In determining the environmental significance of exposures to pollutant mixtures, special consideration must be given to the concentrations and durations of exposures. The exposure conditions used in some studies may provide useful scientific information but they deviate significantly from ambient exposure conditions restricting the usefulness of the data for direct evaluation of environmental problems.

2.5.2.3 Diagnosis of Effects

The factors to be considered when attempting to diagnose a probable cause of injury were discussed in Chapter 2.3 as were detailed descriptions of ozone and PAN injury. The initial report of O_3–SO_2 synergism on tobacco described the injury from the mixture as tiny white flecks scattered randomly over the upper leaf surface and stated that the symptoms resembled O_3 injury (Menser and Heggestad 1966). In other species, the injury was characterized as stipple or small discrete interveinal necrotic and/or chlorotic lesions on the upper leaf surface (Tingey et al. 1973 d). An intense red pigmentation, stippling, and silvering of the interveinal areas were observed on the lower leaf surface. The similarity between the symptoms induced by O_3 or the mixture of O_3 and SO_2 have been observed in numerous species (e. g., Tingey et al. 1973 d, Olszyk and Tibbitts 1982, Reinert and Nelson 1980). In bean, the symptoms ranged between tan interveinal necrosis associated with bifacial injury to silvering of the lower leaf surface, depending on temperature (C. A. Miller and D. D. Davis 1981). However, in white bean O_3 induced a bronze flecking on mature leaves, while SO_2 caused a bifacial necrosis (Hofstra and Ormrod 1977). In contrast, the symptoms of the combined gases appeared as a yellow interveinal chlorosis, becoming apparent several days later than the ozone symptom would. In soybean, characteristic ozone injury appeared following exposure to the combined gases, but the symptoms appeared several days later than on plants exposed to ozone alone (Hofstra and Ormrod 1977). The undersurface of newly developed petunia leaves became glazed after exposure to a combination of O_3 and SO_2, which was similar to the description of PAN injury (Lewis and Brennan 1978). The injury followed a similar distribution with leaf development as with PAN. In trembling aspen, the foliar symptoms caused by the two-pollutant combination were similar to those induced by the individual pollutants (Karnosky 1976). The foliar symptoms of chlorotic dwarf of white pine were similar whether induced by ozone or the mixture of ozone and sulfur dioxide (Dochinger et al. 1970).

The general similarity between the symptoms induced by ozone and the mixture of ozone and sulfur dioxide may explain why the "oxidant type" of foliar injury may occur without the occurrence of elevated ozone concentrations. This similarity may also complicate the field diagnosis and make it difficult to assign the cause to a specific pollutant based on visible injury alone.

2.5.2.4 Factors Influencing Plant Response

Various external and internal growth factors would be expected to influence the magnitude and possibly the type of plant response to pollutant mixtures, just as they do for the individual pollutants. There have been only a few studies concerning the influence of plant or environmental factors as they influence plant response to pollutant mixtures. However, it is anticipated that developmental or environmental factors that alter pollutant uptake, pollutant scavenging, or homeostatic processes would also be expected to influence the response to pollutant combinations.

2.5.2.4.1 Concentration and Duration

The magnitude of plant responses to pollutant combinations tends to increase with increasing concentration of each gas, as is true when the gases are applied individually. However, the type of response is not consistent with concentration. Menser and Heggestad (1966) observed foliar injury on tobacco below the injury threshold of each gas. In contrast, MacDowall and Cole (1971) found the synergistic response below the injury threshold for SO_2 but not below the O_3 injury threshold. Synergism was most pronounced at the injury threshold doses for each pollutant. Jacobson and Colavito (1976) showed that the extent of synergistic or antagonistic responses in tobacco and bean plants depended not only on the species, but also on the concentrations of each gas. Studies with soybean (Heagle and Johnston 1979) showed that plants could display both synergistic and antagonistic responses, depending on the extent of foliar injury; antagonistic responses predominated when the injury was extensive. Similar results were obtained with pinto bean (Matsushima 1971). Studies on grape and apple (Shertz et al.

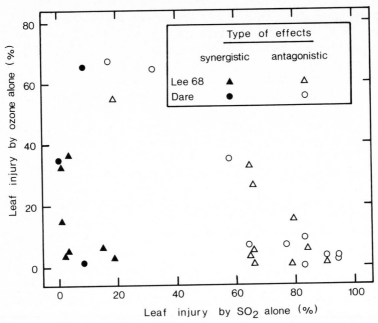

Fig. 2.15. The relationship between the foliar injury intensity of two soybean cultivars as affected by various concentrations of O₃ and SO₂ and the nature of their interactive effects. Each symbol shows the coordinates for mean foliar injury, from O₃ and SO₂ singly, at which significant injury responses occurred in the corresponding mixture. (Heagle and Johnston 1979)

1980 a, b) confirmed the observation that plants typically displayed an antagonistic response to the mixture when the injury from the individual gases was extensive. The relationship between the injury intensity from the individual pollutants and the type of response (antagonistic or synergistic) is shown in Fig. 2.15.

The extent of foliar injury from the individual gases is important to consider when attempting to determine if the responses to the pollutant combinations is greater than the summed effects of the individual ones. Synergism (greater than additive injury) occurred when the injury to soybean foliage from the individual pollutants was less than 45% (Heagle and Johnston 1979); antagonism resulted when the injury from the single gases exceeded 50 to 70%. In some cases the summed effects of the individual gases exceeded 100%, which represents a mathematical but not necessarily a biological antagonism. There have been only a few observations of antagonism when the injury amounts were small. The observations of antagonism are more common in growth than foliar injury responses.

2.5.2.4.2 Significance of External Growth Factors

As with O₃, various external growth factors can influence the severity of injury from exposures to combined pollutants. However, there have been too few reports concerning the influence of external growth factors to generalize about the responses. Through their influence on plant morphology and physiology, the external growth factors can influence pollutant uptake and its reaction within the plant directly, during and following exposure, and indirectly, prior to exposure.

Increasing light intensity during exposure to a mixture of O_3 and SO_2 increased the severity of injury to spinach (Yamazoe and Mayumi 1977). Plants placed in the dark following exposure in the light were as severely injured as plants left in the light. Plants pretreated in the light or dark and subsequently exposed in the light were equally responsive to the pollutant combination.

Pinto bean exposed to a mixture of O_3 and SO_2 displayed the maximum injury at 15 °C; the injury decreased with increasing temperature (C. A. Miller and D. D. Davis 1981). The individual gases induced similar amounts of injury at both 15 ° and 32 °C. The symptom expression changed with temperature. At 15 °C the injury symptom was similar to SO_2, while at 32 °C the injury was similar to O_3.

Ozone uptake and injury tend to increase with increasing relative humidity during exposure. A similar trend was observed in studies using mixtures of O_3 and SO_2. Carlson (1979) exposed young sugar maple, American ash, and black oak (*Quercus velutina*) plants to 0.05 ppm ozone and/or 0.5 ppm SO_2 at various relative humidities and light intensities and determined their effects on photosynthesis and visible injury. In general, the impact of the combined pollutants on photosynthesis and injury intensity was reduced at low relative humidities and low light intensities. However, the environmental conditions moderated the responses of the three tree species to the individual and combined pollutants differentially. Studies with three petunia cultivars (Elkiey and Ormrod 1979 c) showed that exposures to pollutant combinations at 50% RH caused a rapid increase in leaf resistance, suggesting stomatal closure, but at 90% RH there was only a slight increase.

2.5.2.4.3 Significance of Internal Growth Factors

The stage of leaf development and plant ontogeny have been shown to influence plant response to O_3 and PAN (cf. Chap. 2.4.3). There have been only limited studies concerning the influence of leaf or plant development on the plant sensitivity to mixtures of O_3 and SO_2. At this time, there is no clear trend. In tobacco, the mature leaves (Menser and Heggestad 1966) were most sensitive, while the middle-aged leaves of petunia cultivars were most sensitive to injury from the pollutant combination (Elkiey and Ormrod 1979c). The midshoot leaves of grape and apple were the most sensitive to the mixture (Shertz et al. 1980 a, b). Expanding and expanded pea leaves reacted differentially when exposed to lower concentrations of O_3 and SO_2, but at higher concentrations they responded similarly (Olszyk and Tibbitts 1982).

Extensive efforts have been made to determine the relative sensitivity ranking of plants to O_3 (see Chap. 2.4.3.2) and SO_2 (van Haut and Stratmann 1970). In comparison, there have been only a few studies on the responses of a range of species to mixtures of O_3 and SO_2. Individual studies have shown that species vary in their sensitivity to combined pollutant exposures. However, there has not been a concerted attempt to develop listings of relative sensitivity to mixtures as there has been for the individual gases. Studies of herbaceous agricultural crops showed that broccoli and cabbage were least sensitive, while tobacco and radish were the most sensitive of the six species tested at three different concentrations of O_3 + SO_2 (Tingey et al. 1973 d). Karnosky (1981) studied the relative sensitivity of 32 urban-tree cultivars, representing 11 species, to O_3 and SO_2 singly and in combination. Some cultivars of *Acer platanoides, Acer rubrum, Acer saccharum, Fagus sylvatica, Fraxinus pennsylvanica*, and *Ginkgo biloba* were tolerant to all three exposure combinations, while cultivars of *Fraxinus americana* and *Platanus acerifolia* were sensitive to all three exposures. However, all species and cultivars did not display the same sensitivity rankings to the individual and combined pollutants. For example, *Acer rubrum* Bowhall was highly sensitive to the combined pollutants and displayed intermediate or resistant reactions to O_3 and SO_2, respectively. In contrast, *Quercus robur* was resistant to the combination and SO_2, but sensitive to O_3. The limited studies have not shown a clear similarity in the sensitivity to the combined and individual pollutants over a range of species and cultivars.

Not only do species display differential sensitivity to mixtures of O_3 and SO_2, but cultivars and clones within species have shown differential tolerance. Several species and cultivars of tobacco known to vary in sensitivity to ozone were tested for their sensitivity to mixtures of SO_2/O_3 (Menser and Hodges 1970, Grosso et al. 1971, Hodges et al. 1971, MacDowall and Cole 1971, Menser et al. 1973). Most types displayed different relative sensitivity rankings to ozone, the mixture of ozone plus sulfur dioxide and photochemical smog. In general, the effect of the combined gases was greater than for the individual pollutants.

Differential plant response in beans (33 cultivars) was studied using exposure to O_3 and SO_2 singly and in combination (Beckerson et al. 1979). The O_3 injury among the cultivars ranged from 0 to 68% while, from the combination, injury ranged from 0 to 30%. Only two of the cultivars exhibited more than 10% injury from SO_2. The injury response between plants exposed to O_3 or SO_2 was not significant, but the correlation coefficients for injury among the cultivars from the combination and either gas alone was highly significant, suggesting that the plant response to the mixture was the result of the reactions of both gases. In 29 of the cultivars, injury symptoms from the combined pollutants developed more slowly than the symptoms from O_3 alone. Not only was the rate of injury development different among cultivars, but also the nature of the response was different. After 5 days of exposure, most of the cultivars showed more injury from O_3 alone than from the combination, but for the remaining cultivars injury was similar in both treatments.

The relative sensitivity ranking for three petunia cultivars was similar for O_3, SO_2, and their mixture (Elkiey and Ormrod 1979 b). However, for Elatior begonia cultivars, the plant response to O_3 and the combination were similar for some cultivars but different for others (Reinert and Nelson 1980).

Six grass species, including 18 cultivars, exposed to mixtures of O_3, SO_2, and NO_2 (each et 0.15 ppm) displayed variable responses: the various cultivars of Kentucky bluegrass generally displayed a greater than additive response. The other five species displayed additive responses except for two cultivars which exhibited an antagonistic response (Elkiey and Ormrod 1980 b).

Karnosky (1976) studied the effects of O_3 and SO_2 singly and in combination on the foliar injury of five trembling aspen clones. The clones exhibited different patterns of sensitivity to the individual gases, but exposure to the combined pollutants injured more ramets per clone and the percentage of leaves injured increased. There were variations in injury intensity among the clones.

The inheritance of differential tolerance to combined pollutants has been studied for two tree species. Repeatability studies with eastern white pine demonstrated that tolerance-sensitivity to the mixture of O_3 and SO_2 was under strong genetic control (Houston and Stairs 1973). Plant responses to the effects of the combined pollutants correlated better with field observations than the responses from either pollutant individually (Houston 1974). High success has been obtained in the field selection of trees displaying differential sensitivity to low or moderately high ambient levels of these pollutants. In trembling aspen, repeatability estimates illustrated that the responses to the individual and combined pollutants was under strong genetic control (Karnosky 1977).

2.5.3 Ozone and Other Pollutants

There have been only a few reports concerning the effects of ozone in combination with other pollutants (either gases or heavy metals), making it difficult to draw general conclusions. However, for the gases the responses range from antagonistic to synergistic, while for the heavy metals the responses tend to be synergistic. Substantially more research will be needed to determine if these general conclusions are correct, especially for a range of plant species and environmental conditions.

2.5.3.1 Ozone and Nitrogen Dioxide

Matsushima (1971) showed that tomato and pepper exposed to either simultaneous or alternating mixtures of O_3 (0.4 ppm) and NO_2 (1.5 ppm) developed less injury than that caused by either gas alone. The addition of a nonphytotoxic level of NO_2 to O_3 increased foliar injury in rice, but decreased it in corn (Yamazoe and Mayumi 1977). The addition of SO_2 to the $NO_2 + O_3$ mixture tended to decrease injury below that caused by the two-gas mixture. The sensitivity ranking of 12 clones of eastern white pine was the same following exposure to either ambient polluted air or to a combination of O_3 and NO_2 in the greenhouse (Nicholson and Skelly 1977).

2.5.3.2 Ozone and Hydrogen Sulfide

The principal concern for this interaction was that H_2S is released from geothermal operations and synfuel plants which may be located near areas impacted with photochemical oxidants. Initial studies on the effects of H_2S on photosynthesis and leaf conductance showed that a low concentration of H_2S (0.74 ppm) stimulated both leaf conductance and photosynthesis, but the addition of O_3 (0.072 ppm) significantly depressed both responses (Coyne and Bingham 1978). Higher concentrations of H_2S inhibited both photosynthesis and leaf conductance, while ozone intensified the effects. The effects of various concentrations of H_2S (0.30 to 7.0 ppm) in various combinations with O_3 (0.046 to 0.127 ppm) on leaf injury and plant growth and yield were studied (J. P. Bennett et al. 1980). The amount of foliar injury was similar in plots receiving H_2S or H_2S plus O_3. Predicted yield losses at a H_2S dose of 450 ppm showed that H_2S alone significantly reduced yield and the addition of O_3 caused a greater reduction in plant growth and yield.

2.5.3.3 Ozone and Heavy Metals

Zinc and cadmium reacted synergistically with ozone in producing visible injury and chlorophyll loss in garden cress (*Lepidium sativum*) and lettuce (Czuba and Ormrod 1974). The combination of cadmium and O_3 induced the earlier development of necrosis and chlorosis and the injury was observed at lower O_3 and cadmium levels than for the individual treatments (Czuba and Ormrod 1981). In tobacco and pinto bean, Cd (0.1 ppm) added to the soil interacted with O_3 (0.3 to 0.5 ppm) to produce an antagonistic response when visible injury, photosynthesis, and transpiration were measured (Faensen-Thiebes 1981). Cadmium and nickel (concentrations 1, 10, 100 μmol) in the nutrient solution interacted to reduce root and shoot growth of pea (Ormrod 1977). An ozone exposure increased the effects, but the increase was less than additive. However, low concentrations of cadmium and nickel tended to enhance O_3 phytotoxicity. The interaction of cadmium and ozone was influenced by both concentration and the environmental conditions. Tomato plants grown at 0.25 and 0.75 mg Cd l^{-1} developed only slight foliar injury when exposed to ozone (0.20 ppm for 3 h) under cloudy skies and the Cd had no significant effect (Harkov et al. 1979). In full sun, there was extensive ozone injury and the joint response was synergistic. The changes in the cellular ultrastructure of pea leaves by exposure to ozone (0.50 ppm) were increased when plants were grown in nutrient solutions containing 100 μmol $NiSO_4$ (Mitchell et al. 1979).

The limited amount of published data indicates that heavy metals can increase the phytotoxic reactions of ozone. At present, it is not possible to assess the risk from the joint action of gaseous and heavy metal pollutants to vegetation in industrial areas, along heavily traveled highways, or by the use of sludge as a fertilizer on crop lands.

2.5.4 PAN and Other Pollutants

As part of the photochemical oxidant complex, PAN co-occurs with ozone in the ambient air. In general, the plant response to combinations of PAN and O_3 at concentrations approximating ambient conditions were less than additive. Using a high PAN concentration (180 ppb), Kohut et al. (1976) showed that it interacted with O_3 synergistically when producing foliar injury on hybrid poplar. Sequential exposures produced more injury than exposures from the individual pollutants (Kress 1972). Using a lower PAN concentration (50 ppb), the combination of PAN and O_3 induced less injury on ponderosa pine primary needles than did O_3 alone (D. D. Davis 1977). In pinto bean, the response to the combined pollutants was additive or synergistic on the adaxial leaf surface, but antagonistic on the abaxial surface (Kohut and D. D. Davis 1978). Simultaneous exposures to PAN and O_3 induced significantly less injury to petunia and kidney bean than either pollutant alone (Nouchi et al. 1984). In alternating exposures, O_3 alone induced as much injury as O_3 followed by PAN; PAN alone induced as much injury in petunia as a PAN exposure followed by O_3 (Nouchi et al. 1984). Combinations of PAN and O_3 generally produced less than additive effects on the growth of four tomato cultivars (Temple 1982).

Studies concerning the effects of PAN plus other pollutants that are not produced photochemically are very limited. In the laboratory, Matsushima (1971) exposed pinto bean, pepper (*Capsicum frutescens*), and tomato to mixtures of SO_2 (1.5 to 2.1 ppm) and PAN (0.27 to 0.40 ppm) and found that foliar injury ranged from additive to greater than additive. However, when the plants were exposed to alternating combinations of the individual gases, the injury was less than additive when compared to the injury caused by the same total exposure duration from the individual gases. The alternating exposure also caused significantly more injury than the simultaneous exposure.

2.5.5 Ozone, Sulfur Dioxide, and Acid Precipitation – the Cause of Forest Decline in Central Europe?

Since the 1970's, in large areas of Central Europe there has been a dramatic decline in the health of forests, including disintegration of the stands. Initially, the term Waldsterben, or forest decline, was primarily coined from emotional reactions shortly after extensive forest injury was observed, however, the threat to the forests has now been clearly established. Based on a series of facts and circumstantial evidence, primary cause of this injury appears to be ozone; in most locations it probably acts in combination with sulfur dioxide and acidic precipitation.

In so-called clean air locations, principally in southern Germany, injury symptoms have been observed mainly on white fir (*Abies alba*) during the last 10 years and since 1980 on Norway spruce (*Picea abies*). However, the symptoms were either different or only partially comparable to known epidemics and disease symptoms on forest trees (Rehbock 1982, Schütt 1982b, Rehfuess 1983, Prinz et al. 1982, BML/LAI 1982, Bosch et al. 1983). A partly new type of injury syndrome, whose origin is not understood, is found on the individual plant species.

2.5.5.1 Symptoms, Intensity, and Distribution of Injury

The type of injury on leaves, twigs, crowns, and roots varies among the tree species, locations, and regions. Fundamentals of the principal syndrome will be described only for fir and spruce, the most threatened conifers, although other species, such as Scots pine (*Pinus sylvestris*), Douglas fir (*Pseudotsuga menziesii*), European beech (*Fagus sylvatica*), and oak species also exhibit increased injury symptoms. For example, in individual cases with beech, the demarcation between the normal variation in the health of the stand is difficult to determine (Mülder 1983).

2.5.5.1.1 Fir

Reports dating back several hundred years discuss the injury and death of firs, a species sensitive to various stresses, in Germany, Czechoslovakia, and Italy. According to Málek (1981), the so-called fir death in central Europe was the consequence of explosive injury outbreaks that periodically occurred, almost simultaneously at several locations. In Bohemia the interval between disease periods has become progressively shorter since the 1600's and within the last 50 to 100 years it has developed into a condition of permanent poor health and tree death. The injury is particularly intense following dry years, winters with low temperatures, or late frosts (Wagner 1981). Since 1960 the injury intensity has increased. Current symptoms occur first and exhibit the most distinct impact on older trees (BML/LAI 1982, SR-U 1983).

The needles, beginning with the older ones, show a light chlorosis, occasionally a dull discoloration, yellowing, and premature abscission. Typically, there is a progressive opening of the crown proceeding from bottom to top; the needles on the outermost tips of the crown remain green the longest. The apical growth stops prematurely, consequently the firs display the characteristic "stork's nest" at an earlier stage of tree development. Parallel with the progressive needle loss, there is frequently an increased production of epicormic branches. With increasing injury, the "stork's nest" loses its contour and epicormic branches, indicating a disturbance in the water supply of the crown, and begins to die, reducing the tree's survival chances (Mülder 1983). In injured but not yet dead crowns, the bark may become detached mainly near the base of the crown. At this stage, there are increased infections from bark beetles, *Armillaria*, and mistletoe. An additional characteristic is the pathological spread of watercore in the roots and lower trunk, where it may, finger-like, invade the sapwood (Schütt 1981a, Brill et al. 1981, Lang 1981, Blaschke 1981a). A higher proportion of the fine roots dies in diseased firs and spruces; also, the regenerative capacity of the fine root system is reduced, as is mycorrhizal formation (Hüttermann 1983, Blaschke 1981a, b).

There is only partial correspondence between the symptoms of the current fir decline and "periodic re-occurring enigmatic disease of fir". The new symptoms of the current decline include (Wagner 1981, Seitschek 1981, 1982, BML/LAI 1982):

- persistence for approximately 10 years with uncertain tendencies;
- occurrence over the fir's northern range but less intensely in the Bavarian Alps and no injury in the southern portion of the range (Kramer 1982);
- rapid disease progress on individual trees;
- frequently irreversible injury with a high proportion of tree mortality;
- injury in different types of forest canopies and structures and at different types of sites.

In the firs' natural range, the earlier disease outbreaks occurred primarily on the eastern and northwestern edges of stands. However, in the current phase of the disease, trees in the natural range, growing in sites with optimum conditions, are affected (Seitschek 1982). Severe disease outbreaks have occurred in the eastern and northeastern areas of the Bavarian subalpine mountains, Bayerischer Wald, Oberpfälzer Wald, Frankenwald, Fichtelgebirge, as well as in the northern and middle Schwarzwald, where more than 80% of the trees are diseased. When the disease progress rate is considered, and the fact that 10- to 15-year-old firs are already affected (Schütt 1982 b), then one must accept that a dramatic deterioration in the condition of the fir stands has occurred. The firs in Eastern Bavaria up to the Czechoslovakian border are threatened with complete extinction. There have been reports of the fir decline in Poland (Sierpinski 1981), Czechoslovakia (Málek 1981), Switzerland (Flühler and Bosshard 1982), and Italy (Moriondo and Covassi 1981), but the symptoms deviate to some extent among the various reports.

In the mountains, firs are ecologically and silviculturally valuable at the middle elevations. They root deeply in the forest subsoil and are well anchored against storms, thereby contributing significantly to stand stability, particularly when it includes shallow-rooted species such as the spruce. Concurrently, the firs extract nutrients from the deeper soil horizons, making nutrients available to the rest of the forest. The mixed alpine forest composed of fir, spruce, and beech is the most productive and valuable in Europe (Bonnemann and Röhrig 1971).

2.5.5.1.2 Spruce

Only the most obvious injury symptoms, as observed in various regions, will be described. Increased needle fall spreading from top to bottom and erratically from the exterior to the interior causes a thinning of the crown over wide areas. Frequently there is a distinct dieback, particularly in the higher elevations of the Harz Mountains (BML/LAI 1982). The injury usually occurs on the crowns of dominant trees which project above the general canopy level. The stand thereby loses the dominant members of its ecological support system; long-term structural destabilization is the probable consequence (Franz 1983). In severely impacted mature trees that would normally display a substantial vertical growth the yellowing and casting of needles frequently begins beyond the first three or four needle whirls behind the tips (Mülder 1983), thus the lower portion of the crown appears to have better needle retention. This symptom appears to be similar to the initial injury phase established for the decline and death of spruce experiencing chronic exposures from air pollution (Pelz and Materna 1964). The needle-casting starts on the side branches at the transition between the upper and middle crown and spreads to the lower and upper portions of the crown (sub-top-dying).

A symptom more frequently found in southern Germany deviates from the above descriptions; the characteristic thinning of the crown proceeds from bottom to top and interior to exterior. The older trees, in general, display the so-called Lametta syndrome; the more or less denuded branches of the second order droop, givin the trees a fringed or comb-like appearance.

The widely distributed yellowing of all needle ages is conspicuous, primarily on the upper sides of light-exposed branches; there is also a previously unobserved chlorosis on the tips of the older needles: "Goldspitzigkeit" (Rehfuess 1983, Zöttl and Mies 1983, Bosch et al. 1983). In addition to needle yellowing and loss, needle and branch anomalies were noted. Increased concentrations of abscisic acid and abscisic acid conjugates, as well as changes in the composition of this metabolic group, indicated a perturbation in phytohormone relations as observed in herbaceous plants subjected to various stresses (Schütt 1983). Resin drops frequently occur inside and just below the crown area. Moreover, breaks occur in the stem bark within the crown and on the underside of the branches. Root deformation and loss is found in young spruce trees.

The spruce injury is most intense on the heights and ridge levels of the subalpine mountains, particularly in the Bayerischer Wald, Oberpfälzer Wald, Fichtelgebirge, Frankenwald, Schwarzwald, Odenwald, Taunus, Spessart, and Harz (BML/LAI 1982). The injury is intensified at lower elevations (Zöttl and Mies 1983, Wentzel 1983). Increased injury has occurred on the spruce in Nordrhein-Westfalen, since the fall of 1982 (Prinz et al. 1982, Knabe 1983). By and large, trees in the west and southwest are particularly impacted. Spruce in its natural habitat as well as in areas outside its natural range is injured. This wide occurrence of injury is alarming because until the latter half of the 1970's spruce was considered to be healthy and vigorous – except in areas heavily impacted with SO_2. Spruce covers approximately 40% of the forested area in the Federal Republic of Germany and is economically the most important tree species. In the summer of 1983, it was estimated that about 40% of the area occupied by spruce stands was injured (BML/LAI 1983). Since then, this percentage has increased. At many locations, primarily in the higher elevations where fir are threatened with extinction, the spruce are not healthy and there is no other comparable species with similar economic and ecologic characteristics available to replace it.

2.5.5.2 Investigations into the Cause of the Forest Injury

Given the current extent of forest injury and the progress rate of the disease, it appears that extensive portions of the forests in Central Europe are endangered. As required for implementation of effective relief measures, intensive research into the probable causes has been initiated, e. g., natural conditions, silvicultural influences, and phytotoxic air pollutants.

2.5.5.2.1 Natural and Silvicultural Influences as Factors in the Cause of Forest Injury and Tree Death

Climate extremes and weather conditions are among the most likely natural causes that should be considered. In Central Europe, numerous observations have established that tree deaths may be caused by drought periods, cold years, a sudden drop in temperature during the winter, early or late frosts, or ice storms, acting at times either alone or in combination (Wagner 1981, Seitschek 1981). Se-

vere injury occurs following extended droughts but only at particularly dry locations such as on shallow soils over a very permeable rock layer. Because forest
injury occurs at well-watered sites and is widely distributed, drought periods must
be excluded as the sole cause of injury (Schütt 1977, BML/LAI 1982). Drought
can weaken trees, disrupt water relations (Schuck 1981), and change their predisposition to biotic factors and air pollutants (Blaschke 1981 a, b, von Schönborn
and Weber 1981). Drought may directly influence the spread of the pathological
disorder "watercore" in diseased firs (Schütt 1981 a, Brill et al. 1981, Bauch
1983).

Air pollution may also reduce the trees' resistance to cold stress (Wentzel
1965, Huttunen 1980, Keller 1981). However, several factors suggest that frost is
not the primary injury cause; frost episodes did not occur at all injured sites and
all frost-exposed locations did not exhibit increased injury (BML/LAI 1982).

There have been extensive discussions to determine if and to what extent the
growing of spruce as a monoculture beyond its natural range has contributed to
the current problem. Also, to determine if previous forest use practices have influenced the injury. The influence of stand structure can be excluded as the primary cause because both fir and spruce are injured independently of stocking
rate and canopy structure. Injury occurs in both pure and mixed species stands
and also in well-managed and thinned forests. Conversely, there are, for example
in the foothills and the edges of the Alps and on the upper Neckar, poorly thinned
single-storied stands of old-growth firs that are at present healthy (BML/LAI
1982, Mülder 1983). Modifications in forest structure and management are not
practical relief measures.

Spruce, particularly, has succeeded on sites outside its natural range. The injury to fir and spruce at different sites argues against unsuitable sites as the cause
of the forest decline. There are natural stands of diseased spruce, for example at
the higher elevations in the interior of the Bayerischer Wald, where the soils have
exhibited an acidic reaction for an extended period and, in part, have a large humus
accumulation. However, spruce on basic and, sometimes, on soils well buffered
with carbonate such as the Parabraunerden (a soil belonging to the suborder
Udolf), in the alpine foothills on gravel plains and on the calcareous soils on the
alpine rivers (Mies and Zöttl 1982) are also injured. Trees growing in areas receiving continual fertilization are injured, while on highly reduced and consequently
acid soils pure stands of healthy spruce are found. Liming the soils has not significantly reduced the needle loss (Aldinger 1983). There is not a close relationship
between the injury intensity and the substrate characteristics or the stand nutrition (BML/LAI 1982, FVF 1984). Injury to fir is especially severe on sites with
an adequate supply of nitrogen but pronounced Ca and Mg deficiencies (Evers
1981, Rehfuess et al. 1982, Zöttl and Mies 1983).

There is broad agreement that monoculturing conifers, particularly spruce, on
weakly buffered substrates with low cation exchange capacity increased the acidification of the upper soil layers, formation of a litter layer, and podzolization;
consequently, the upper soil layers are depleted in soluble nutrients (Rehfuess
1981 b, Rehbock 1982, F. Scheffer and Schachtschabel 1982). Nevertheless, the
widespread fear of a dramatic soil deterioration when conifers are monocultured
has not been established (BML/LAI 1982). Mülder (1983) has indicated that nat-

ural monocultures were as widely distributed as labile subclimax and stable climax forests.

Etiological studies have attempted to determine if animal pests or plant pathogens (viruses, mycoplasms, ricketsiae, bacteria, or fungi) are the cause, or combined causes, of forest decline. Based on previous knowledge and the lack of typical distribution patterns that normally occur with pests and disease incidents, animal pests, pathogenic fungi, and bacteria were excluded as primary causes (Eichorn 1981, Roll-Hansen and Roll-Hansen 1981, Brill et al. 1981). The locally increased infections from bark beetles and *Armillaria*, two typically weak parasites, resulted from a previous weakening of the host trees. Bacterial infections followed physiological disturbances and injury to the fine roots, including the mycorrhiza (Blaschke 1981a), which, in fir, led to watercore formation in the roots and trunk (Schütt 1981a, Brill et al. 1981). The possible significance of root and stem nematodes are the primary cause of death in firs and spruces or animals as possible vectors for viruses, mycoplasms, or ricketsiae are unanswered (Schwenke 1982). The prevailing opinion is that animal pests and phytopathogens are generally of secondary significance for the forest injury (Möhring 1983). However, there are contrary views (GVSt 1982, Kandler 1983, Frenzel 1983).

2.5.5.2.2 Air Pollutants as the Cause of the Injury

Drought, frost, unfavorable edaphic conditions, poor site selection, failures in the structural composition of the forests and their uses are just as unsuccessful in explaining the sudden occurrence and dramatic deterioration in the health of the forest stands in Central Europe as infections from animal pests or pathogenic agents are (FVF 1984). In individual cases, the various abiotic and biotic influences are certainly responsible for injury and as single or combined factors they contribute to the expression and the extent of injury. However, based on current knowledge, they are excluded as the primary cause of the current phenomenon of Waldsterben or forest decline. Consequently, there must be an unknown allogenic factor that satisfies the following requirements:

- distributed over several regions;
- sudden occurrence, in high concentrations that are active over short durations, especially on the western sides of the mountain ranges;
- a high biological activity.

With present knowledge, these conditions are only satisfied by specific air pollutants. However, a new category of pollution response must be present. The symptoms deviate partially from injury previously shown to be caused by acute or chronic exposures. By and large, there is a new syndrome occurring in the forests remote from emission sources, a circumstance that is similar to previous occurrences with photochemical oxidants in the United States (Berry and Ripperton 1963, P. R. Miller et al. 1963), that were only later linked to air pollution.

Given the degree of phytotoxicity, the concentration levels, and spatial distribution, the ambient pollution complex of photochemical oxidant, SO_2, and acid precipitation should be considered as the most likely causes (Guderian 1982, Guderian and Rabe 1982, Arndt et al. 1982b, Prinz et al. 1982).

Previous concern was expressed regarding heavy metals, i. e., that they could present significant local risks for humans and animals as the metals moved through the food chain (Krause 1974, Guderian et al. 1977, UBA 1977, Prinz et al. 1979). The metals certainly demand increased consideration in light of the continuing need to protect the forests. In forest ecosystems the long-term input of heavy metals is a risk, particularly for the soil-dwelling organisms and for the release of nutrients from the litter. There is no evidence that heavy metal-containing compounds alone or in combination are responsible for the forest decline (Zöttl and Mies 1983, Keller 1970). Accordingly, only photochemical oxidants and acid-containing sulfur compounds, the decisive portion of "acidic precipitation", remain as the potential cause of the injury.

While photochemical oxidants can directly effect only the aboveground portions of vegetation, acidic precipitation can also cause indirect effects through the soil, a concept proposed particularly by Ulrich (1981a,b, 1982). From the development of his Ecosystem Theory of forest decline, it follows that the natural coupling of biomass production with acid use and biomass decomposition with acid production in the forest is disrupted as a consequence of acid deposition and the interference from man and the processes are no longer in balance. The increased atmospheric input and acid production within weakly buffered soils ("pedogenic acidification") has led to the rapid decrease in the soil reaction to the buffer range of aluminium, a pH between 3 and 4; consequently phytotoxic concentrations of metal ions are liberated. According to Ulrich, the forest decline is based on a series of processes, anthropogenic soil acidification, leaching of alkaline and alkaline-earth ions, disruption of litter decomposition, promotion of the podzolic processes, root injury from the toxic effects of Al and Mn ions, a reduction in the mass of fine roots, disturbance in the water relations of trees, leaf injury, crown drying, increased storm injury, and infections from pests and ultimately the death of trees. There is no disagreement about potential consequences of these effects but they are seen as long-term. The "destabilization" of the forest ecosystem, postulated by Ulrich, may result from acidic precipitation alone when material import and export are no longer in equilibrium, in contrast to natural ecosystems not influenced by man. In the vicinity of individual pollution sources, following long-term exposure to elevated SO_2 concentrations, distinct reductions in soil pH have been established (Rusnow 1919, Ewert 1924, Guderian and Stratmann 1962).

Wieler (1905) proposed that vegetation injury in the vicinity of SO_2 sources, with low stack heights, occurred primarily through changes in the soil. However, according to our knowledge there is no clear evidence to support this hypothesis. The hills near the iron-ore roasting facility at Biersdorf, Germany, were exposed to elevated SO_2 concentrations (up to 10 ppm) for about 100 years, with the consequence that most of the vegetation near the plant was killed or severely injured (Guderian and Stratmann 1962, 1968). Within a few years after the emissions terminated, birch was revegetating the slope and a few years later the previously denuded hills were completely reforestred. Also shortly following the termination of emissions, vigorous (meter-long) shoot growth occurred from the bases of oaks in the transition zone on the edge of the smoke-denuded areas, which argues against the release of injurious concentrations of aluminium and heavy metals into the soil. Thus, for several reasons the forest decline phenomenon is not explained by indirect effects through the soil.

The sudden appearance of injury and the rapid acute course of the disease implies a high injury potential. There is no proof in the literature that complex soil systems, at markedly different sites, in areas remote from industrial areas, can be so changed that within a few years the death of forests is the consequence. To be sure, there is a reduction in soil pH (Ulrich et al. 1979, Butzke 1981, Evers 1983, Reichmann and Streitz 1983, Wittman and Fetzer 1983), however, neither needle nor bark analyses nor studies of site conditions have provided evidence of Al or Mn toxicity on diseased fir or spruce (Berchtold et al. 1981, Evers 1981, Reiter et al. 1983, Bauch and Schröder 1982, Rehfuess 1981 a, Zöttl 1983, Bosch et al. 1983). The occurrence of injury on fir and spruce growing on fluvents, a soil rich in carbonates (rendzina), for example on the steep limestone slopes of the upper Neckar, the Kalkalpen, and Schwäbische Alb, as well as lithic pystrochrepts soils, argues against Al and Mn toxicity (Rehfuess 1981 a, BML/LAI 1982). To be sure, in most of the southern Schwarzwald, the injured fir stands occur on acidic, rocky, brown soil types with a limited A_h-horizon (Mies and Zöttl 1982). Following an analysis of the results from various countries, Zöttl (1983) concluded that Ulrich's postulated (1980, 1981 c) Al toxicity to firs and spruces, in acidophilic conifer ecosystems, should be considered very unlikely. Also, disruption of litter decomposition causing a significant enhancement of podzolization cannot be unequivocally established (Abrahamsen 1980, Rehfuess 1981 a). Finally the compilation of silvicultural and dendrochronological measurements argue against soil problems being the primary cause of forest decline (Wentzel 1983). However, the long-term pollution-dependent input of materials must be considered as a potentially serious danger to forest ecosystems. A series of inferences suggest that the atmospheric input of acids on weakly buffered sites has already influenced the availability of nutrient cation ions in the soil and their uptake through the plant (Zöttl and Mies 1983, Bosch et al. 1983).

The concept that soil acidification and its consequences, resulting from either atmospheric input of acids or through the monoculture of conifers, is the sole cause of the previously described forest injury and decline cannot be supported. Consequently, one must acknowledge that the direct effects of air pollutants on the aboveground plant organs could play a key role in the disease.

Since Stöckhardt (1871) established the phytotoxic characteristics of SO_2 more than 100 years ago, it has been considered to be an important air pollutant, based on its extensive distribution. Decades of SO_2 exposure have induced injury, principally on conifer groves in industrial areas and their surrounding regions such as the Ruhr Area (Federal Republic of Germany), the area around Bitterfeld/Leipzig, the Erzgebirge (both in the German Democratic Republic and Czechoslovakia), the Riesen-Gebirge (Czechoslovakia), and in Silesia (Poland). From this perspective, SO_2 has been considered as a possible cause of the forest injury. However, several observations argue against the concept that SO_2 is the sole cause of forest decline, i. e., the occurrence of symptoms different from those typical of SO_2, several tree species from various taxa and differential SO_2 resistance levels were impacted nearly simultaneously, the rapid progress of the disease including the death of individuals and complete stands, and the ambient measurements of SO_2 concentrations. During the past few years, there have been some distinct increases in the peak SO_2 concentrations, especially in the so-called clean air areas, particularly in eastern Bavaria near the borders with the German Democratic Republic and Czechoslovakia. Ambient measurements showed that the yearly mean SO_2 concentrations ranged from 10 to 36 µg m^{-3} (0.0038 to 0.0137 ppm), but on a regional basis the mean concentrations exhibited only a very slight increase (UBA 1982, Landesanstalt für Umweltschutz, Baden-Württemberg 1981). The available dose-response relationships, established after decades of research, suggest that long-term mean SO_2 concentrations in this range pose only a slight risk to fir and spruce stands growing on extremely unfavorable sites (Guderian and Stratmann 1968, VDI 1978b, IUFRO 1979, 1981, Wentzel 1981). The fact that the lichen populations in areas of southern Germany exhibiting forest decline show little or no injury argues against the occurrence of sulfur dioxide in phytotoxic concentrations (Prinz et al. 1982) (cf. Chap. 2.2.6). When areas such as eastern Bavaria, which experience high acutely injurious SO_2 concentrations, are ignored, it is apparent that SO_2 is not the sole cause of the new type of forest injury. The possible participation of SO_2, acting in concert with other pollutants, will be discussed.

Based on their distribution and phytotoxicity, photochemical oxidants should be considered as the initial cause of injury. The preliminary observations of Darley (1963/1964) established that ozone and PAN induced foliar injury on annual plants in Germany. In 1967, subsequent studies with tobacco plants established that phytotoxic concentrations of ozone occurred at several locations in the Federal Republic of Germany (Knabe et al. 1973). Photochemical oxidant injury has been reported in other European countries (see Chap. 2.3.2.2). Guderian et al. (1984) in their report Luftqualitätskriterien für Photooxidantien (Air Quality Criteria for Photochemical Oxidants) for the Umweltbundesamt, Berlin, concluded that the established effects on plants and the existing measurements of ambient concentrations overlap, indicating that phytotoxic concentrations of photochemical oxidants that can endanger vegetation are widely distributed in the Federal Republic of Germany. A comparison of the available ambient ozone measurements (see Chap. 1.3.2.2.1) and dose-response relations for vegetation (for example Table 2.24) supports this conclusion. The measured peak ozone concentrations [maximum half-hour mean values of up to 0.338 ppm (664 µg m^{-3})], as well as mean values for longer time intervals, indicate that ozone concentrations frequently exceed the tolerance level for vegetation. Obviously the maximum peak concentrations generally occur in congested areas. However, when evaluating risks to vegetation, it should be remembered that anthropogenic ozone concentrations, in areas remote from pollution-emitting sources, may remain at elevated levels for longer periods because the low NO concentrations will only partially decompose the ozone. This ozone-frequency distribution yields high arithmetic mean concentrations (Prinz et al. 1982). Also the extended ozone-exposure time series reduce the duration of pollution-free periods that are important for plant recovery from pollution episodes (see Chap. 2.4.1).

As a probable cause of the new type of forest injury, ozone satisfies the three conditions mentioned earlier in this section: distribution over large regions, a high biological activity over a narrow range of concentrations, and ambient concentrations that fluctuate more with secondary pollutants than for primary ones. Repeatedly in the past few years, meteorological conditions in Europe have favored significant photochemical oxidant formation (1.4.3). In the time period 1966 to 1978, the emissions of nitrogen oxides, photochemical oxidant precursors, have increased in the Federal Republic of Germany from 2 to 3 million metric tons years^{-1} (UBA 1981). In contrast, the SO_2 concentrations have remained approximately the same. Also by way of comparison, vertical profile measurements found high ozone impact several hundred meters above the ground corresponding with the particularly intense injury at these elevations (see 1.3.2.2.1). In this context, there are unanswered questions about the differential ozone resistance of individual european plant species (see Chap. 2.4.3.2).

Neither the direct effects of acidic precipitation on aboveground plant parts nor its indirect effects through the soil explain the new type of forest injury. Supporting this conclusion are decades of experience in areas exposed to high SO_2 concentrations and the fact that SO_2 emissions, which control the acidity of rain, have remained approximately constant at 3.5 million metric tons year^{-1} between 1966 to 1978 (UBA 1981).

In principle, acidic precipitation can directly injure the cuticle or, following its penetration, be biologically active within the leaf interior. The expression of effect is the consequence of the effective dose, that is to say the product of H^+-ion concentration and exposure duration as influenced by differences in plant morphology and wettability of the plant organs (Evans 1980). The xerophytic conifer needles are significantly less susceptible to acidic precipitation (Haines et al. 1980) than the tender leaves of many angiosperms (Evans et al. 1977, Shriner 1978). There have been no previous reports of cuticular injury on vegetation from areas in Central Europe exhibiting the forest decline. Scanning electron microscope studies on injured spruce needles found no significant cuticular injury, but perhaps a premature loss of the wax plugs from the stomatal pore (Rehfuess 1983). If cuticular erosion or injury occurs on conifer needles, it would be expected to develop first on the young, still developing needles, as found by Wood and Bormann (1976). The small loss in epicuticular waxes on the needles of Scots pine (Fowler et al. 1980) in an area exposed to SO_2 are more probably the result of stress-induced changes in wax formation than the erosion of the cuticular waxes from the needle surface.

The nutrient supply (Ca, Mg, and, in some incidents, zinc) of the trees apparently plays a central role in the forest injury problem. Deficiencies of these elements, especially Mg, are widely distributed in injured forest stands (Zöttl et al. 1977, Altherr and Evers 1977, Prinz et al. 1982). For example, in 4-year-old needles from diseased spruce trees, Zöttl and Mies (1983) found a Mg content of <200 ppm which was significantly less than the amount found in needles from visibly healthy trees. A similar relation was found by Zech and Popp (1983), the recent needles in the upper crown of diseased spruce trees contained 250 to 270 ppm Mg compared to 530 to 590 ppm in control samples from healthy trees. Visible Mg deficiencies develop on spruce when the Mg content of the needle tissue is less than approximately 350 ppm (Zech 1968), leading Zech and Popp (1983) to conclude that a Mg deficiency was the principal cause of the chlorotic needle tip form of the disease causing fir and spruce death on the acid granitic and phyllonitic soils of northeastern Bavaria. There have been three suggestions given for the cause of the deficiencies of Mg and other nutrients: a limited supply from the soil, restrictions in their uptake from the soil because of injury to the fine root system, and leaching of the minerals from the foliar organs.

The first evidence of a Mg deficiency was found by soil and leaf analyses collected from a fertilizer study (established in 1958/1959) on beech growing on middle Buntsandstein (Lower Triassic) in the Odenwald (Altherr and Evers 1974, 1977), a result that initially was not widely noticed. During the last few years it has been established that Mg and Ca deficiencies are widely distributed on acid soils, leading to the conclusion that the atmospheric input of acidic compounds promoted the leaching of alkaline and alkaline-earth cations; also Mg deficiencies on vegetation are especially intense on Mg-poor substrates (Rehfuess 1981a, Zech and Popp 1983, Zöttl and Mies 1983, Bosch et al. 1983). The increased mobilization of Mg from the sorption complex of acidic silicate soils has been found in the Solling area, which has experienced heavy pollution exposures (Ulrich 1981 a) and also from experiments with artificial acidic rain (Abrahamsen 1980). The extent to which the Mg deficiency results from or is intensified by losses to the fine root network (Hüttermann 1983) is unanswered. While Bauch and Schröder (1982) found an extremely low concentration of Ca and Mg in the xylem and cortical cells from the fine roots of diseased firs and spruces, Reiter et al. (1983) found no significant differences between healthy and diseased trees. Because the diseased trees generally have an adequate supply of nitrogen, phosphorus, and potassium, a general reduction in root biomass or root activity is not likely.

The leaching of nutrients from foliar organs has only recently been considered as the possible cause of forest injury, although there is extensive literature on leaching (Tukey 1980). In addition to macro- and micronutrients, numerous organic substances such as sugars, amino acids, and phytohormones are leached. The nutrients, potassium, calcium, magnesium, and manganese, that occur in higher concentrations in the "apparent free space" can be leached in large amounts. Increased losses from the "inner space", i.e., symplastic transport, results from injury to the plasmalemma, tonoplast, and other biomembranes.

Artificial rain studies found that decreasing the pH increased the foliar leaching of potassium, calcium, and magnesium from sugar maple (*Acer saccharum*) and bean leaves prior to the appearance of tissue injury (Wood and Bormann 1975). Also, studies with Norway spruce (*Picea abies*) and eastern white pine (*Pinus strobus*), using artificial rain, showed an increased leaching from the needles with a decreasing pH, however, the content of the corresponding nutrients remained unchanged in the needles (Wood and Bormann 1976). In preliminary experiments, Prinz et al. (1982) found an increased leaching of Mg by artificial rain from spruce exposed to 0.15 ppm (300 µg m^{-3}) ozone for 4 weeks; the amount leached was pH-dependent. The observations of Evers (1981) supported the concept of nutrient leaching from needles. In June, shortly after the new growth reached full development, there were only slight differences in the calcium and magnesium contents among trees from different vigor (injury) classes, however, distinct differences were observed in the fall.

The extent to which foliar leaching in the injured forests is influenced by ozone (Arndt et al. 1982 b) and acidic precipitation (Rehfuess 1983) has not been determined. Certainly it will depend on the actual exposure intensity from the individual air pollutants. Given the available data concerning the mode of action of these compounds, it appears likely that acidic precipitation is primarily active in the "apparent free space", while ozone increases the permeability of biomembranes, permitting the increased loss of nutrients and other substances from the cell interior. The combination of ozone and acidic precipitation therefore poses an important threat to the nutrient relations of plants.

As discussed above, the increased forest injury could result from ozone or SO_2 alone, or the combined effects of both. However, the injury intensity and the rate of disease progress, including the death of trees, are not explainable over large areas without considering a disruption in nutrient relations.

Analyses and evaluations of the available ambient air monitoring information, and biological and soils data suggest the following causes and consequences from the disease progress in the injured forests in Central European areas remote from pollution sources:

- The available data support the hypothesis that the new type of forest injury occurring in areas of Central Europe, remote from pollution sources, is caused by photochemical oxidants, SO_2, and acid precipitation. The direct effects of ozone and sulfur dioxide as single and combined agents – in some regions in combination with acidic precipitation impair the photosynthetic performance and water relations of forest trees with the consequence that vertical, radial, and especially root growth are reduced, as is the formation of mycorrhiza and the resistance of trees to biotic and abiotic stress factors.
- Disturbances in nutrient relations, particularly Mg and Ca, enhanced the direct effects and increased susceptibility of trees to air pollutants and also climatic stresses and disease agents.
- The deficiencies in these nutrients result, according to current knowledge, from an insufficient supply in the soil, a reduced uptake as a consequence of injury to the fine roots, and/or especially the leaching of the nutrients from the leaves as influenced by ozone and acidic precipitation.
- There is little evidence for the hypothesis that acidic precipitation, in combination with natural soil acidification under coniferous monocultures growing outside their natural range, releases phytotoxic concentrations of Al. Also, there is no evidence that exposure to heavy metals causes the forest injury described.
- In the long term, however, acidic precipitation and heavy metal-containing compounds pose a potential serious threat to soil productivity and therefore the health of the forests.

2.5.6 Summary

Plant response to pollutant mixtures can range from antagonistic to synergistic, depending on the species, the specific pollutant mixtures, and concentrations, among other things. Our understanding of how environmental conditions modify these reactions is very limited. Based on available data, it is not possible to assess the influences of pollutant sequence, meteorological factors, cultural practices, or many other factors in relation to the joint action of pollutants on vegetation. The available data indicate that pollutant interactions do occur and they occur in the same concentration range as the compounds occur in the atmosphere. These observations are significant for environmental management, therefore better information concerning combined effects is needed. In pollutant-impacted areas, the ambient concentrations of the individual pollutants are generally low, so that at least an additive plant response to pollutant mixtures should be considered. This is especially true for subtle injury following chronic exposures to low concentrations. When establishing air quality criteria, one should consider not only the effects of individual pollutants, but also the possible effects of pollutant mixtures.

Chapter 2.6 Dose-Response Relationships

In this chapter, experimental results concerning the quantitative relationship between the atmospheric concentrations of ozone and PAN and effects on higher plants will be described, documented, and evaluated. The goal is to establish the limits of impact by photochemical oxidants on vegetation as a basis for risk prognosis and to pave the way for remedial and mitigative actions. The principal concern is the most accurate identification possible of the lowest effective dose (i. e., concentration and exposure duration).

The derivation of such threshold values should be clear and verifiable because they have implications for the affected plant species and for the pollution source. Previous chapters concerning the nature and mechanism of injury, methods of detection and evaluation, as well as the decisive factors controlling plant responses, provide a significant basis for ascertaining the relationships between pollutant exposure and effect. Consequently, only a limited discussion of special methods is required, including the selection of experimental results as a basis for deriving air quality criteria and methods for summarizing individual values.

Only those experimental studies were considered that included the type and expression of injury on individual plant species or cultivars in relation to pollutant exposure, including concentration and exposure duration. Moreover, the measured responses must provide direct information concerning the impact on the useful value of the plants or permit the assessment of the indirect reduction in their useful values (see Chap. 2.3.3). The compilation of individual values for air quality criteria is based on the work of Jacobson (1977a), who reviewed numerous reports from controlled exposure studies to develop the concepts for determining limiting values and from the derivation of dose values for definite injury levels (Heck and Brandt 1977, Linzon et al. 1975).

2.6.1 Dose-Effect Values from Gas Exposure Studies and Surveys in Pollutant Impacted Areas

The previous presentations of the relationships between pollutant exposure and effects have shown how strongly the genetic resistance level can vary in relation to plant development and external factors – edaphic and climatic. Under these conditions, it is necessary to utilize data from several biological approaches, especially when they involve a complex pollutant mixture such as photochemical oxidants.

Results from the following types of pollutant studies were evaluated: fumigations in controlled environment chambers, greenhouse exposure chambers, field exposure chambers, with and without pollutant exclusion (air filtration) systems, and field studies. In reference to the presentations in Chapter 2.4.1 and 2.4.2, there is the general question of the applicability of the experimental results from such studies which deviate to varying degrees from natural conditions. Problems in the areas of cause and effect can also result from surveys of pollutant impacted areas and the observed effects cannot, unequivocally, be related to a specific cause or origin (Guderian 1978). Conducting studies under various conditions can reduce these risks. The majority of the published data from experimental and epidemiological studies were evaluated within the context of the criteria concerning pollutant exposure and its effects on the useful value of the impacted plants. These evaluations were used for the derivation of the maximum acceptable ozone concentrations (see Table 2.24), the attainment of which should provide vegetation, in its various functions, reasonable protection from ozone effects. Because of space limitations, it was not possible to reiterate all studies which contained dose-response values. In the tabular material concerning ozone (Tables 2.16, 2.17, 2.20), only examples are presented. The examples were selected to illustrate the range and types of results that can be used to support the derivation of air quality criteria. A more detailed listing of dose-response values is contained in the publications prepared for the Umweltbundesamt, Berlin and The European Community, Brussels (Guderian and Rabe 1982, Guderian et al. 1983) and the summary publications by Linzon et al. (1975) and EPA (1978a).

2.6.1.1 Dose-Effect Values from Pollutant Exposures

The importance of photochemical oxidants as phytotoxic air pollutants has received extensive experimental investigations because of their complex composition, containing several biologically active components, and wide distribution. The majority of these investigations were conducted with pot-grown plants in either the greenhouse or controlled-environment chambers. These methods with controlled pollutant exposure and varying degrees of climatic control have the advantage of greater reproducibility; they are particularly well suited for investigations of cause and effect and for hypothesis testing (cf. Chap. 2.3.4).

The pollutant exposures were conducted in chambers with varying deviations from natural growth conditions. Using artificial light sources with a composition similar to daylight, representative plant reactions, at least for short term exposures at higher concentrations, can be established.

Foliar injury is a principal response criterion that is frequently, but not always, associated with growth reductions. Also, growth reductions may occur with no visible foliar injury symptoms (cf. Chaps. 2.2.5 and 2.3.3). In addition to foliar symptoms, there are other parameters which may be used to evaluate impacts, e.g., effects on the growth of leaves, shoots and roots, yield, pollen germination, pollen tube growth, photosynthate production, respiration, or rhizobium-induced nodulation in legumes.

Table 2.16. Effects of defined amounts of ozone on plants; results from short-term exposures (≤ 1 d)

Plant species	Concen-tration (ppm)	Exposure duration (h)	Type and intensity of effect	References
28 different agricultural and horticultural species	0.25	1	Acute foliar injury on 21 of 28 exposed plant species	Heck and Tingey (1971)
Radish *(Raphanus sativus)* Tobacco *(Nicotiana tabacum)*, Bel W3 and White Gold	0.075	2	Acute leaf injury	
In addition to radish and tobacco also tomato *(Lycopersicon esculentum)*, smooth bromegrass *(Bromus inermis)*, oats *(Avena sativa)*, Lettuce *(Lactuca sativa)*	0.15	2	Acute leaf injury	
White clover *(Trifolium repens)* cv. Tillman	0.30	2	Reduction in shoot dry weight 17%; root dry weight 33% and nodule number 48%	Kochhar et al. (1980)
Tall fescue *(Festuca arundinacea)* cv. Kentucky 3	0.30	2 (3×)	Reduction in shoot dry weight, 22%; root dry weight 27%	
9 different turfgrass species and/or cultivars (3-year-old stand)	0.20	2	Foliar injury on 5 of 9 grass species and cultivars	G.A. Richards et al. (1980)
Kentucky bluegrass *(Poa pratensis)*, 7 cultivars (3-year-old stand)	0.10	3.5 h/day for 5 days	Foliar injury on 6 of 7 cultivars	
19 different turfgrass species and/or cultivars	0.30	3	Foliar injury on 14 of 19 grass species and cultivars	
Soybean *(Glycine max)* 4 cultivars	0.15–0.30 0.30–0.45	1.5 1.5	Threshold for foliar injury Threshold for reduction of shoot growth	Heagle (1979a)
Snap bean *(Phaseolus vulgaris)*	0.30	2 × 1.5	Up to 71% foliar injury to primary leaves and 36% on other leaves; plant dry weight reduced 10% and pod dry weight 12%	Blum and Heck (1980)
	0.60	2 × 1.5	Up to 95% foliar injury to primary leaves and 68% on other leaves; plant dry weight reduced 25% and pod dry weight 41%	
Pinto bean *(Phaseelus vulgaris)*	0.05 0.10	24 12	Significant reduction in leaf growth Significant reduction in leaf growth	Evans (1973)
Tobacco *(Nicotiana tabacum)*, Bel W3	0.10	5.5	In vivo and in vitro exposure reduced pollen germination 45% and pollen tube growth 50%	Feder (1968)

Table 2.16 (continued)

Plant species	Concentration (ppm)	Exposure duration (h)	Type and intensity of effect			References
Grapevine (*Vitis labrusca*), 2 cultivars	0.40	4		Ives	Delaware	Shertz et al. (1980a)
			Leaf injury	27%	1%	
			Reduced shoot growth	40%	17%	
			Premature leaf drop	5%	–	
	0.08	4	Leaf injury	67%	56%	
			Reduced shoot growth	60%	33%	
			Premature leaf drop	15%	–	
18 different coniferous tree species	0.10	8	Needle injury on Austrian pine (*Pinus nigra*), jack pine (*Pinus banksiana*), and Virginia pine (*Pinus virginiana*)			D. D. Davis and Wood (1972)
	0.25	4	Needle injury on Austrian pine (*Pinus nigra*), jack pine (*Pinus banksiana*), and European larch (*Larix decidua*)			
	0.25	8	Needle injury on all 8 coniferous species			

For ozone, exposure durations ≤ 1 day (Table 2.16) and ≥ 1 day (Table 2.17), will differentiate between experimental results. Short-term studies are concerned primarily with a single exposure on a single day; the longer-lasting exposures encompass exposure periods greater than a single day to several months or cover the complete vegetation period. The injurious concentrations vary over a wide range, depending on individual plant species, cultivars, developmental stage, and external growth conditions and also duration of exposure. For ozone, the lowest limit for injury follows several hours of exposure to a concentration range of 0.02 to 0.05 ppm (Jacobson 1977 a).

In addition to the authors cited in Table 2.16, the dose-response values from the research of the following authors was considered. Adedipe et al. (1972 a, 1973 a), Adedipe and Ormrod (1974), D. D. Davis (1977), D. D. Davis and Kress (1974), Feder and Campbell (1968), Feder et al. (1969 a), Houston (1974), Jacobson and Colavito (1976), Khatamian et al. (1973), Letchworth and Blum (1977), MacDowall and Cole (1971), Manning et al. (1972), V. L. Miller et al. (1974), O'Connor et al. (1975), Ormrod (1977), Ormrod et al. (1971), Pell and Brennan (1973), Proctor and Ormrod (1977), Shertz et al. (1980 b), Tingey and Blum (1973), Tingey et al. (1973a), Tingey and Reinert (1975), Townsend and Dochinger (1974).

In addition to the authors cited in Table 2.17, the dose-response data for exposure periods greater than 1 day from the following articles were evaluated:

Table 2.17. Effects of defined amounts of ozone on plants; results from chronic exposures (≥ 1 d)

Plant species	Concen-tration (ppm)	Exposure duration (h)	Type and intensity of effect	References
Winter wheat (*Triticum aestivum*)	0.10	54 days, 7 h daily	Seed yield reduced: container grown 10%, field grown 16%,	Heagle et al. (1979e)
	0.13	54 days, 7 h daily	seed yield reduced: container grown 27%, field grown 33%	
Corn (*Zea mays*)	0.03	60 days, 5.5 h daily	No effect on pollen germination	Mumford et al. (1972)
	0.06	60 days, 5.5 h daily	Pollen germination reduced about 50%	
	0.12	60 days, 5.5 h daily	Pollen germination reduced about 60%	
Corn (*Zea mays*) Golden	0.05	64 days, 6 h daily	14% leaf injury, seed weight reduced about 9%	Heagle et al. (1972)
	0.10	64 days, 6 h daily	25% leaf injury, seed dry weight reduced about 45%	
Orchardgrass (*Dactylis glomerata*), Perennial Ryegrass (*Lolium perenne*)	0.09	5 weeks, 5 days/week, 4 h daily	Some leaf chlorosis, shoot dry weight reduced 14 to 21%	Horsman et al. (1980)
Tall fescue (*Festuca* sp.)	0.08	6 weeks	Leaf dry weight reduced about 17%, shoot dry weight reduced about 15%, and tillering about 8%	Johnston et al. (1980)
Crimson Clover (*Trifolium incarnatum*) and	0.03	6 weeks, 8 h daily	Yield reduced <10%, not significant	J.P. Bennett and Runeckles (1977)
Italian Ryegrass (*Lolium multiflorum*) as monocultures and as a mixture	0.09	6 weeks, 8 h daily	Reductions in monocultures Crimson Clover: leaf area, 50%; yield (dry weight), 36%; Italian Ryegrass: leaf area, 35%; yield (dry weight), 36%. Reductions in the mixture of Crimson Clover and Italian ryegrass: leaf area 26%, yield (dry weight) 22%	
Alfalfa (*Medicago sativa*)	0.05	68 days, 7 h daily	Shoot dry weight reduced 30 and 50% for the 1st and 2nd harvests, respectively	Neely et al. (1977)

Plant species	Concen-tration (ppm)	Exposure duration (h)	Type and intensity of effect	1977	1978	References
Potato (*Solanum tuberosum*) 2 cultivars	0.20	6 × 3 h biweekly	*Norland* Reduction in tuber number	19%	21%	Pell et al. (1980)
			Reduction in tuber weight	30%	20%	
			Kennebeck Reduction in tuber number	40%	32%	
			Reduction in tuber weight	54%	30%	

Table 2.17 (continued)

Plant species	Concen- tration (ppm)	Exposure duration (h)	Type and intensity of effect	References
Soybean (Glycine max) Corsoy	0.064 0.079 0.094	55 days, 9 h daily	Seed yield reduced about 31% Seed yield reduced about 45% Seed yield reduced about 56%	Kress and Miller (1981)
Soybean (Glycine max), 4 cultivars	0.06–0.09 0.09–0.12	10 days, 6 h daily 10 days, 6 h daily	Visible injury on all 4 cultivars, primarily chlorosis Shoot growth of 2 cultivars reduced	Heagle (1979a)
Pinto bean (Phaseolus vulgaris)	0.06	40 days, 5 days/ week	Height growth reduced about 26%, shoot and root dry weights 48 and 50%, respectively, rhizobium nodule number and weight reduced about 45 and 74%, respectively	Manning (1978)
Radish (Raphanus sativus)	0.05	5 weeks, 5 days/ week, 8 h daily	Plant dry weight reduced about 31%, root fresh weight and leaf fresh weight about 54 and 20%, respectively	Tingey et al. (1971)
Spinach (Spinacia oleracea), 11 cultivars, values are the mean of 11 cultivars	0.06 0.10 0.13	37 days, 7 h daily 37 days, 7 h daily 37 days, 7 h daily	Leaf necrosis 6%, reductions in fresh weight, container grown 4%, field grown 18% Leaf necrosis 38%, reductions in fresh weight, container grown 25%, field grown 37% Leaf necrosis 65%, reductions in fresh weight, container grown 65%, field grown 69%	Heagle et al. (1979b)
Cotton (Gossypium hirsutum), Alcala SJ2	0.25 0.25	15 weeks, 2 × weekly for 6 h daily 22 weeks, 2 × weekly for 6 h daily	Reduced weight for open bolls about 60%, seeds about 60%, and ginned fibers about 62% Reduced weight for open bolls about 59%, seeds about 60%, and ginned fibers about 59%	Oshima et al. (1979)
Petunia (Petunia hybrida)	0.05–0.07	53 days, 24 h daily	Flower fresh weight reduced about 30%	Craker (1972)
Carnation (Dianthus caryophyllus)	0.05–0.09	90 days, 24 h daily	Leaf tip necrosis; reduced vegetative growth, and flower- ing reduced about 50%	Feder (1970a)
Geranium (Pelargonium hortorum)	0.07–0.10	90 days, 9.5 h daily	Leaf chlorosis and leaf drop; reduced vegetative growth, flowering reduced 50%, and decreased flower retention	

Table 2.17 (continued)

Plant species	Concen-tration (ppm)	Exposure duration (h)	Type and intensity of effect	References
Ponderosa pine (*Pinus ponderosa*)	0.15	10 days, 9 h daily	Reduction in apparent photosynthesis about 4%	P. R. Miller et al. (1969)
	0.15	20 days, 9 h daily	Reduction in apparent photosynthesis about 25%	
	0.15	60 days, 9 h daily	Reduction in apparent photosynthesis about 34%	
	0.30	10 days, 9 h daily	Reduction in apparent photosynthesis about 12%	
	0.30	20 days, 9 h daily	Reduction in apparent photosynthesis about 50%	
	0.30	30 days, 9 h daily	Reduction in apparent photosynthesis about 72%	
	0.45	30 days, 9 h daily	Reduction in apparent photosynthesis about 85%	
Sugar maple (*Acer saccarum*)	0.05	2 days, 7–11 h daily	Net photosynthesis reduced about 30%, 21%, and 0%, for the three cultivars, respectively	Carlson (1979)
Black oak (*Quercus velutina*)	0.05	1 week, 7–11 h daily	Net photosynthesis reduced about 48%, 27%, and 0%, for the three cultivars, respectively	
White ash (*Fraxinus americana*)	0.05	3 weeks, 7–11 h daily	Net photosynthesis reduced about 34%, 55%, and 6%, for the three cultivars, respectively	
Conifer species	0.10	20 weeks, 6 h daily	The most severe foliar injury occurred on ponderosa pine (20%), lodgepole pine (16%), and shore pine (13%), significant growth reduction occurred in ponderosa pine and eastern white pine	Wilhour and Neely (1977)
Yellow poplar (*Liriodendron tulipifera*)	0.10	6 weeks, 12 h daily	Reduction in relative growth rate about 37%, relative leaf area growth rate about 34%, and net photosynthesis about 17%	Jensen (1981b)
Eastern cottonwood (*Populus deltoides*)	0.10	6 weeks, 12 h daily	Reduction in relative growth rate about 59%, relative leaf area growth rate about 52%, and net photosynthesis about 58%	

Barnes (1972a), Craker and Feder (1972), Dochinger and Seliskar (1970), Engle and Gabelman (1967), Feder and Campbell (1968), Heagle et al. (1974), Heagle et al. (1979c), Henderson and Reinert (1979), Hoffman et al. (1973), Jensen (1973), Maas et al. (1973), Manning and Vardaro (1974a, b), Ogata and Maas (1973), Oshima (1973), Oshima et al. (1975), Price and Treshow (1972), Reinert and Nelson (1979), Shannon and Mulchi (1974), Tingey et al. (1973c), Tingey and Reinert (1975), Wammsley et al. (1980).

Table 2.18. Effects of defined amounts of PAN of plants; results from controlled exposures

Plant species	Concentration (ppm)	Exposure duration (h)	Type and intensity of effect	References
Petunia *(Petunia hybrida)*	100	5	Severe foliar injury	Stephens et al. (1961)
Pinto bean *(Phaseolus vulgaris)*	100	5	Moderate foliar injury	
Petunia, tomato	14	4	Foliar injury (field observations)	O.C. Taylor (1969)
Petunia *(Petunia hybrida)*	140	1	33% foliar injury	
Pinto bean *(Phaseolus vulgaris)*	140 40 20	1 4 8	55% foliar injury 90% foliar injury 44% foliar injury	
Petunia, tomato, dwarf meadowgrass, romaine lettuce	15–20	4	"May be severely damaged ... in the polluted atmosphere"	O.C. Taylor and McLean (1970)
Petunia *(Petunia hybrida)*	200 100 50	1/2 1 2	74 ⎫ 97 ⎬ injury index, scale 116 ⎭ 0–500	Drummond (1972)
Petunia *(Petunia hybrida),* 6 cultivars	120 250 500	1 1 1	A trace to slight foliar injury on 5 of 6 cultivars A trace to moderate foliar injury on all 6 cultivars Slight to moderate foliar injury on all 6 cultivars	Feder et al. (1969a)
Petunia *(Petunia hybrida)* cv. White Cascade	150	1.5	24% foliar injury	Pell and Gardner (1979)
Petunia *(Petunia hybrida)*	32 14 7	1 3 8	Leaf injury threshold concentration for each combination of concentration and duration	Nouchi (1979)
Pinto bean *(Phaseolus vulgaris)*	150	3	21% foliar injury	Pell (1976)
Bean *(Phaseolus vulgaris),* cv. Provender	80	0.5	Disruption in plant water relations	Starkey et al. (1981)
Pinto bean *(Phaseolus vulgaris)*	50	4	Mild foliar injury, primarily abaxial bronzing with light bifacial necrosis	Kohut and Davis (1978)
Chard, spinach, japanese radish, kidney bean, tomato, lettuce	100	6–10	Foliar injury	Nouchi et al. (1975)
Petunia *(Petunia hybrida)*	30	6–10	Foliar injury	
Lettuce *(Lactuca sativa)* cv. Empire	20	2 × weekly, 4 h each from seedling to maturity	Reduction in leaf fresh weight	O.C. Taylor et al.(1981) (cit.in Temple 1982)

Table 2.18 (continued)

Plant species	Concen-tration (ppb)	Exposure duration (h)	Type and intensity of effect	References
Lettuce *(Lactuca sativa)* cv. Empire	40	2 × weekly, 4 h each from seed-ling to maturity	Reduction in leaf fresh weight	
Lettuce *(Lactuca sativa)* cv. Empire	50	1 × weekly, 4 h, 3 to 4 weeks	15% foliar injury, no effect on growth	Temple (1982)
Tomato *(Lycopersicon esculentum)*	50	1 × weekly, 4 h, 3 weeks	Significant reductions in plant height, leaf and root dry weights, but no foliar injury	
Maple, ash, locust, and oak spp.	200–300	8	Differentially severe injury depending on the age-dependent susceptibility	Drummond (1971)
Little leaf nettle *(Urtica urens)*	50 50	5 × 7 h 9 × 2 h	20% foliar injury 12% foliar injury	Posthumus (1977)
Annual bluegrass *(Poa annua)*	50 50	6 × 7 h 9 × 2 h	3% foliar injury 2% foliar injury	

The limited number of experimental PAN exposures are presented in Table 2.18, and include exposure periods longer than 24 h. The lowest PAN concentration that caused foliar injury on petunia was 7 ppb for 8 h (Nouchi 1979).

2.6.1.2 Dose-Effect Values from Field Studies and Pollutant-Impacted Areas

Surveys of dose-effect relationships in field studies or pollutant-impacted areas have significant advantages. In contrast to the extrapolation problems from controlled pollutant exposures, the experimental results are representative of the location and may be generalized to areas with similar soil and climatic conditions. Also, the frequent use of response data relating to reductions in useful values of plants as response criteria in these studies is advantageous. Nevertheless, there are difficulties, namely the identification and evaluation of the injurious components and confounding factors, the determination of the pollutant dose and the establishment of the damage intensity (Table 2.19).

To determine pollutant exposure, ozone is usually measured; total oxidants are measured less often. A lack of reliable monitoring techniques that can relate total oxidants concentrations to vegetation effects contributes to these limited measurements. The exposure dose can be determined more or less exactly from information on the ozone frequency distribution at specific concentration ranges (e. g., Table 2.20). Lefohn and Benedict (1982) developed an "integrated Expo-

Table 2.19. Effects of defined amounts of oxidants (ozone) on plants; results from studies in polluted areas

Plant species	Concentration (ppm)	Exposure duration	Type and Intensity of effect	References
Corn *(Zea mays)*, 2 cultivars	<0.08 0.08–0.16 <0.06	68 days from germination to harvest 467 h = 42.4% 366 h = 33.3% 266 h = 24.3% of light period from 6.00 h–21.00 h	On Monarch Advance, the ozone-sensitive cultivar the yield (dry weight) was reduced about 58%; also there was no foliar injury, premature chlorosis and senescence; on Bonanza, the ozone-resistant cultivar, the yield (dry weight) was reduced about 44% with slight foliar injury	Thompson et al. (1976a)
Potato *(Solanum tuberosum)*	≥0.05	326–444 h in 2 years	Yields reduced about 34–50% (2 years for 2 cultivars) 20–26% (1 year for 2 cultivars)	Heggestad (1973)
Alfalfa *(Medicago sativa)*, 2 cultivars	>0.08 >0.12 >0.16	Monthly mean, proportion of light period ~ 50% ~ 30% ~ 20%	Mean reduction in forage 33 to 42% and reductions in crude fiber, β-carotene, and vitamin C in both cultivars	Thompson et al. (1976b)
Bush bean *(Phaseolus vulgaris)*, 4 cultivars	≥0.10 ≥0.05	From seedling to harvest, 60 days, 1 to 40 h ~ 250–480 h, in Beltsville, Maryland, during the growth periods 1972–1976, 2 plantings each year	Yield reductions between open-top chambers with and without air filtration ranged between 5 and 27% and premature leaf drop on oxidant sensitive cultivars during the growth periods of 1972 to 1976	Heggestad et al. (1980)
Bush bean *(Phaseolus vulgaris)*, cv. Tendergreen	0.041 (daily mean) 0.017–0.09 (variation about daily mean)[a]	43 days	Reduction in the number and weight of marketable beans, about 24 to 26% compared to the controls which received 60–70% less oxidant exposure	MacLean and Schneider (1976)
Tobacco *(Nicotiana tabacum)* Bel W3	0.02–0.03	6–8 h	Trace of foliar injury	Heck and Dunning (1967)
Orange *(Citrus sinensis)*	>0.10	148 h monthly average, from March to October; 254 h monthly average from July to September	Yield reduced 54%	Thompson and Taylor (1969)

Table 2.19 (continued)

Plant species	Concentration (ppm)	Exposure duration	Type and Intensity of effect	References
Navel Orange	0.00–0.69[b] 0.02–0.67[c]	8 months from blossom to harvest	Increased leaf and fruit drop; significant reduction in fruit yield in the absence of visual injury[d]	Thompson et al. (1972)
Eastern white pine (*Pinus strobus*)	>0.03	10–15 h continuously	Chlorotic fleck on new needles	Linzon (1973)
Ponderosa pine (*Pinus ponderosa*)	>0.15	2 h daily over 2 months	Chlorotic decline	P. R. Miller et al. (1969)

[a] Oxidant concentration
[b] Measured peak oxidant concentrations
[c] Measured maximum hourly average of oxidant for individual months
[d] In experiments units with filtered ambient air and comparable O_3 concentrations, there was less damage observed; additional evidence that other compounds of photochemical smog, i.e., PAN and NO_x, increase the ozone toxicity

sure Index" that expressed the frequency distribution as a single value that the authors believed could be related to plant response.

The precision with which effects are established depends on the type of field study. Investigations in open-top (for example, MacLean and Schneider 1976, Heggestad et al. 1980) or closed-top (for example, Thompson and Taylor 1969) chamber systems with regulated pollutant addition systems or filtered and unfiltered ambient air exposures allow a precise determination of effects, i. e., growth or yield. However, with surveys of cultivated or native plants (for example, Brennan and Rhoads 1976, Mosley et al. 1978) without similar chambers, one should not expect comparable results because a "control" is lacking.

The data in Tables 2.19 and 2.21 were developed using various experimental designs and ozone monitoring techniques. Different measures of growth and yield have been used; it is not clear if these different response measures are equally sensitive to ozone. The ozone exposure methods included: ambient air, ambient air with supplemental ozone additions, and ozone additions to charcoal-filtered air. However, most ozone exposures utilized constant ozone addition for fixed time periods, producing artificial exposure regimes. A wide range of ozone concentrations and exposure durations were used. Although the published reports usually contain the concentration, duration, and sometimes exposure frequency, the number of exposures, the diurnal timing of exposures, and the plant development stages are not given. In some studies the plant received "square wave" (constant ozone concentrations for a fixed time period) exposures interspersed with pollutant-free periods of a few hours to several days, while in other studies variable concentrations were used during the ozone exposure periods. In most of these studies, the ozone exposures were characterized by a mean concentration, how-

Table 2.20. An example of selected ozone frequency distributions from several sites in California. (EPA 1979)

County	Hourly ozone concentrations (L) $\mu g\ m^{-3}$ percentile								Integrated[a] exp. index ($\mu g\ m^{-3}$ h)
	10	30	50	70	90	95	99	Max	
Riverside	10	10	20	39	176	255	372	725	208,808
Kern	10	20	39	59	137	176	216	294	50,070
Fresno	10	10	20	59	118	137	196	294	20,031
Merced	10	20	39	59	118	137	196	274	19,447
Sacramento	10	20	20	59	98	137	216	431	35,082

[a] Calculated according to the method of Lefohn and Benedict (1982)

Table 2.21. Percent yield reduction as a function of ozone concentration[a]. (Heck et al. 1982)

Crop	Linear function[b]	Predicted % yield reduction at two O_3 concentrations[d] ($\pm SE$)	
		0.06 ppm	0.10 ppm
Corn (Coker 16)	$Y = -2.7 + 108x$	3.83 ± 1.0	8.1 ± 2.1
Soybean (Corsoy)	$Y = -15.5 + 621x$	21.7 ± 1.4	46.5 ± 2.9
Soybean – 1977 (Davis)	$Y = -11.1 + 443x$	15.5 ± 2.4	33.2 ± 5.1
Soybean – 1978 (Davis)	$Y = -8.8 + 353x$	12.3 ± 3.6	26.4 ± 7.7
Kidney bean (Calif. Light Red)	$Y = -4.8 + 193x$	6.8 ± 1.3	14.5 ± 2.8
Lettuce, Head (Empire)	$Y = -16.3 + 652x$	22.8 ± 1.7	48.9 ± 3.6
Peanut (NC-6)	$Y = -17.8 + 711x^c$	24.9 ± 0.7	53.3 ± 1.5
Spinach (America)	$Y = -13.2 + 527x$	18.5 ± 3.0	39.6 ± 6.4
Spinach (Hybrid 7)	$Y = -13.0 + 517x$	18.1 ± 3.7	38.9 ± 8.0
Spinach (Viroflay)	$Y = -14.8 + 594x$	20.7 ± 3.1	44.4 ± 6.6
Spinach (Winter Bloomsdale)	$Y = -14.9 + 599x$	20.9 ± 3.4	44.8 ± 7.3
Turnip (Just Right)	$Y = -22.1 + 887x^c$	31.0 ± 2.1	66.4 ± 4.5
Turnip (Purple Top White Globe)	$Y = -20.5 + 817x^c$	28.6 ± 2.4	61.4 ± 5.0
Turnip (Shogoin)	$Y = -20.5 + 818x$	28.6 ± 2.6	61.4 ± 5.5
Turnip (Tokyo Cross)	$Y = -19.1 + 763x^c$	26.8 ± 4.6	57.4 ± 9.8
Wheat (Blueboy II)	$Y = -7.4 + 290x$	10.4 ± 1.4	22.3 ± 3.0
Wheat (Coker 47–27)	$Y = -10.1 + 405x$	14.2 ± 1.2	30.4 ± 2.6
Wheat (Holly)	$Y = -7.5 + 304x^c$	10.6 ± 1.5	22.6 ± 3.3
Wheat (Oasis)	$Y = -6.2 + 250x$	8.7 ± 1.7	18.6 ± 3.7

[a] This includes all data sets. The O_3 is expressed as the seasonal 7-h-day^{-1} mean concentration

[b] $Y = \dfrac{100 b_1}{a} [0.025 - x]$

where Y = predicted percent yield reduction, b_1 = the regression coefficient from the yield model, a = the predicted yield at a seasonal 7-h-day^{-1} mean O_3 concentration of 0.025 ppm, and x = seasonal 7-h-day^{-1} mean O_3 concentration

[c] These models must be considered with caution. More complex models account for significantly more of the variation than these simple linear models. Compare the predicted percent yield reductions for the plateau-linear model at 0.06 ppm to get an idea of the differences

[d] Values may differ slightly from those predicted by the equations due to a rounding of numbers in the functions shown. The 0.06 ppm seasonal 7-h-day^{-1} mean O_3 concentration is the maximum mean concentration expected in many parts of the US when the 1-h^{-1} standard of 0.12 ppm is just met; the 0.10 ppm was arbitrarily used as a maximum 7-h-day^{-1} mean O_3 concentration for comparison

ever, other exposure parameters have been used (Jacobson 1982). The use of a mean concentration is not reliable because it assumes that all products of time and concentration produce equal effects, an assumption which has not been verified by research (see Chap. 2.4.1). Also, there is no direct, verified method to relate mean ozone concentrations or the results of most experimental exposures to ambient pollutant regimes. Jacobson (1982) summarized the experimental approaches to determine ozone effects on crop productivity and concluded that the studies are rarely performed or described in terms that can be compared to ambient air quality standards. Despite these confounding factors and limitations, the data in Tables 2.19 and 2.21 are the best available to describe ozone effects on plant growth and to establish limiting values to protect plants from growth and yield losses.

The data in Table 2.19 are only examples as in Tables 2.16 and 2.17. Also, the work of the following authors was considered: Bell and Cox (1975), Bisessar and Temple (1977), Brasher et al. (1973), Brennan and Rhoads (1976), Cameron et al. (1970), Heagle et al. (1973), Howell et al. (1976), Mosley et al. (1978), Thompson and Kats (1970), Thompson et al. (1969), Thompson and Taylor (1969), Weaver and Jackson (1968).

2.6.2 Methods to Derive Limiting Factors

The level of certainty for limiting values depends on the amount and quality of data and the models used to evaluate it (Jacobson 1977a). The method for deriving limiting values is based on these criteria. The models should describe, with reasonable accuracy, the relationship between exposure and effect. Limiting values are generally useful for establishing air-quality criteria and they also allow the evaluation of the phytotoxic risk potential from measured exposure concentrations and may assist in remedial or mitigative actions.

With present knowledge, the following two models are suited for the stated role: (1) the concept of determining limiting values; and (2) the determination of specific doses that cause specific injury levels. Both models are suited for short-term effects from acute injurious concentrations. However, for intermittent and long duration exposures at low concentrations, both methods are deficient. At present, the data are lacking to develop and use such models.

2.6.2.1 Determination of Limiting Values

The concept of limiting values elaborated by Jacobson (1977a) during his presentation to the VDI-Congress, *Ozon und seine Begleitsubstanzen im photochemischen Smog* at Düsseldorf (1976) resembles the models of McCune (1969) for fluorides and Larsen and Heck (1976) for ozone and sulfur dioxide, singly. The objective of these models is to determine the boundary conditions between doses that are probably injurious and those that are not.

Limiting values for ozone and PAN (Figs. 2.16–2.18) were developed from studies conducted in the USA that contained definite information concerning pol-

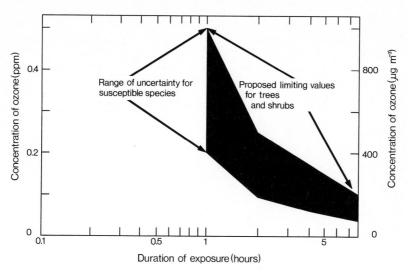

Fig. 2.16. Proposed limiting values for foliar injury on trees and shrubs by ozone. (Jacobson 1977a) For further explanation see text

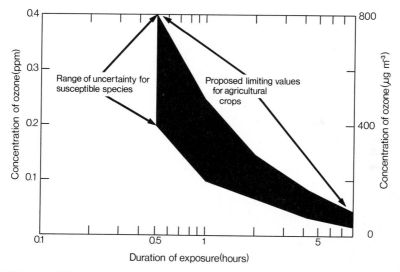

Fig. 2.17. Proposed limiting values for foliar injury on agricultural crops by ozone. (Jacobson 1977a) For further explanation see text

lutant concentration and duration of exposure. Each study was represented by a point in the coordinate system representing the lowest concentration and duration of exposure used in the study that induced foliar injury symptoms. The data sets used did not necessarily establish an injury threshold which would prevent foliar injury if obtained, because the data sets reflected the limitations in measurement and experimental techniques. In addition, it is possible to establish only the

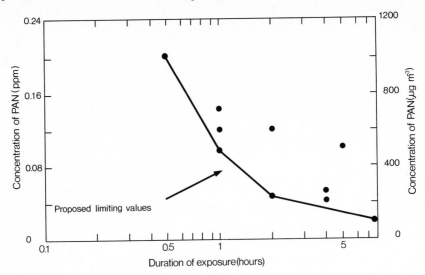

Fig. 2.18. Dose-response relationships and proposed limiting values for PAN-induced foliar injury (Jacobson 1977 a). For further explanation see text

injuriousness of an exposure but not the harmlessness. Despite extensive investigations, it is still possible that combinations of conditions that were previously thought to be harmless can now be shown to cause injury. It follows that the derivation and the subsequent observance of limiting values will produce a high, but not absolute, guarantee of protecting vegetation; a small risk will remain.

As a response criterion, Jacobson (1977 a) used visible foliar injury including necrotic lesions, chlorosis, or pigmentation. The limiting values are based on dose-response relationships for visible injury on the assumption that if visible foliar injury is prevented, other adverse effects are unlikely. Although this assumption is not always valid, it provides the only basis for current estimates of limiting values.

With the previously discussed criteria, individual data sets were used to derive limiting values for reducing ozone-induced foliar injury to trees and shrubs (Fig. 2.16). The presentation is based on data from: Berry (1971, 1974), Berry and Ripperton (1963), Costonis and Sinclair (1969 b), D. D. Davis and Coppolino (1974 a, b), D. D. Davis and Wood (1972, 1973 b), Drummond (1972), Hibben (1969 a, b), Hill et al. (1961), Jensen (1973), Jensen and Dochinger (1974), Jensen and Masters (1975), Ledbedder et al. (1959), Karnosky (1976), Kohut (1975), Kress (1972), Linzon (1967), P. R. Miller et al. (1969), P. M. Miller and Rich (1968), B. L. Richards et al. (1958), Santamour (1969), Wilhour (1970 b), Wood and Coppolino (1972).

In Figs. 2.16 and 2.17, the shaded area gives the range for the lowest ozone doses causing foliar injury to a range of plant species. Doses above and to the right of the shaded area are likely to cause foliar injury on susceptible plant species. It is unlikely that ozone doses below and to the left of the shaded area will cause foliar injury. However, certain highly susceptible clones that were selected

Table 2.22. The range in limiting values for foliar injury from ozone on trees, shrubs, and agricultural crops. (Jacobson 1977a)

Duration	Trees and shrubs		Agricultural crops	
h	$\mu g\ m^{-3}$	ppm	$\mu g\ m^{-3}$	ppm
0.5			400–800	0.2 –0.4
1.0	400–1,000	0.2 –0.5	200–500	0.1 –0.25
2.0	200– 500	0.1 –0.25		
4.0	120– 340	0.06–0.17	75–180	0.04–0.09

for their ozone sensitivity may show injury at ozone doses below the curves. Data for these plants were not included because they may not be representative of the sensitivity of native and cultivated plants.

In comparison to the woody plants, there is considerably more dose-response information available for agricultural crops (cf. Table 2.16). In Fig. 2.17, the shaded area, representing the uncertainty range between probably harmless and injurious ozone doses, is especially large because (1) the large number of plant species tested exhibited considerable differences in their relative sensitivity and (2) there were substantial differences in external growth conditions among the various studies. The uncertainty area includes the concentration range producing 5% injury on agricultural crops as reported by Heck and Tingey (1971) and corresponds closely to the response threshold published by Linzon et al. (1975).

For long-term or repeated exposures, the data are not sufficient to determine if the curves in Figs. 2.16 and 2.17 depart from their asymptotic approach to parallel the abscissa. Interpretations of the data for repeated or long term exposures are difficult and limiting values at concentrations less than 0.05 ppm (100 μg m^{-3}) are not practical because of inaccuracies in the older air-monitoring equipment (Jacobson 1977a). Until additional information is obtained, Jacobson (1977a) assumed that limiting values for exposures up to 8 h would also protect plants from foliar injury from long-term or repeated exposures. Considering the mode of action of ozone, and also the possible growth reductions in the absence of visible foliar injury, such protection is not probable; however, at this time no other approaches are practical.

The range of limiting values for exposures up to 4 h listed in Table 2.22 were derived from Figs. 2.16 and 2.17. According to Jacobson (1977a), limiting values for exposure times >4 h had little significance because of the difficulties in measuring low concentrations and, at longer durations, the limiting values were in the same range as the background (natural) ozone concentrations.

It is disproportionately more difficult to evaluate the potential hazards from PAN than from ozone. PAN, as mentioned earlier, is the most important phytotoxic component of the family of peroxy compounds formed within the atmosphere; however, it is not the most phytotoxic. The studies of O. C. Taylor (1969) showed that peroxypropionyl nitrate (PPN) was more injurious than PAN and peroxybutyryl nitrate (PBN) was more injurious than PPN. The ambient concentrations of the PAN analogs are thought to be too small to be injurious by themselves, but they may possibly increase the phytotoxicity of PAN (O.C. Taylor 1969).

For PAN, the dose-response data are too limited to derive separate limiting values for trees and agricultural crops (Jacobson 1977a). Also, no range of uncertainty can be given for the doses likely to be injurious or harmless. Each coordinate in Fig. 2.18, as with ozone, corresponds to the lowest PAN concentration and exposure duration producing foliar injury as reported in: Drummond (1972), Feder et al. (1969a), Jaffe (1967), Kohut (1975), Kress (1972), R.G. Pearson et al. (1974), Pell (1976), Starkey (1975), Wood and Drummond (1974), Wood et al. (1969). The solid line in Fig. 2.18 provides a first approximation of limiting values for foliar injury from PAN. To the right and above the line, there is an appreciable risk of foliar injury, but for doses below and to the left there is low risk. Posthumus and Tonneijck (1982) studied the dose-response relationship of little leaf nettle (*Urtica urens*) under conditions of maximum sensitivity. They found a similar type curve as Jacobson (1977a) (Fig. 2.18) but the line was displaced to the left, indicating that their species was more sensitive.

Jacobson (1977a) derived the following concentrations for reducing the probability of PAN-induced foliar injury:

1,000 μg m^{-3} (200 ppb) for 0.5 h exposure duration
500 μg m^{-3} (100 ppb) for 1.0 h exposure duration
175 μg m^{-3} (35 ppb) for 4.0 h exposure duration

The level of protection to vegetation that would be achieved by the attainment of the air quality criteria proposed by Jacobson (1977a) cannot be determined at this time. The uncertainty results not only from limited data, but also because the data were derived from controlled exposures. It is thought that plants exposed to PAN in controlled exposures are less sensitive than when exposed under field conditions. Such differences have been observed (Stephens et al. 1961, O.C. Taylor 1969) and confirmed by others (Floor and Posthumus 1977, Temple 1982). In controlled PAN exposures (25 to 50 ppb, for one 4-h exposure per week for 3 weeks), no symptoms were observed on four tomato cultivars (Temple 1982); however, under field conditions, sensitive plants such as tomato and petunia displayed foliar injury following a 4-h exposure to 14 ppb PAN (O.C. Taylor 1969). In contrast, Nouchi (1979), using controlled exposures, found low injury thresholds for petunia at 32, 14, and 7 ppb for 1,3, and 8 h exposure durations respectively.

Available data on ambient concentrations of PAN are limited, restricting the determination of possible risk to vegetation. Ambient air monitoring at Riverside, California, during 1980 for the hours of 8.00 to 20.00 found monthly maximum concentrations of 8.1 ppb – July, 9.4 ppb – September, 8.8 ppb – October (Temple and O.C. Taylor 1983). Hourly average concentrations greater than 30 ppb occurred 23 times between April and October. "Potentially phytotoxic episodes," defined as PAN concentrations exceeding 15 ppb for 5 h in the morning or 25 to 30 ppb for 4 h in the afternoon, occurred at Riverside 27 times during 1980. Along with this, PAN-induced foliar injury was observed numerous times on sensitive plant species. Based on this, concentrations for "potentially phytotoxic episodes" and limited measurements of atmospheric concentrations, it appears that PAN may endanger vegetation in only a few regions (Temple and O. C. Taylor 1983). The ambient PAN measurements for Europe are from a few to several times less than the California concentrations (see 1.3.2.2 and 1.3.2.2.2), and as a single atmospheric component should pose only a limited risk to vegetation (Posthumus and Tonneijck 1982; Tonneijck 1983a). However, it should be realized that these concentrations represent limited data from sporadic monitoring and are scarcely more than chance occurrences. Since Darley (1963/64) first identified PAN injury on annual bluegrass (*Poa annua*) in the Federal Republic of Germany, repeated observations of PAN injury to many plant species in The Netherlands have been made (ten Houten 1966, Posthumus 1977), verifying the occurrence of phytotoxic PAN concentrations.

2.6.2.2 Dose Values for Definite Injury Levels

The dose-response function proposed by O'Gara (1922) has been shown to be suitable for deriving dose values that cause a defined amount of injury from short term exposures to acute injurious concentrations (cf. Chap. 2.4.1). Heck and Tingey (1971) modified the O'Gara function to predict ozone injury; using this function, Linzon et al. (1975) and Heck and Brandt (1977) developed curves for definite injury levels on plants. These curves, illustrated in Fig. 2.19, show time-concentration combinations that cause 5% injury levels. The equation for the 5% foliar injury is: c (concentration in ppm) $= 0.16 + 0.12\,t^{-1}$ (t in h) (Linzon et al. 1975). The concentration for 33% foliar injury on plants is: $c = 0.21 + 0.21\,t^{-1}$ (Heck and

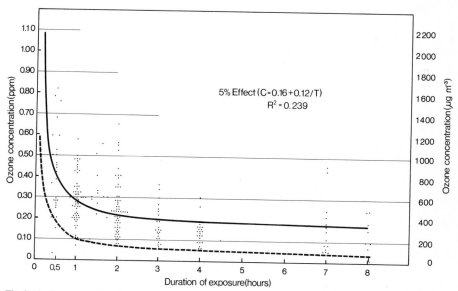

Fig. 2.19. Concentration-time graph for foliar ozone injury developed from a single short-term acute exposure to injurious concentrations using the O'Gara function: $c = a + b/t$. *Solid* line mean 5% injury level for all the species studied; *dashed line* threshold injury level for sensitive plants. (Linzon et al. 1975)

Table 2.23. Ozone concentrations for short-term exposures that will produce 5% detectable injury on plants of various resistance levels grown under sensitive conditions. (Heck and Brandt 1977)

Exposure duration	Resistance level					
	Sensitive		Intermediate		Less sensitive	
h	$\mu g\, m^{-3}$	ppm	$\mu g\, m^{-3}$	ppm	$\mu g\, m^{-3}$	ppm
0.5	400	0.20	600	0.30	1,000	0.50
1.0	200	0.10	400	0.20	600	0.30
2.0	140	0.07	300	0.15	500	0.25
4.0	100	0.05	240	0.12	460	0.23
8.0	60	0.03	200	0.10	400	0.20

Brandt 1977). Larsen and Heck (1976) reviewed foliar injury data from numerous plant species and concluded: (1) that a constant percentage of foliar injury was caused by a pollutant concentration that was negatively related to the exposure duration raised to an exponent and (2) for a given exposure duration, the amount of foliar injury as a function of concentration was approximated by a lognormal frequency distribution. They combined these characteristics into a mathematical model to estimate the amount of foliar injury as a function of concentration and exposure duration for a single exposure. Pratt and Krupa (1981) found a similar response in soybean.

The solid line (Fig. 2.19) shows dose values for 5% injury on all the plants studied; the dashed line shows the threshold concentrations for sensitive plant species under conditions of high susceptibility. The individual points in the concentration-time coordinate system illustrate corresponding studies that established injury at specific ozone doses. In deriving this relationship, results from short-term controlled exposures using more than 100 plant species and cultivars were considered. The individual values used to derive the relationships (Fig. 2.19) are shown in Table 2.16.

Neither the graphical representation of ozone dose based on the O'Gara function nor the limiting values (Sect. 2.6.2.1) proposed by Jacobson (1977a) were developed to consider the differential resistance levels of individual species or cultivars. In consideration of the requirement to provide the largest possible set of decision criteria, limiting values for air quality criteria were derived for several resistance groupings (cf. Chap. 2.4.3.2). Heck and Brandt (1977) used a modification of the O'Gara function (Heck and Tingey 1971) to evaluate the results of ozone concentration and exposure duration on injury development from short term controlled pollutant exposure to derive the dose-response values for sensitive, intermediate, and resistant plant species (Table 2.23).

2.6.2.3 Ozone – Maximum Acceptable Concentrations for the Protection of Vegetation

The ozone concentrations and exposure durations shown in Table 2.23 will induce foliar injury following a single exposure of plants grown under conditions

Table 2.24. Proposed maximum acceptable ozone concentrations for protection of vegetation

Exposure duration	Resistance level					
	Sensitive		Intermediate		Less sensitive	
h	$\mu g\, m^{-3}$	ppm	$\mu g\, m^{-3}$	ppm	$\mu g\, m^{-3}$	ppm
0.5	300	0.150	500	0.25	1,000	0.50
1.0	150	0.075	350	0.18	500	0.25
2.0	120	0.060	250	0.13	400	0.20
4.0	100	0.050	200	0.10	350	0.18

favorable for injury development. When one considers that with repeated exposures a summation or intensification of individual effects can occur and that in the complete absence of visible symptoms growth reductions are possible, then it is obvious that even if the concentration values in Table 2.23 are obtained, a considerable risk to vegetation remains. Similar conclusions were reached by Prinz and Brandt (1978) in their study for the European Community. The attainment of the concentrations and durations shown in Table 2.24 should guarantee a practical far-reaching protection of vegetation in its various functions, for ozone acting alone.

The numerical values in Table 2.24 are based on the limiting values proposed by Jacobson (1977a), and the dose-response values for definite injury levels developed by Heck and Brandt (1977) in which they used research from North America. The question remains, if and to what extent these values can be extrapolated to Middle and Western European conditions. However, it is known that plant response is dependent on the pollutant exposure and internal and external resistance factors. In deriving the dose-response values from various experiments, it should be remembered that various edaphic and climatic conditions were used. Given this wide spectrum of conditions, it is assumed that the prevailing conditions of Middle and Western Europe are also included. From this perspective, there is no valid reason why the dose-response values developed in North America cannot be used in Europe as an initial basis for remedial and mitigative actions.

It is not possible to answer the question of the significance of differences in genetically controlled resistance levels between European and North American vegetation. As shown in Chapter 2.4.3.2, large resistance differences exist not only among individual species but also among cultivars and plants with different seed origins. From available data, it is not possible to determine if the resistance levels of the native and cultivated plants of Europe partially or completely overlap with the tested species from North America. However, recent studies in Holland showed that numerous European bean cultivars were as sensitive or more sensitive to ozone than a sensitive North American cultivar (Tonneijck 1983b). Given the small difference between clearly phytotoxic ozone concentrations and the natural ozone background, it appears, at least for native plants and forest stands with natural reproduction, that the assumption is justified that the resistance increases with concentrations of the natural ozone background. Plants from areas with a high ozone background would have a higher ozone tolerance. If the hypothesis is correct, then the air quality criteria determined for North American vegetation will be adequate for vegetation of other countries only if it has experienced similar natural background concentrations.

An additional uncertainty in establishing dose-response values from North American findings is that dose-response relationships have been established for only the early development stages of long-lived perennial plant cultures. For example, the exposure doses that will injure forest trees in the mature stages cannot be determined with certainty based on the available field data (cf. Chap. 2.4.3.1), although, for some species, there is a relationship between the relative sensitivity of seedlings and older trees.

These important questions for deriving limiting values can only be clarified from corresponding experimental studies. Investigations of oxidant effects are necessary, especially under the relatively high SO_2 and acidic precipitation burden in Europe and other countries. For example, the decline in forest conditions in Central Europe is possibly the result of the combined effects of ozone, sulfur dioxide, and acid precipitation on forest ecosystems (see Chap. 2.5.5).

References

Abeles AL, Abeles FB (1972) Biochemical pathway of stress-induced ethylene. Plant Physiol 50:496–498

Abeles FB, Heggestad HE (1973) Ethylene: An urban air pollutant. J Air Pollut Control Assoc 23:517–521

Abrahamsen G (1980) Acid precipitation, plant nutrients, and forest growth. Proc Int Conf Ecol Impact of Acid Precipitation, Sandefjord, Norway, pp 58–63

Adams RM, Crocker TD (1982) Economically relevant response estimation and the value of information: The case of acid deposition. In: The economics of acid rain. Ann Arbor Sci, Ann Arbor

Adams RM, Crocker TD, Thanabivulchai N (1982) An economic assessment of air pollution damage to selected annual crops in southern California. J Environ Econ Manage 9:42–58

Adedipe NO, Ormrod DP (1972) Hormonal regulation of ozone phytotoxicity in *Raphanus sativus*. Z Pflanzenphysiol 68:254–258

Adedipe NO, Ormrod DP (1974) Ozone-induced growth suppression in radish plants in relation to pre- and post-fumigation temperatures. Z Pflanzenphysiol 71:281–287

Adedipe NO, Barrett RE, Ormrod DP (1972a) Phytotoxicity and growth responses of ornamental bedding plants to ozone and sulfur dioxide. J Am Soc Hortic Sci 97:341–345

Adedipe NO, Hofstra G, Ormrod DP (1972b) Effects of sulfur nutrition on phytotoxicity and growth responses of bean plants to ozone. Can J Bot 50:1789–1793

Adedipe NO, Fletcher RA, Ormrod DP (1973a) Ozone lesions in relation to senescence of attached and detached leaves of tobacco. Atmos Environ 7:357–361

Adedipe NO, Khatamian H, Ormrod DP (1973b) Stomatal regulation of ozone phytotoxicity in tomato. Z Pflanzenphysiol 68:323–328

Aldaz L (1969) Flux measurement of atmospheric ozone over land and water. J. Geophys Res 74:6943–6946

Aldinger E (1983) Gesundheitszustand von Nadelholzbeständen auf gedüngten und ungedüngten Standorten im Buntsandstein-Schwarzwald. Allg. Forstz 38:794–796

Altherr E, Evers FH (1974) Unerwarteter Düngungserfolg bei Magnesiummangel in einem jungen Buchenbestand auf mittlerem Buntsandstein des Odenwaldes. Allg Forst- Jagdz 145:121–125

Altherr E, Evers FH (1977) Nachweis eines Magnesium-Düngeeffekts in einem Buchenbestand auf mittlerem Buntsandstein des Odenwaldes. Allg Forst- Jagdz 148:45–48

Anderson FK (1966) Air pollution damage to vegetation in Georgetown Canyon, Idaho. MS thesis, Univ Utah, 102 p. Cited in Treshow M (1968): Phytopathology 58:1108–1113

Anderson LE, Ng T-CL, Park K-EY (1974) Inactivation of pea leaf chloroplastic and cytoplasmic glucose 6-phosphate dehydrogenases by light and dithiothreitol. Plant Physiol 53:835–839

Arndt U, Lindner G (1981) Zur Problematik phytotoxischer Ozonkonzentrationen im südwestdeutschen Raum. Staub Reinhalt Luft 41:349–352

Arndt U, Nobel W, Bünau H von (1982a) Wirkungskataster für Luftverunreinigungen Baden-Württemberg. Ulmer, Stuttgart

Arndt U, Seufert G, Nobel W (1982b) Die Beteiligung von Ozon an der Komplexkrankheit der Tanne (*Abies alba* Mill.) – eine prüfenswerte Hypothese. Staub Reinhalt Luft 42:243–246

Arndt U, Obländer W, König E, Maier W (1983) Auftreten und Wirkungen gasförmiger Luftverunreinigungen in Waldgebieten Baden-Württembergs. VDI – Ber 500:249–255

Ashenden TW, Mansfield TA (1977) Influence of wind speed on the sensitivity of ryegrass to SO_2. J Exp Bot 28:729–735

Asher JE (1956) Observations and theory on x-disease or needle dieback. Unpubl Rep Arrowhead
    Ranger District, San Bernardino Nat Forest, 1956. Cited in Hill AC, Heggestad HE, Linzon SN
    (1970) Ozone, pp B1–B6. In: Jacobson JS, Hill AC (eds) Recognition of air pollution injury to vege-
    tation: a pictorial atlas. Air Pollut Control Assoc, Pittsburgh, Pa
Ashmore MR, Bell JNB, Reily CL (1978) A survey of ozone levels in the British Isles using indicator
    plants. Nature (London) 276(5690):813–815
Ashmore MR, Bell JNB, Reily CL (1980a) The distribution of phytotoxic ozone in the British Isles.
    Environ Pollut Ser B 1:195–216
Ashmore MR, Bell JNB, Dalpra C, Runeckles VC (1980b) Visible injury to crop species by ozone in
    the United Kingdom. Environ Pollut Ser A 21:209–215
Athanassious R (1980) Ozone effects on radish (*Raphanus sativus* L. cv. Cherry Belle): gradient of ul-
    trastructural changes. Z Pflanzenphysiol 97:227–232
Aycock MK (1972) Combining ability estimates for weather fleck in Nicotiana tabacum L. Crop Sci
    12:672–674
Barnes RL (1972a) Effects of chronic exposure to ozone on photosynthesis and respiration of pines.
    Environ Pollut 3:133–138
Barnes RL (1972b) Effects of chronic exposure to ozone on soluble sugar and ascorbic acid contents
    of pine seedlings. Can J Bot 50:215–219
Barnes RL, Berry CR (1969) Seasonal changes of carbohydrates and ascorbic acid in white pine and
    possible relation to tipburn sensitivity. US For Serv, Southeast For Exp Stn, Note SE-124, US Dep
    Agric, Ashville, North Carolina, 4 pp
Bauch J (1983) Biological alterations in the stem and root of fir and spruce due to pollution influence.
    In: Ulrich B, Pankrath J (eds) Accumulating of air pollutants in forest ecosystems. Reidel, Dord-
    recht, pp 377–386
Bauch J, Schröder W (1982) Zellulärer Nachweis einiger Elemente in den Feinwurzeln gesunder und
    erkrankter Tannen (*Abies alba* Mill.) und Fichten [*Picea abies* (L.) Karst.]. Forstwiss Centralbl.
    101:285–294
Beckerson DW, Hofstra G (1979a) Stomatal responses of white bean to O_3 and SO_2 singly or in com-
    bination. Atmos Environ 13:533–535
Beckerson DW, Hofstra G (1979b) Response of leaf diffusive resistance of radish, cucumber, and soy-
    bean to O_3 and SO_2 singly or in combination. Atmos Environ 13:1263–1268
Beckerson DW, Hofstra G (1979c) Effect of sulphur dioxide and ozone singly or in combination on
    leaf chlorophyll, RNA, and protein in white bean. Can J Bot 57:1940–1945
Beckerson DW, Hofstra G (1980) Effects of sulphur dioxide and ozone, singly or in combination, on
    membrane permeability. Can J Bot 58:451–457
Beckerson DW, Hofstra G, Wukasch R (1979) The relative sensitivities of 33 bean cultivars to ozone
    and sulfur dioxide singly or in combination in controlled exposures and to oxidants in the field.
    Plant Dis Rep 63:478–482
Bell JNB, Clough WS (1973) Depression of yield in ryegrass exposed to sulfur dioxide. Nature (Lon-
    don) 241(5384):47–49
Bell JNB, Cox RA (1975) Atmospheric ozone and plant damage in the United Kingdom. Environ Pol-
    lut 8:163–170
Bell JNB, Mudd CH (1976) Sulphur dioxide resistance in plants. A case study of *Lolium perenne*. In:
    Mansfield TA (ed) Effects of air pollution on plants. Soc Exp Biol Semin Ser 1. Cambridge Univ
    Press, Cambridge, pp 87–104
Bennett JH (1969) Effects of ozone on leaf metabolism. PhD thesis, Univ Utah, Salt Lake City,
    103 pp
Bennett JH, Hill AC (1974) Acute inhibition of apparent photosynthesis by phytotoxic air pollutants.
    In: Dugger M (ed) Air pollution effects on plant growth. ACS Symp Ser 3. Am Chem Soc, Wash-
    ington DC, pp 115–127
Bennett JH, Hill AC (1975) Interaction of air pollutants with canopies of vegetation. In: Mudd JB,
    Kozlowski TT (eds) Responses of plants to air pollution. Academic Press, London New York,
    pp 276–306
Bennett JH, Hill AC, Gates DM (1973) A model for gaseous pollutant sorption by leaves. J Air Pollut
    Control Assoc 23:957–962
Bennett JH, Heggestad HE, McNulty IB (1977) Ozone and leaf physiology. Proc 4th Annu Plant
    Growth Regul Work Group. Agway, Syracuse New York, pp 323–330

Bennett JH, Lee EH, Heggestad HH (1978) Apparent photosynthesis and leaf stomatal diffusion in EDU treated ozone-sensitive bean plants. Proc 5th Annu Meet Plant Growth Regul Work Group. Agway, Syracuse New York, pp 242–246

Bennett JP, Runeckles VC (1977) Effects of low levels of ozone on plant competition. J Appl Ecol 14:877–880

Bennett JP, Resh HM, Runeckles VC (1974) Apparent stimulations of plant growth by air pollutants. Can J Bot 52:35–41

Bennett JP, Barnes K, Shinn JH (1980) Interactive effects of H_2S and O_3 on the yield of snap beans (*Phaseolus vulgaris* L.). Environ Exp Bot 20:107–114

Berchthold R, Alcubilla M, Evers FH, Rehfuess KE (1981) Standortkundliche Studien zum Tannensterben. Nadel- und bastanalytischer Vergleich zwischen befallenen und gesunden Bäumen. Forstwiss Centralbl 100:236–253

Berry CR (1970) A plant fumigation chamber suitable for forestry studies. Phytopathology 60:1613–1615

Berry CR (1971) Relative sensitivity of red, jack, and white pine seedlings to ozone and sulfur dioxide. Phytopathology 61:231–232

Berry CR (1974) Age of pine seedlings with primary needles affects sensitivity to ozone and sulfur dioxide. Phytopathology 64:207–209

Berry CR, Ripperton LA (1963) Ozone, a possible cause of white pine emergence tipburn. Phytopathology 53:552–557

Bialobok St, Karolewski P, Oleksyn J (1980) Sensitivity of scots pine needles from mother trees and their progenies to the action of SO_2, O_3, a mixture of these gases, NO_2 and HF. Arbor Kornickie 25:289–303

Bisessar S (1982) Effect of ozone, antioxidant protection, and early blight on potato in the field. J Am Soc Hortic Sci 107:597–599

Bisessar S, Temple PJ (1977) Reduced ozone injury on virus-infected tobacco in the field. Plant Dis Rep 61:961–963

Black VJ, Ormrod DP, Unsworth MH (1982) Effects of low concentration of ozone, singly, and in combination with sulphur dioxide on net photosynthesis rates of *Vicia faba* L. J Exp Bot 33:1302–1311

Blanchard RO, Baas J, Cotter H van (1979) Oxidant damage to eastern white pine in New Hampshire. Plant Dis Rep 63:177–182

Blaschke H (1981 a) Schadbild und Ätiologie des Tannensterbens. II. Mykorrhizastatus und pathogene Vorgänge im Feinwurzelbereich als Symptome des Tannensterbens. Eur J For Pathol 11:375–379

Blaschke H (1981 b) Veränderungen bei der Feinwurzelentwicklung in Weißtannenbeständen. Forstwiss Centralbl 100:190–195

Blessin CW, Garcia WJ, Cavins JF, Inglett GE (1979) Effect of atmospheric ozone on physical and chemical characteristics of three field corn hybrids. Abstr Cereal Foods World 24:456

Blum U, Heck WW (1980) Effects of acute ozone exposure on snap bean at various stages of its life cycle. Environ Exp Bot 20:73–85

Blum U, Tingey DT (1977) A study of the potential ways in which ozone could reduce root growth and nodulation of soybean. Atmos Environ 11:737–739

Blum U, Heagle AS, Burn JC (1980) Effects of O_3 on the yield of tall fescue-ladino clover pasture. Abstr Bull Ecol Soc 61:125

BMI (Bundesminister des Innern) (1978) Ökosystemforschung im Hinblick auf Umweltpolitik und Entwicklungsplanung, Bonn

BML/LAI (1982) Waldschäden durch Luftverunreinigungen. Schriftenreihe des Bundesministeriums für Landwirtschaft und Forsten, Reihe A. Angew Wiss 273. Landwirtschaftsverlag, Münster-Hiltrup

BML/LAI (Bundesministerium für Ernährung, Landwirtschaft und Forsten) (1983) Neuartige Waldschäden durch Baumarten. BML-Informationen Nr. 43 vom 24. 10. 1983

Bobrov RA (1952) The effect of smog on the anatomy of oat leaves. Phytopathology 42:558–563

Bobrov RA (1955a) The leaf structure of the *Poa annua* with observations on its smog sensitivity in Los Angeles County. Am J Bot 42:467–474

Bobrov RA (1955b) Use of plants as biological indicators of smog in the air of Los Angeles County. Science 121:510–511

Bobrov-Glater RA (1956) Smog damage to ferns in the Los Angeles area. Phytopathology 46:696–698

Bonnemann A, Röhrig E (1971) Waldbau auf ökologischer Grundlage, 4. Aufl, Bd 1. Der Wald als Vegetationstyp und seine Bedeutung für den Menschen. Parey, Berlin Hamburg

Bosch E, Pfannkuch E, Baum U, Rehfuess KE (1983) Über die Erkrankung der Fichte (*Picea abies* Karst.) in den Hochlagen des Bayerischen Waldes. Forstwiss Centralbl 102:167–185

Botkin DB, Smith WH, Carlson RW, Smith TL (1972) Effects of ozone on white pine saplings: Variation in inhibition and recovery of net photosynthesis. Environ Pollut 3:273–289

Bradshaw AD (1971) Plant evolution in extreme environments. In: Creed R (ed) Ecological genetics and evolution. Oxford, Blackwell Scientific Publications, pp 20–50

Bradshaw AD (1972) Some of the evolutionary consequences of being a plant. Evol Biol 5

Bradshaw AD (1976) Pollution and evolution. In: Mansfied TA (ed) Effects of air pollution on plants. Soc Exp Biol. Semin Ser 1. Cambridge Univ Press, Cambridge, pp 135–159

Brandt CS (1962) Effects of air pollution on plants. In: Stern AC (ed) Air pollution, vol I. Academic Press, London New York, pp 255–281

Brandt CS, Heck WW (1968) Effects of air pollutants on vegetation. In: Stern AC (ed) Air pollution, vol I. Academic Press, London New York, pp 401–443

Brandt CJ, Rhoads RW (1972) Effects of limestone dust accumulation on composition of a forest community. Environ Pollut 3:217–225

Brasher EP, Fieldhouse DJ, Sasser M (1973) Ozone injury in potato variety trials. Plant Dis Rep 57:542–544

Brennan E (1975) Exclusion as mechanism of ozone-resistance in virus-infected plants. Phytopathology 659:1054–1055

Brennan E, Leone IA (1968) The response of plants to sulfur dioxide or ozone polluted air supplied at varying flow rates. Phytopathology 58:1661–1664

Brennan E, Leone IA (1970) Interaction of tobacco mosaic virus and ozone in *Nicotiana sylvestris*. J Air Pollut Control Assoc 20:470

Brennan E, Leone IA (1972) Crysanthemum response to sulfur dioxide and ozone. Plant Dis Rep 56:85–87

Brennan E, Rhoads A (1976) Response of field-grown bean cultivars to atmospheric oxidant in New Jersey. Plant Dis Rep 60:941–945

Brennan E, Leone IA, Halisky PM (1969) Response of forage legumes to ozone fumigations. Phytopathology 59:1458–1459

Brewer RF, Guillemet FB, Creveling RK (1961) Influence of N-P-K fertilization on incidence and severity of oxidant injury to mangels and spinach. Soil Sci 92:298–301

Briggs D (1972) Population differentiation in *Marchantia polymorpha* L. in various lead pollution levels. Nature (London) 238:166–167

Brill H, Bock E, Bauch J (1981) Über die Bedeutung von Mikroorganismen im Holz von *Abies alba* Mill. für das Tannensterben. Forstwiss Centralbl 100:195–206

Brown DH (1980) Notes on the instability of extracted chlorophyll and a reported effect of ozone on lichen algae. The Lichenologist 12:151–154

Brown DH, Smirnoff N (1978) Observations on the effect of ozone on *Cladonia rangiformis*. Lichenologist 10:91–94

Burleson GR, Murray TM, Pollard M (1975) Inactivation of viruses and bacteria by ozone, with and without sonication. Appl Microbiol 29:340–344

Butler LK, Tibbitts TW (1979a) Variation in ozone sensitivity and symptom expression among cultivars of *Phaseolus vulgaris* L. J Am Soc Hortic Sci 104:208–210

Butler LK, Tibbitts TW (1979b) Stomatal mechanisms determining genetic resistance to ozone in *Phaseolus vulgaris* L. J Am Soc Hortic Sci 104:213–216

Butler LK, Tibbitts TW, Bliss FA (1979) Inheritance of resistance to ozone in *Phaseolus vulgaris* L. J Am Soc Hortic Sci 104:211–213

Butzke H (1981) Versauern unsere Wälder? Erste Ergebnisse der Überprüfung 20 Jahre alter pH-Wert-Messungen in Waldböden Nordrhein-Westfalens. Forst- Holzwirt 36:542–548

Bystrom BG, Glater RB, Scott FM, Bowler FSC (1968) Leaf surface of *Beta vulgaris* – electron microscope study. Bot Gaz 129:133–138

California Department of Food and Agriculture (1977)

Cameron JW (1975) Inheritance in sweet corn for resistance to acute injury. J Am Soc Hortic Sci 100:577–579

Cameron JW, Taylor OC (1973) Injury to sweet corn inbreds and hybrids by air pollutants in the field and by ozone treatments in the greenhouse. J Environ Qual 2:387–389

Cameron JW, Johnson H Jr, Taylor OC, Otto HW (1970) Differential susceptibility of sweet corn hybrids to field injury by air pollution. HortScience 5:217–219

Campbell TA, Devine TE, Howell RK (1977) Diallel analyses of resistance to air pollutants in alfalfa. Crop Sci 17:664–665

Carlson RW (1979) Reduction in the photosynthetic rate of *Acer*, *Quercus*, and *Fraxinus* species caused by sulphur dioxide and ozone. Environ Pollut 18:159–170

Carnahan JE, Jenner EL, Wat EKW (1978) Prevention of ozone injury to plants by a new protectant chemical. Phytopathology 68:1225–1229

Carney AW, Stephenson GR, Ormrod DP, Ashton GC (1973) Ozone-herbicide interactions on crop plants. Weed Sci 21:508–511

Cathey HM, Heggestad HE (1972) Reduction of ozone damage to *Petunia hybrida* Vilm. by use of growth regulating chemicals and tolerant cultivars. J Am Soc Hortic Sci 97:685–700

Cathey HM, Heggestad HE (1973) Effects of growth retardants and fumigations with ozone and sulfur dioxide on growth and flowering of *Euphorbia pulcherrima* Willd. J Am Soc Hortic Sci 98:3–7

Cathey HM, Heggestad HE (1982 a) Ozone and sulfur dioxide sensitivity of petunia: Modification by ethylenediurea. J Am Soc Hortic Sci 107:1028–1035

Cathey HM, Heggestad HE (1982 b) Ozone sensitivity of herbaceous plants: Modification by ethylenediurea. J Am Soc Hortic Sci 107:1035–1042

Cathey HM, Heggestad HE (1982c) Ozone sensitivity of woody plants: Modification by ethylenediurea. J Am Soc Hortic Sci 107:1042–1045

Chang CW (1971 a) Effect of ozone on ribosomes in pinto bean leaves. Phytochemistry 10:2863–2868

Chang CW (1971 b) Effect of ozone on sulfhydryl groups of ribosomes in pinto bean leaves. Relationship with ribosome dissociation. Biochem Biophys Res Commun 44:1429–1435

Chang CW, Heggestad HE (1974) Effect of ozone on photosystem II in *Spinacia oleracea* chloroplasts. Phytochemistry 13:871–873

Chimiklis PE, Heath RL (1972) Effluxes of K^+ and H^+ from *Chlorella sorokiana* as affected by ozone. Plant Physiol (Suppl) 49:3

Chimiklis PE, Heath RL (1975) Ozone-induced loss of intracellular potassium ion from *Chlorella sorokiniana*. Plant Physiol 56:723–727

Claussen T (1980) Die Bedeutung biologischer Indikatoren für die Luftqualitätsüberwachung. Umwelt 6:601–606

Clayberg CD (1971) Screening tomatoes for ozone resistance. HortScience 6:396–397

Clayberg CD (1972) Searching for ozone resistance to tomato varieties. Front Plant Sci 24:5–6

Cobb FW Jr, Stark RW (1970) Decline and mortality of smog-injured ponderosa pine. J For 68:147–149

Comeau G, LeBlanc F (1971) Influence de l'ozone et l'anhydride sulfureux sur la regeneration des feuilles de *Funaria hygrometrica*. Hedw Nat Can 98:347–358

Constantinidou HA, Kozlowski TT (1979 a) Effects of sulfur dioxide and ozone on *Ulmus americana* seedlings. I. Visible injury and growth. Can J Bot 57:170–175

Constantinidou HA, Kozlowski TT (1979 b) Effects of sulfur dioxide and ozone on *Ulmus americana* seedlings. II. Carbohydrates, proteins, and lipids. Can J Bot 57:176–184

Costonis AC (1968) The relationships of ozone, *Lophodermium pinastri* and *Pullularia pullulans* to needle blight of eastern white pine. PhD thesis, Cornell Univ, Ithaca, New York, 176 pp

Costonis AC (1973) Injury to eastern white pine by sulfur dioxide and ozone alone and in mixture. Eur J For Pathol 3:50–55

Costonis AC, Sinclair WA (1969 a) Relationships of atmospheric ozone to needle blight of eastern white pine. Phytopathology 59:1566–1574

Costonis AC, Sinclair WA (1969 b) Ozone injury to *Pinus strobus*. J Air Pollut Control Assoc 19:867–872

Coulson CL, Heath RL (1974) Inhibition of the photosynthetic capacity of isolated chloroplasts by ozone. Plant Physiol 53:32–38

Coulson CL, Heath RL (1975) The interaction of peroxyacetyl nitrate (PAN) with the electron flow of isolated chloroplasts. Atmos Enrivon 9:231–238

Coyne PI, Bingham GE (1978) Photosynthesis and stomatal light responses in snap beans exposed to hydrogen sulfide and ozone. J Air Pollut Contr Assoc 28:1119–1123

Coyne PI, Bingham GE (1981) Comparative ozone dose response of gas exchange in a ponderosa pine stand exposed to long-term fumigations. J Air Pollut Control Assoc 31:38–41

Craker LE (1971a) Ethylene production from ozone injured plants. Environ Pollut 1:299–304

Craker LE (1971b) Effects of mineral nutrients on ozone susceptibility of Lemna minor. Can J Bot 49:1411–1414

Craker LE (1972) Decline and recovery of petunia flower development from ozone stress. HortScience 7:484

Craker LE, Feder WA (1972) Development of the inflorescence in petunia, geranium, an poinsettia under ozone stress. HortScience 7:59–60

Craker LE, Starbuck JS (1972) Metabolic changes associated with ozone injury of bean leaves. Can J Plant Sci 52:589–597

Craker LE, Starbuck JS (1973) Leaf age and air pollutant susceptibility: Uptake of ozone and sulfur dioxide. Environ Research 6:91–94

Craker LE, Berube JL, Fredrickson PB (1974) Community monitoring of air pollution with plants. Atmos Environ 8:845–853

Curtis CR, Howell RK, Kremer DF (1976) Soybean peroxidases from ozone injury. Environ Pollut 11:189–194

Curtis LR, Edgington LF, Littlejohns DJ (1975) Oxathiin chemicals for control of bronzing of white beans. Can J Plant Sci 55:151–156

Czuba W, Ormrod DP (1974) Effects of cadmium and zinc on ozone-induced phytotoxicity in cress and lettuce. Can J Bot 52:645–649

Czuba M, Ormrod DP (1981) Cadmium concentrations in cress shoots in relation to cadmium-enhanced ozone phytotoxicity. Environ Pollut Ser A 25:67–76

Dahlsten DL (1978) The role of bark beetles in oxidant stressed forests. In: Proc simulation modeling of oxidant air pollution effects on mixed conifer forests and possible role of models in timber management planning for southern California National Forests. Statewide Air Pollut Res Cent, Univ California, Riverside, p 15

Daines RH, Leone IA, Brennan E (1960) Air pollution as it affects agriculture in New Jersey. N J Agric Exp Stn Bull 794:1–14

Darley EF (1963/64) Feststellungen von Ozon- und PAN-Wirkung an Pflanzen in der Bundesrepublik Deutschland, Essen (unpublished)

Dass HC, Weaver GM (1968) Modification of ozone damage to Phaseolus vulgaris by antioxidants, thiols, and sulfhydryl reagents. Can J Plant Sci 48:569–574

Dass HC, Weaver GM (1972) Enzymatic changes in intact leaves of Phaseolus vulgaris following ozone fumigation. Atmos Environ 6:759–763

Davis DD (1970) The influence of ozone on conifers. Ph D thesis, Pennsylvania State Univ, Univ Park, Pa, 93 pp

Davis DD (1977) Response of ponderosa pine primary needles to separate and simultaneous ozone and PAN exposures. Plant Dis Rep 61:640–644

Davis DD, Coppolino JB (1974a) Relationship between age and ozone sensitivity of current needles of ponderosa pine. Plant Dis Rep 58:660–663

Davis DD, Coppolino JB (1974b) Relative ozone susceptibility of selected woody ornamentals. HortScience 9:537–539

Davis DD, Kress L (1974) The relative susceptibility of ten bean varieties to ozone. Plant Dis Rep 56:14–16

Davis DD, Wilhour RG (1976) Susceptibility of woody plants to sulfur dioxide and photochemical oxidants. A literature review. Corvallis Environ Res Lab . US Environ Protect Agency, EPA-660/3-76-102

Davis DD, Wood FA (1972) The relative susceptibility of eighteen coniferous species to ozone. Phytopathology 62:14–19

Davis DD, Wood FA (1973a) The influences of environmental factors on the sensitivity of Virginia pine to ozone. Phytopathology 63:371–376

Davis DD, Wood FA (1973b) The influence of plant age on the sensitivity of Virginia pine to ozone. Phytopathology 63:381–388

Davis I (1959) The survival and mutability of *Escherichia coli* in aqueous solutions of ozone. Ph D thesis, Univ Pennsylvania, Philadelphia, Pa

Dean CE (1972) Stomata density and size as related to ozone-induced weather fleck in tobacco. Crop Sci 12:547–548

Dean CE, Davis DR (1967) Ozone and soil moisture in relation to the occurrence of weather fleck on Florida cigar-wrapper tobacco in 1966. Plant Dis Rep 51:72–75

Dean CE, Davis DR (1972) Ozone air pollution and weather fleck of tobacco. Univ Florida, Gainesville, Agric Exp Stn, Cincinnati, S-218

Decker RK, Postlethwait SM (1960) The maturation of the trifoliate leaf of *Glycine max*. Proc Indiana Acad Sci 70:66–72

Dijak M, Ormrod DP (1982) Some physiological and anatomical characteristics associated with differential ozone sensitivity among pea cultivars. Environ Exp Bot 22:395–402

Dochinger LS (1968) The impact of air pollution on eastern white pine: The chlorotic dwarf disease. J Air Pollut Contr Assoc 18:814–816

Dochinger LS, Seliskar CE (1965) Results from grafting chlorotic dwarf of eastern white pine. Phytopathology 55:404–407

Dochinger LS, Seliskar CE (1970) Air pollution and the chlorotic dwarf disease of eastern white pine. For Sci 16:46–55

Dochinger LS, Townsend AM (1979) Effects of roadside deicer salts and ozone on red maple progenies. Environ Pollut 19:229–237

Dochinger LS, Seliskar CE, Bender FW (1965) Etiology of chlorotic dwarf of eastern white pine. Phytopathology 55:1055

Dochinger LS, Bender FW, Fox FL, Heck WW (1970) Chlorotic dwarf of eastern white pine caused by an ozone and sulfur dioxide interaction. Nature (London) 225:476

Dolzmann P, Ullrich H (1966) Einige Beobachtungen über Beziehungen zwischen Chloroplasten und Mitochondrien im Palisadenparenchym von *Phaseolus vulgaris*. Z Pflanzenphysiol 55:165–180

Donaubauer E (1966) Durch Industrieabgase bedingte Sekundärschäden am Wald. Mitt Forstl Bundesversuchsanst Mariabrunn 73:101–110

Drummond DB (1971) Influence of high concentrations of peroxyacetylnitrate on woody plants. Abstr Phytopathology 61:128

Drummond DB (1972) The effect of peroxyacetyl nitrate on petunia (*Petunia hybrida* Vilm.). Cent Air Environ Stud Publ 260-72. Pennsylvania State Univ, Univ Park, 72 pp

Dubeau H, Chung YS (1979) Ozone response in wild type and radiation sensitive mutants of *Saccharomyces cerevisiae*. Mol Gen Genet 176:393–398

Duchelle SF, Skelly JM, Kress LW (1980) The impact of photochemical oxidant air pollution on biomass development of native vegetation and symptom expression of *Asclepias* spp. Abstr Proc Potomac Div Am Phytopathol Soc Morgantown, West Virginia

Dugger WM Jr, Palmer RL (1969) Carbohydrate metabolism in leaves of rough lemon as influenced by ozone. In: Chapman HD (ed) Proc 1st Int Citrus Symp, vol II. University of California, Riverside, pp 711–715

Dugger WM Jr, Ting IP (1968) The effect of peroxyacetyl nitrate on plants: Photoreductive reactions and susceptibility of bean plants to PAN. Phytopathology 56:1102–1107

Dugger WM, Ting IP (1970a) Physiological and biochemical effect of air pollution oxidants on plants. Recent Adv Phytochem 3:31–58

Dugger WM, Ting IP (1970b) Air pollution oxidants – their effects on metabolic processes in plants. Annu Rev Plant Physiol 21:215–234

Dugger WM Jr, Taylor OC, Cardiff E, Thompson CR (1962a) Relationship between carbohydrate content and susceptibility of pinto bean plants to ozone damage. Proc Am Soc Hortic Sci 81:304–314

Dugger WM Jr, Taylor OC, Cardiff E, Thompson CR (1962b) Stomatal action in plants as related to damage from photochemical oxidants. Plant Physiol 37:487–491

Dugger WM Jr, Koukol J, Reed WD, Palmer RL (1963a) Effect of peroxyacetyl nitrate on $^{14}CO_2$ fixation by spinach chloroplasts on pinto bean plants. Plant Physiol 38:468–472

Dugger WM Jr, Taylor OC, Klein WH, Shropshire W (1963b) Action spectrum of peroxyacetyl nitrate damage to bean plants. Nature (London) 198:75–76

Dugger WM, Koukol J, Palmer RL (1965) Effect of ozone on permeability of leaves to sugar and subsequent metabolism. Plant Physiol Suppl XX 40

Dugger WM, Koukol J, Palmer RL (1966) Physiological and biochemical effects of atmospheric oxidants on plants. J Air Pollut Control Assoc 16:467–471

Dunn DB (1959) Some effects or air pollution on *Lupinus* in the Los Angeles area. Ecology 40:621–625

Dunning JA, Heck WW (1973) Response of pinto bean and tobacco to ozone as conditioned by light intensity and/or humidity. Environ Sci Technol 7:824–826

Dunning JA, Heck WW (1977) Response of bean and tobacco to ozone: Effect of light intensity, temperature, and relative humidity. J Air Pollut Control Assoc 27:882–886

Dunning JA, Heck WW, Tingey DT (1974) Foliar sensitivity of pinto bean and soybean to ozone as affected by temperature, potassium nutrition, and ozone dose. Water Air Soil Pollut 3:305–313

Eastmond M, Skärby L (eds) (1979) Report from the workshop "Ozone Effects on Vegetation in Europe", 18–21 April 1979 Swed Water Air Pollut Res Inst (IVL), Göteborg, Sweden

Eichhorn O (1981) Zoologische Aspekte zum Tannensterben. Forstwiss Centralbl 100:270–274

Elford WG, Ende J van den (1942) An investigation on the merits of ozone as an aerial disinfectant. J Hyg 42:240–265

Elfving DC, Gilbert MD, Edgerton LF, Wilde MH, Lisk DJ (1976) Antioxidant and antitranspirant protection of apple foliage against ozone injury. Bull Environ Contamin Toxicol 15:336–341

Elkiey T, Ormrod DP (1979a) Ozone and sulphur dioxide effects on leaf water potential on petunia. Z Pflanzenphysiol 91:177–181

Elkiey T, Ormrod DP (1979b) Petunia cultivar sensitivity to ozone and sulphur dioxide. Sci Hortic 11:269–280

Elkiey T, Ormrod DP (1979c) Leaf diffusion resistance responses of three petunia cultivars to ozone and/or sulfur dioxide. J Air Pollut Control Assoc 29:622–625

Elkiey T, Ormrod DP (1979d) Ozone and/or sulfur dioxide effects on tissue permeability of petunia leaves. Atmos Environ 13:1165–1168

Elkiey T, Ormrod DP (1980a) Sorption of ozone and sulfur dioxide by petunia leaves. J Environ Qual 9:93–95

Elkiey T, Ormrod DP (1980b) Response of turfgrass cultivars to ozone, sulfur dioxide, nitrogen dioxide, or their mixtures. J Am Soc Hortic Sci 105:664–668

Elkiey T, Ormrod DP (1981a) Absorption of ozone, sulfur dioxide, and nitrogen dioxide by petunia plants. Environ Exp Bot 21:63–70

Elkiey T, Ormrod DP (1981b) Sorption of O_3, SO_2, NO_2, or their mixture by nine *Poa pratensis* cultivars of differing pollutant sensitivity. Atmos Environ 15:1739–1743

Elkiey T, Ormrod DP (1981c) Sulphate, total sulphur, and nitrogen accumulation by petunia leaves exposed to ozone, sulphur dioxide, and nitrogen dioxide. Environ Pollut Ser A 22:233–241

Elkiey T, Ormrod DP, Pelletier RL (1979) Stomatal and leaf surface features as related to the ozone sensitivity of petunia cultivars. J Am Soc Hortic Sci 104:510–514

Endress AG, Suarez SJ, Taylor OC (1980) Peroxidase activity in plant leaves exposed to gaseous HCl or ozone. Environ Pollut Ser A 22:47–58

Engle RL, Gabelman WH (1966) Inheritance and mechanism for resistance to ozone damage in onion, *Allium cepa* L. Proc Am Soc Hortic Sci 89:423–430

Engle RL, Gabelman WH (1967) The effects of low levels of ozone on pinto beans, *Phaseolus vulgaris* L. Proc Am Soc Hortic Sci 91:304–309

Ensing J, Hofstra G (1982) Impact of air pollutant ozone on acetylene reduction and shoot growth of red clover. Can J Plant Pathol 4:237–242

EPA (US Environmental Protection Agency) (1976) The photochemical oxidants. In: Diagnosing vegetation injury caused by air pollution. Appl Sci Assoc Inc, EPA Contr 68-02-1344

EPA (US Environmental Protection Agency) (1978a) Effects of photochemical oxidants on vegetation and certain microorganisms. In: Air quality criteria for ozone and other photochemical oxidants. EPA-600/8-78-004, pp 253–293

EPA (US Environmental Protection Agency) (1978b) Ecosystems. In: Air quality criteria for ozone and other photochemical oxidants. EPA-600/8-78-004, pp 294–328

EPA (US Environmental Protection Agency) (1979) Air quality data (1978) Annual statistics including summaries with reference to standards. EPA-450/4-79-037

Epstein SS, Bishop Y (1977) Protection by antioxidants against the toxicity of ozone to microbial systems. Environ Res 14:187–193

Evans LS (1973) Bean leaf growth response to moderate ozone levels. Environ Pollut 4:17–26

Evans LS (1980) Foliar responses that may determine plant injury by simulated acid rain. In: Toribara TY, Miller MW, Morrow PE (eds) Polluted air. Proc 12th Rochester Int Conf Environ Toxicity, Rochester, New York, May 21–23, 1979. Plenum Press, New York, pp 239–258

Evans LS, Miller PR (1972) Comparative needle anatomy and relative ozone sensitivity of four pine species. Can J Bot 50:1067–1071

Evans LS, Ting IP (1973) Ozone-induced membrane permeability changes. Am J Bot 60:155–162

Evans LS, Ting IP (1974) Ozone sensitivity of leaves: Relationship to leaf water content, gas transfer resistance, and anatomical characteristics. Am J Bot 61:592–597

Evans LS, Gmur NF, DaCosta F (1977) Leaf surface and histological perturbations of leaves of *Phaseolus vulgaris* and *Helianthus annuus* after exposures to simulated rain. Am J Bot 64:903–913

Evers FH (1981) Ergebnisse ernährungskundlicher Erhebungen zur Tannenerkrankung in Baden-Württemberg. Forstwiss Centralbl 100:253–265

Evers FH (1983) Orientierende Untersuchungen langfristiger Bodenreaktionsänderungen in südwestdeutschen Düngungs-Versuchsflächen. Forst Holzwirtsch 39:317–320

Eversman S (1982) Quantifying changes in ultrastructural features in lichens treated with air pollutants. Bot Soc Am Abstr Annu Meet, 8–12 August 1982, Pennsylvania State Univ, Univ Park, Pa, Miscellaneous Ser Publ No 162:3

Ewert R (1924) Rauchkranke Böden. Angew Bot 97–104

Faensen-Thiebes A (1981) Wirkungen von Ozon und Cadmium auf die CO_2-Assimilation und Transport von *Nicotiana tabaccum* L. und *Phaseolus vulgaris* L. Dissertation, Tech Univ, Berlin

Feder WA (1968) Reduction in tobacco pollen germination and tube elongation induced by low levels of ozone. Science 160:1122

Feder WA (1970) Plant response to chronic exposures to low levels of oxidant type air pollution. Environ Pollut 1:73–79

Feder WA (1977) Adverse effects of chronic low level ozone exposure to tomato fruit set and yield. Cottrell Centennial Symp. Air pollution and its impact on agriculture, January 13–14, 1977. Calif State College, Stanislaus

Feder WA, Campbell FJ (1968) Influence of low levels of ozone on flowering of carnations. Phytopathology 58:1038–1039

Feder WA, Sullivan F (1969) Differential susceptibility of pollen grains to ozone injury. Phytopathology 59:399

Feder WA, Fox FL, Heck WW, Campbell FJ (1969 a) Varietal responses of Petunia to several air pollutants. Plant Dis Rep 53:506–610

Feder WA, Sullivan F, Perkins I (1969 b) The effect of chronic exposure to low levels of ozone upon growth and development of Geranium cultivar Olympic Red. Phytopathology 59:1026

Ferry BW, Baddeley MS, Hawksworth DL (eds) (1973) Air pollution and lichens. Athlone, London

Fetner RH (1958) Chromosome breakage in *Vicia faba* by ozone. Nature (London) 181:504–505

Fetner RH (1962) Ozone-induced chromosome breakage in human cell cultures. Nature (London) 194:793–794

Fletcher RA, Adedipe NO, Ormrod DP (1972) Abscisic acid protects bean leaves from ozone-induced phytotoxicity. Can J Bot 50:2389–2391

Floor H, Posthumus AC (1977) Biologische Erfassung von Ozon- und PAN-Immissionen in den Niederlanden 1973, 1974 und 1975. VDI Ber 270:183–190

Flueckiger W, Flueckiger-Keller H, Oertli JJ (1978) Der Einfluß verkehrsbedingter Luftverunreinigungen auf die Peroxydaseaktivität, das ATP-Bildungsvermögen isolierter Chloroplasten und das Längenwachstum von Mais. Z Pflanzenkrankh Pflanzenschutz 85:41–47

Flueckiger W, Oertli JJ, Flueckiger-Keller H, Braun S (1979) Premature senescence in plants along a motorway. Environ Pollut 19:171–176

Flühler H, Bosshard W (1982) Waldschäden und Immissionsbelastungen im Walliser Rhonetal. Neue Zürcher Ztg 16. 6. 1982

Foster KW, Timm H, Labanauskas CK, Oshima RJ (1983) Effects of ozone and sulfur dioxide on tuber yield and quality of potatoes. J Environ Qual 12:75–80

Fowler D, Cape JN, Nicholso JA, Kinwaird JW, Patersow JS (1980) The influence of a polluted atmosphere on cuticle degradation in Scots pine (*Pinus silvestris*). In: Drabløs D, Tallos A (eds) Proc Int Conf Ecol Impact Acid Precip. Sanderfjord, Norway

Fox FM, Caldwell MM (1978) Competitive interaction in plant populations exposed to supplementary ultraviolet-B radiation. Oecologia 36:173–190

Franz F (1983) Auswirkungen der Walderkrankungen auf Struktur und Wuchsleistung von Waldbeständen. Tagungsber Symp Saurer Regen – Waldschäden, Jülich 27.–28. Januar 1983, pp 22–23

Frederick PE (1973) Lipid and viability changes in *Chlorella* subjected to ozone stress. M A thesis, Univ
California, Riverside

Frederick PE, Heath RL (1975) Ozone-induced fatty acid viability changes in *Chlorella*. Plant Physiol
55:15–19

Freebairn HT (1963) Uptake and movement of l-C^{14} ascorbic acid in bean plants. Plant Physiol
16:517–522

Freebairn HT, Taylor AC (1960) Prevention of plant damage from airborne oxidizing agents. Proc Am
Soc Hortic Sci 76:693–699

Freeman BA, Miller BE, Mudd JB (1979) Reaction of ozone with human erythrocytes. In: Lee SD,
Mudd JF (eds) Assessing toxic effects of environmental pollutants. Ann Arbor Science, Ann Arbor, pp 151–171

Frenzel B (1983) Beobachtungen eines Botanikers zur Koniferenerkrankung. Allg Forstz 38:743–747

Fridovich I (1975) Oxygen: Boon and bane. Am Sci 62:54–59

Fridovich I (1978) The biology of oxygen radicals. Science 201:875–880

Furukawa A, Kadota M (1975) Effect of ozone on photosynthesis and respiration in poplar leaves. Environ Control Biol 13:1–7

Furukawa A, Totsuka T (1979) Effects of NO$_2$, SO$_2$, and O$_3$ alone and in combinations on net photosynthesis in sunflower. Environ Control Biol 17:161–166

FVF (Forstliche Versuchs- und Forschungsanstalt Baden-Württemberg) (1984) Hauptergebnisse der
Datenanalyse der großräumigen terrestrischen Waldschadensinventur Baden-Württemberg 1983,
Freiburg

Gardner JD, Ormrod DP (1976) Response of the Rieger begonia to ozone and sulphur dioxide. Sci
Hortic 5:171–181

Garland JA (1976) Dry deposition of SO$_2$ and other gases. In: Atmosphere surface exchange of par-
ticulates and gaseous pollutants. Proc Symp, Richland, Washington, Sept 4–6, 1974. ERDA Tech
Inf Cent, Oak Ridge, Tennessee, pp 212–227

Garland JA, Penkett SA (1976) Absorption of peroxyacetyl nitrate and ozone by natural surfaces. Atmos Environ 10:1127–1131

Gäumann E (1951) Pflanzliche Infektionslehre. Birkhäuser, Basel

Gesalman CM, Davis DD (1978) Ozone susceptibility of ten azalea cultivars as related to stomatal frequency or conductance. J Am Soc Hortic Sci 103:489–491

Giese AC, Christensen E (1954) Effects of ozone on organisms. Physiol Zool 27:101–115

Gilbert MD, Elfving DC, Lisk DJ (1977) Protection of plants against ozone injury using the antioxidant N-(1,3-dimethylbutyl)-N'-phenyl-p-phenylene-diamine. Bull Environ Contamin Toxicol
18:783–785

Glater RB, Solberg RA, Scott FM (1962) A developmental study of the leaves of *Nicotiana glutinosa* as
related to their smog-sensitivity. Am J Bot 49:954–970

Goetz A, Tsuneishi N (1959) A bacteriological irritation analogue for aerosols. Arch Ind Health
20:167–180

Goldstein BD (1977) Cellular effects of ozone. Rev Environ Health 2:177–202

Gordon WC, Ordin L (1972) Phosphorylated and nucleotide sugar metabolism in relation to cell wall
production in *Avena* coleoptiles treated with fluoride and peroxyacetyl nitrates. Plant Physiol
49:542–545

Goren AA, Donagi AE (1980) Assessment of atmospheric ozone levels in Israel through foliar injury
to Bel-W$_3$ tobacco plants. Oecologia 44:418–421

Greef JA De, Verbelen JP (1973) Physiological stress and crystallites in leaf plastids of *Phaseolus vulgaris* L. Ann Bot (London) [New Ser] 37:593–596

Greenwood P, Greenhalgh A, Baker C, Unsworth M (1982) A computer-controlled system for exposing field crops to gaseous air pollutants. Atmos Environ 16:2261–2266

Gross RE, Dugger WM (1969) Responses of *Chlamydomonas reinhardtii* to peroxyacetylnitrate. Environ Res 2:256–266

Grosso JJ, Menser HA, Hodges GA, McKinney HH (1971) Effects of air pollutants on *Nicotiana* cultivars and species used for virus studies. Phytopathology 61:945–950

Grzywacz A, Wazny J (1973) The impact of industrial air pollutants on the occurrence of several important pathogenic fungi of forest trees in Poland. Eur J For Pathol 3:129–141

Guderian R (1966) Reaktionen von Pflanzengemeinschaften des Feldfutterbaues auf Schwefeldioxid-einwirkungen. Schriftenr Landesanst Immissions-Bodennutzungsschutz Landes Nordrhein-West-falen Essen 4:80–100

Guderian R (1969) Obstbau in Gebieten mit Schwefeldioxid-Immissionen. Erwerbsobstbau 11:110–113

Guderian R (1970) Untersuchungen über quantitative Beziehungen zwischen dem Schwefelgehalt von Pflanzen und dem Schwefeldioxidgehalt der Luft. Z Pflanzenkrankh Pflanzenschutz 77:200–220

Guderian R (1977) Air pollution. Phytotoxicity of acidic gases and its significance in air pollution control. Ecol Stud 22. Springer, Berlin Heidelberg New York

Guderian R (1978) Wirkungen sauerstoffhaltiger Schwefelverbindungen. Einführung. VDI Ber 314:207–217

Guderian R (1982) Diskussionsbeiträge. In: Graf Hatzfeld (ed) Stirbt der Wald? Energiepolitische Vor-aussetzungen und Konsequenzen. Müller, Karlsruhe

Guderian R, Haut H van (1970) Nachweis von Schwefeldioxid-Wirkungen an Pflanzen. Staub Reinhalt Luft 30:17–26

Guderian R, Küppers K (1980a) Problems in determining dose-response relationships as a basis for ambient pollutant standards. Proceedings of the symposium on the effects of airborne pollution on vegetation, August 20–24, Warsaw (Poland), United Nations, Econ Commiss Eur, pp 196–212

Guderian R, Küppers K (1980b) Responses of plant communities to air pollution. Proceedings of the symposium on effects of air pollutants on mediterranean and temperate forest ecosystems, June 22–27. Pac Southwest For Range Exp Stn. Berkeley, Gen Tech Rep PSW-43, pp 187–199

Guderian R, Rabe R (1982) Effects of photochemical oxidants on plants. Final Rep. Prep Commiss Eur Commun, Contr No U/81/519

Guderian R, Reidl K (1982) Höhere Pflanzen als Indikatoren für Immissionsbelastungen im terrestri-schen Bereich. Decheniana Beih 26:6–22

Guderian R, Stratmann H (1962) Freilandversuche zur Ermittlung von Schwefeldioxidwirkungen auf die Vegetation. Teil I. Übersicht zur Versuchsmethodik und Versuchsauswertung. Forschungsber Landes Nordrhein-Westfalen Nr 1118 Westdeutscher Verlag, Köln Opladen

Guderian R, Stratmann H (1968) Freilandversuche zur Ermittlung von Schwefeldioxidwirkungen auf die Vegetation. Teil III. Grenzwerte schädlicher SO_2-Immissionen für Obst- und Forstkulturen so-wie für landwirtschaftliche und gärtnerische Pflanzenarten. Forschungsber Landes Nordrhein-Westfalen Nr 1920. Westdeutscher Verlag, Köln Opladen

Guderian R, Haut H van, Stratmann H (1960) Probleme der Erfassung und Beurteilung von Wirkun-gen gasförmiger Luftverunreinigungen auf die Vegetation. Z Pflanzenkrankh Pflanzenschutz 67:257–264

Guderian R, Krause GHM, Kaiser H (1977) Untersuchungen zur Kombinationswirkung von Schwe-feldioxid und schwermetallhaltigen Stäuben auf Pflanzen. Schriftenr Landesanst Immissionsschutz Landes Nordrhein-Westfalen Essen 40:23–30

Guderian R, Rabe R, Tingey DT (1983) Wirkungen von Photooxidantien auf Pflanzen. In: Umwelt-bundesamt (ed) Luftqualitätskriterien für photochemische Oxidantien. Berichte 5/83. Schmidt, Berlin

Guderian R, Vogels K (1982) Recognition of air pollution effects on plants. Proc 5th Int Clean Air Congr, October 1980, Buenos Aires, pp 773–778

Guicherit P (1975) Fotochemische smogvorming in Nederland. Rapp G 646 Inst Milieuhyg Gezond-heidstech TNO, Afdeling Buitenlucht, Schoemakerstr 97, Delft

GVST (Gesamtverband des deutschen Steinkohlenbergbaus) (1982) Saurer Regen und Forstschäden: Eine Dokumentation. Gesamtverb Dtsch Steinkohlenbergbaus, February 1983

Haagen-Smit AJ (1952) Chemistry and physiology of Los Angeles smog. Ind Eng Chem 44:1342–1346

Haagen-Smit AJ, Darley EF, Zaitlin M, Hull H, Noble W (1952) Investigation on injury to plants from air pollution in the Los Angeles area. Plant Physiol 27:18–34

Haas JH (1970) Relation of crop maturity to air pollution incited bronzing of *Phaseolus vulgaris*. Phy-topathology 60:407–410

Haines BL, Stefani M, Hendrix F (1980) Acid rain: Threshold of leaf damage in eight species from a forest succession. Water Air Soil Pollut 14:403–407

Hajdúk J (1961) Quantitative und qualitative Änderungen der Phytozönosen, verursacht durch Exhalationsprodukte von Fabriken. Biologia Bratislawa XVI 6:404–419

Hajdúk J, Ruzicka M (1969) Das Studium der Schäden an Wildpflanzen und Pflanzengesellschaften verursacht durch Luftverunreinigungen. In: Air pollution. Proc 1st Eur Congr Infl Air Pollut Plants Animals, Wageningen, pp 183–192

Halbwachs G, Kisser J (1967) Durch Rauchimmissionen bedingter Zwergwuchs bei Fichte und Birke. Centralbl Gesamte Forstwesen 84:156–173

Hamelin C, Chung YS (1974a) Lethal and mutagenic effects of ozone on *Escherichia coli*. Can J Genet Cytol 16:706

Hamelin C, Chung YS (1974b) Optimal conditions for mutagenesis by ozone in *Escherichia coli* K 12. Mutat Res 24:271–279

Hamelin C, Chung YS (1975) Characterization of mucoid mutants of *Escherichia coli* K 12 isolated after exposure to ozone. J Bacteriol 122:19–24

Hamelin C, Chung YS (1976) Rapid test for assay of ozone sensitivity in *Escherichia coli*. Mol Gen Genet 145:191–194

Hamelin C, Sarkan F, Chung YS (1977) Ozone-induced DNA degradation in different DNA polymerase I mutants of *Escherichia coli* K 12. Biochem Biophys Res Commun 77:220–224

Hanson GP, Thorne L, Jativa CD (1970) Vitamin C – A natural smog resistance mechanism in plants? Lasca Leaves 20:6–7

Hanson GP, Thorne L, Jativa CD (1971) Ozone tolerance of petunia leaves as related to their ascorbic acid concentration. 2nd Int Clean Air Congr Proc Digest, Washington DC

Hanson GP, Addis DH, Thorne L (1974) Smog-tolerant petunias developed at the Los Angeles State and County Arboretum. Lasca Leaves 24:117–119

Hanson GP, Thorne L, Addis DH (1975) The ozone sensitivity of *Petunia hybrida* Vilm. as related to physiological age. J Am Soc Hortic Sci 100:188–190

Hanson GP, Addis DH, Thorne L (1976) Inheritance of photochemical air pollution tolerance in petunias. Can J Genet Cytol 18:579–592

Harding PR Jr (1968) Effect of ozone on *Penicillium* mold decay and sporulation. Plant Dis Rep 52:246–247

Harkov R, Brennan E (1979) An ecophysiological analysis of the response of trees to oxidant pollution. J Air Pollut Control Assoc 29:157–161

Harkov R, Clarke B, Brennan E (1979) Cadmium contamination may modify response of tomato to atmospheric ozone. J Air Pollut Control Assoc 29:1247–1249

Harris MJ, Heath RL (1981) Ozone sensitivity in sweet corn (*Zea mays* L.) plants: A possible relation to water balance. Plant Physiol 68:885–890

Harrison BH, Feder WA (1974) Ultrastructural changes in pollen exposed to ozone. Phytopathology 64:257–258

Harward MR, Treshow M (1971) The impact of ozone on understory plants of the aspen zone. Proc 64th Annu Meet Pollut Control Assoc, Atlantic City, NJ

Harward MR, Treshow M (1975) Impact of ozone on the growth and reproduction of understory plants in the aspen zone of the western USA. Environ Conserv 2:17–24

Haselhoff E, Lindau G (1903) Die Beschädigung der Vegetation durch Rauch. Borntraeger, Leipzig

Hasse HE (1913) The lichen flora of Southern California. Contrib U S Nat Herb 17, p 182

Haut H van (1961) Die Analyse von Schwefeldioxidwirkungen auf Pflanzen im Laboratoriumsversuch. Staub Reinhalt Luft 21:52–56

Haut H van, Stratmann H (1967) Experimentelle Untersuchungen über die Wirkung von Stickstoffdioxid auf Pflanzen. Schriftenr Landesanst Immissions- Bodennutzungsschutz Landes Nordrhein-Westfalen Essen 7:50–70

Haut H van, Stratmann H (1970) Farbtafelatlas über Schwefeldioxid-Wirkungen an Pflanzen. Girardet, Essen

Hawksworth DL, Rose R (1976) Lichens as pollution monitors. Arnold, London

Heagle AS (1973) Interactions between air pollutants and plant parasites. Annu Rev Phytopathol 11:365–388

Heagle AS (1979a) Ranking of soybean cultivars for resistance to ozone using different ozone doses and response measures. Environ Pollut 19:1–10

Heagle AS (1979b) Effects of growth media, fertiliser rate, and hour and season of exposure on sensitivity of four soybean cultivars to ozone. Environ Pollut 18:313–322

Heagle AS, Heck WW (1974) Predisposition of tobacco to oxidant air pollution injury by previous exposure to oxidants. Environ Pollut 7:247–252

Heagle AS, Johnston JW (1979) Variable responses of soybeans to mixtures of ozone and sulfur dioxide. J Air Pollut Control Assoc 29:729–732

Heagle AS, Letchworth MB (1982) Relationships among injury, growth, and yield responses of soybean cultivars exposed to ozone at different light intensities. J Environ Qual 11:690–694

Heagle AS, Key LW (1973 a) Effect of ozone on the wheat stem rust fungus. Phytopathology 63:397–400

Heagle AS, Key LW (1973 b) Effect of *Puccinia graminis* f. sp. *tritici* on ozone injury on wheat. Phytopathology 63:609–613

Heagle AS, Philbeck RB (1978) Exposure techniques. In: Heck WW, Krupa SW, Linzon SN (eds) Handbook of methodology for the assessment of air pollution effects on vegetation. Upper Midwest Sect, Air Pollut Control Assoc, pp 6–1 – 6–18

Heagle AS, Strickland A (1972) Reaction of *Erysiphe graminis* f. sp. *hordei* to low levels of ozone. Phytopathology 62:1144–1148

Heagle AS, Body DE, Pounds EK (1972) Effect of ozone on yield of sweet corn. Phytopathology 62:683–687

Heagle AS, Body DE, Heck WW (1973) An open-top field chamber to assess the impact of air pollution on plants. J Environ Qual 2:365–368

Heagle AS, Body DE, Neely GE (1974) Injury and yield responses of soybean to chronic doses of ozone and sulfur dioxide in the field. Phytopathology 64:132–136

Heagle AS, Philbeck RB, Rogers HH, Letchworth MB (1979 a) Dispensing and monitoring ozone in open-top field chambers for plant-effects studies. Phytopathology 69:15–20

Heagle AS, Philbeck RB, Letchworth MB (1979 b) Injury and yield responses of spinach cultivars to chronic doses of ozone in open-top field chambers. J Environ Qual 9:368–373

Heagle AS, Philbeck RB, Knott WM (1979 c) Thresholds for injury, growth, and yield loss caused by ozone on field corn hybrids. Phytopathology 69:21–26

Heagle AS, Riodan AJ, Heck WW (1979 d) Field methods to assess the impact of air pollutants on crop yield. Pap No 79-46.6, Pres 72nd APCA Annu Meet, Cincinnati, Ohio

Heagle AS, Spencer S, Letchworth MB (1979 e) Yield response of winter wheat to chronic doses of ozone. Can J Bot 57:1999–2005

Heagle AS, Heck WW Rawlings JO, Philbeck RB (1983) Effects of chronic doses of ozone and sulfur dioxide on injury and yield of soybean in open-top field chambers. Crop Sci 23: 1184–1191

Heath RL (1975) Ozone. In: Mudd JB, Kozlowski TT (eds) Responses of plants to air pollution. Academic Press, London New York pp 23–55

Heath RL (1978) The inhibition of respiration and glycolysis in *Chlorella* exposed to ozone. Plant Physiol 61:93

Heath RL, Frederick PE (1977) Ozone-induced alterations of K^+ influx and efflux from *Chlorella sorokiana*. Plant Physiol 59(Suppl):124

Heath RL, Frederick PE (1979) Ozone alteration on membrane permeability in *Chlorella*. I. Permeability of potassium ion as measured by [86]Rubidium tracer. Plant Physiol 64:455–459

Heath RL, Chimiklis PE, Frederick PE (1974) Role of potassium and lipids in ozone injury to plant membranes. In: Dugger WM (eds) Air pollution effects on plant growth. ACS Symp Ser 3. Am Chem Soc, Washington, pp 58–75

Heath RL, Frederick PE, Chimiklis PE (1982) Ozone inhibition of photosynthesis in *Chlorella sorokiana*. Plant Physiol 69:229–233

Heck WW, Brandt CS (1977) Effects on vegetation: Native crops, forests. In: Air pollution, vol II. Academic Press, London New York, pp 157–229

Heck WW, Dunning JA (1967) The effects of ozone on tobacco and pinto bean as conditioned by several ecological factors. J Air Pollut Control Assoc 17:112–114

Heck WW, Heagle AS (1970) Measurement of photochemical air pollution with a sensitive monitoring plant. J Air Pollut Control Assoc 20:97–99

Heck WW, Pires EG (1962) Effect of ethylene on horticultural and agronomic plants. Tex Agric Exp Stn Misc Publ No MP-613:3–12

Heck WW, Tingey DT (1971) Ozone, time-concentration model to predict acute foliar injury. In: Proc 2nd Int Clean Air Congr, Washington DC, pp 249–355

Heck WW, Dunning JA, Hindawi IJ (1965) Interactions of environmental factors on the sensitivity of plants to air pollution. J Air Pollut Control Assoc 15:511–515

Heck WW, Dunning JA, Hindawi IJ (1966) Ozone: Nonlinear relation of dose and injury to plants. Science 151:577–578

Heck WW, Fox FL, Brandt CS, Dunning JA (1969) Tobacco, a sensitive monitor for photochemical air pollution. Nat Air Pollut Control Administr Publ No AP-55, Cincinnati, US Dep Health, Education, and Welfare

Heck WW, Mudd JB, Miller PR (1977) Plants and microorganisms. In: National Research Council (ed) Ozone and other photochemical oxidants. Nat Acad Sci, Washington DC, pp 437–585

Heck WW, Philbeck RB, Dunning JA (1978) A continuous stirred tank reactor (CSTR) system for exposing plants to gaseous air contaminants: Principles, specifications, construction, and operation. Agric Res Serv, US Dep Agric, ARS-S-181, New Orleans

Heck WW, Taylor OC, Adams R, Bingham G, Miller J, Preston E, Weinstein L (1982) Assessment of crop loss from ozone. J Air Pollut Control Assoc 32:353–361

Heggestad HE (1966) Ozone as a tobacco toxicant. J Air Pollut Control Assoc 16:691–694

Heggestad HE (1970) Variation in response of potato cultivars to air pollution. Abstr Phytopathology 60:1015

Heggestad HE (1973) Photochemical air pollution injury to potatoes in the Atlantic coastal states. Am Potato J 50:315–328

Heggestad HE, Bennett RH (1981) Photochemical oxidants potentiate yield losses in snap beans attributable to sulfur dioxide. Science 213:1008–1010

Heggestad HE, Heck WW (1971) Nature, extent, and variation of plant response to air pollutants. Adv Agron 23:111–145

Heggestad HE, Menser HA (1962) Leaf spot-sensitive tobacco strain Bel W3, a biological indicator of the air pollutant ozone. Phytopathology 52:735

Heggestad HE, Middleton JT (1959) Ozone in high concentrations as cause of tobacco leaf injury. Science 129:208–210

Heggestad HE, Burleson FR, Middleton JT, Darley EF (1964) Leaf injury on tobacco varieties resulting from ozone, ozonated hexene-1 and ambient air of metropolitan areas. Int J Air Water Pollut 8:1–10

Heggestad HE, Heagle AS, Bennett JH, Koch EJ (1980) The effects of photochemical oxidants on the yield of snap beans. Atmos Environ 14:317–326

Henderson WR, Reinert RA (1979) Yield response of four fresh market tomato cultivars after acute ozone exposure in the seedling stage. J Am Soc Hortic Sci 104:754–759

Hibben CR (1966) Sensitivity of fungal spores to sulphur dioxide and ozone. Abstr Phytopathology 56:880–881

Hibben CR (1969 a) The distinction between injury to tree leaves by ozone and mesophyll-feeding leafhoppers. For Sci 15:154–157

Hibben CR (1969 b) Ozone toxicity to sugar maple. Phytopathology 59:1423–1428

Hibben CR, Stotzky G (1969) Effects of ozone on the germination of fungus spores. Can J Microbiol 15:1187–1196

Hill AC (1971) Vegetation: A sink for atmospheric pollutants. J Air Pollut Control Assoc 21:341–346

Hill AC, Littlefield N (1969) Ozone. Effect on apparent photosynthesis, rate of transpiration, and stomatal closure in plants. Environ Sci Technol 3:52–56

Hill AC, Pack MR, Treshow M, Downs RJ, Transtrum LG (1961) Plant injury induced by ozone. Phytopathology 51:356–363

Hill AC, Heggestad HE, Linzon SN (1970) Ozone. In: Jacobson JS, Hill AC (eds) Recognition of air pollution injury to vegetation: A pictorial atlas. Air Pollut Control Assoc, Pittsburgh, Pa, pp B1–B22

Hodges GH, Menser HA Jr, Ogden WB (1971) Susceptibility of Wisconsin Havana tobacco cultivars to air pollutants. Agron J 63:107–111

Hoffman GJ, Maas EV, Rawlins SL (1973) Salinity-ozone interactive effects on yield and water relations of pinto bean. J Environ Qual 2:148–152

Hoffman GJ, Maas EV, Rawlins SL (1975) Salinity-ozone interactive effects on alfalfa yield and water relations. J Environ Qual 4:326–331

Hofstra G, Ormrod DP (1977) Ozone and sulphur dioxide interaction in white bean and soybean. Can J Plant Sci 57:1193–1198

Hofstra G, Littlejohns DA, Wukasch RT (1978) The efficacy of the antioxidant ethylene-diurea (EDU) compared to carboxin and benomyl in reducing yield losses from ozone in navy bean. Plant Dis Rep 62:350–352

Hofstra G, Ali A, Wukasch RT, Fletcher RA (1981) Rapid inhibition of root respiration after exposure of bean (*Phaseolus vulgaris* L.) plants to ozone. Atmos Environ 15:483–487

Horsman DC, Wellburn AR (1977) Effect of SO_2 polluted air upon enzyme activity in plants originating from areas with different annual mean atmospheric SO_2 concentration. Environ Pollut 31:33–40

Horsman DC, Nicholls AO, Calder DM (1980) Growth responses of *Dactylis glomerata, Lolium perenne* and *Phalaris aquatica* to chronic ozone exposure. Aust J Plant Physiol 7:511–517

Houston DB (1974) Responses of selected *Pinus strobus* L. clones to fumigations with sulfur dioxide and ozone. Can J For Res 4:65–68

Houston DB, Stairs GR (1973) Genetic control of sulfur dioxide and ozone tolerance in eastern white pine. For Sci 19:267–271

Houten JG ten (1966) Bezwaren van luchtverontreiniging voor de landbouw. Landbouwkd Tijdschr 78:2–13

Howell RK (1970) Influence of air pollution on quantities of caffeic acid isolated from leaves of *Phaseolus vulgaris*. Phytopathology 60:1626–1629

Howell RK (1974) Phenols, ozone and their involvement in pigmentation and physiology of plant injury. In: Dugger M (ed) Air pollution effects on plant growth. ACS Symp Ser 3. Am Chem Soc, Washington DC, pp 94–105

Howell RK, Kremer DR (1973) The chemistry and physiology of pigmentation in leaves injured by air pollution. J Environ Qual 2:434–438

Howell RK, Devine TE, Hanson CH (1971) Resistance of selected alfalfa strains to ozone. Crop Sci 11:114–115

Howell RK, Koch EJ, Rose LP (1976) Field assessment of air pollution induced soybean yield losses. Agron Abstr Am Soc Agron 68th Annu Meet, Madison, Wisconsin, pp 84–85

Huang TR, Aycock MK, Mulchi CL (1975) Heterosis and combining ability estimates for air pollution damage, primarily ozone, in Maryland tobacco. Crop Sci 15:785–789

Hull HM, Went FW (1952) Life processes of plants as affected by air pollution. In: Proc 2nd Nat Air Pollut Symp. Nat Air Pollut Symp, Pasadena, Calif, pp 122–128

Hull HM, Went FW, Yamada N (1954) Fluctuations in sensitivity of the *Avena* test due to air pollutants. Plant Physiol 29:182–187

Hutchinson TC, Havas M (1980) Effects of acid precipitation on terrestrial ecosystems. NATO Conf Ser I. Ecology, vol IV. Plenum Press, New York London

Hüttermann A (1983) Immissionsschäden im Bereich der Wurzeln von Waldbäumen: Frühdiagnose, biochemische und physiologische Untersuchungen. Mitt LÖLF Sonderh: Immissionsbelastungen von Waldökosystemen. Erweiterte Neuaufl 1983 10 a–14 a

Huttunen S (1979) Some experience about biological monitoring of air pollutants by biochemical methods – Finland. In: Eastmond M, Skärby L (eds) Report from the workshop „Ozone effects on vegetation in Europe". Swed Water Air Pollut Inst, Göteborg, p 17

Huttunen S (1980) The integrative effects of airborne pollutants on boreal forest ecosystems. In: United Nations (ed) Symposium on the effects of airborne pollution on vegetation. Warsaw, Poland

IUFRO (International Union of Forestry Research Organizations) (1979, 1981) Resolution on air quality standard for the protection of forests. IUFRO-Fachgruppe S 2.09.00

Jacobson JS (1977 a) The effects of photochemical oxidants on vegetation. VDI-Ber 270:163–173

Jacobson JS (1977 b) Plants as indicators of photochemical oxidants in the USA. VDI-Ber 270:191–196

Jacobson JS (1982) Ozone and the growth and productivity of agricultural crops. In: Unsworth MH, Ormrod DP (eds) Effects of gaseous air pollution in agriculture and horticulture. Butterworth, London, pp 293–304

Jacobson JS, Colavito LJ (1976) The combined effect of sulfur dioxide and ozone on bean and tobacco plants. Environ Exp Bot 16:277–285

Jacobson JS, Feder WA (1974) A regional network for environmental monitoring: Atmospheric oxidant concentrations and foliar injury to tobacco indicator plants in the eastern United States. Mass Agric Exp Stn, Coll Food Nat Resourc. Univ Mass Amherst, Bull No 604

Jacobson JS, Hill AC (1970) Recognition of air pollution injury to vegetation: A pictorial atlas. Air Pollut Control Assoc, Pittsburgh, Pa

Jaffe LS (1967) Effects of photochemical air pollution on vegetation with relation to the air quality requirements. J Air Pollut Control Assoc 17:38–42

Jahns HM, Neumann K (1981) Flechtenwachstum im Frankfurter Raum. Nat Mus 3:333–338

James RL, Cobb FW, Wilcox WW, Rowney DL (1980) Effects of photochemical oxidant injury on susceptibility of ponderosa and jeffrey pines sapwood and freshly cut stumps to *Fomes annosus*. Ecol Epidemiol 70:704–708

Jensen KF (1973) Response of nine forest tree species to chronic ozone fumigation. Plant Dis Rep 57:914–917

Jensen KF (1981 a) Ozone fumigation decreased the root carbohydrate content and dry weight of green ash seedlings. Environ Pollut Ser A 26:147–152

Jensen KF (1981 b) Air pollutants affect the relative growth rate of hardwood seedlings. US Dep Agric For Serv Res Pap NE 470

Jensen KF (1981 c) Growth analysis of hybrid poplar cuttings fumigated with ozone and sulphur dioxide. Environ Pollut Ser A 26:243–250

Jensen KF, Dochinger LS (1974) Responses of hybrid poplar cuttings to chronic and acute levels of ozone. Environ Pollut 6:289–295

Jensen KF, Masters RG (1975) Growth of six woody species fumigated with ozone. Plant Dis Rep 59:760–762

Jensen KF, Dochinger LS, Roberts BR, Townsend AM (1976) Pollution responses. In: Miksche JP (ed) Modern methods in forest genetics. Springer, Berlin Heidelberg New York, pp 189–216

Johnston WJ, Dickens R, Haaland RL (1980) Exposure to ozone affects growth of tall fescue. Highlight Agric Res 27:11

Jones JL (1963) Ozone damage: Protection for plants. Science 140:1317–1318

Juhren M, Noble W, Went FW (1957) The standardization of the *Poa annua* as an indicator of smog concentrations. I. Effects of temperature, photoperiod, and light intensity during growth of the test plants. Plant Physiol 32:576–586

Kadota M, Ohta K (1972) Ozone sensitivity of japanese plant species in summer with special reference to a tentative sensitivity grade list for applying to field survey on ozone injury. Taiki Osen Kenkyu (J Jpn Soc Air Pollut) 7:19–26

Kandler O (1983) Waldsterben: Emissions- oder Epidemie-Hypothese. Naturwiss Rundsch 36:488–490

Karnosky DF (1976) Threshold levels for foliar injury to *Populus tremuloides* by sulfur dioxide and ozone. Can J For Res 6:166–169

Karnosky DF (1977) Evidence for genetic control of response to sulfur dioxide and ozone in *Populus tremuloides*. Can J For Res 7:437–440

Karnosky DF (1978) Testing the air pollution tolerances of shade tree cultivars. J Arboric 4:107–110

Karnosky DF (1981) Chamber and field evaluations of air pollution tolerances of urban trees. J Arboric 7:99–105

Keen NT, Taylor OC (1975) Ozone injury in soybeans. Isoflavonoid accumulation is related to necrosis. Plant Physiol 55:731–733

Keitel A, Arndt U (1983) Ozoninduzierte Turgeszenzverluste bei Tabak (*Nicotiana tabacum* var. Bel W3) – ein Hinweis auf schnelle Permeabilitätsveränderungen der Zellmembranen. Angew Bot 57:193–204

Kelleher TJ, Feder WA (1978) Phytotoxic concentrations of ozone on Nantucket Island: Long range transport from the Middle Atlantic States over the open ocean confirmed by bioassay with ozone-sensitive tobacco plants. Environ Pollut 17:187–194

Keller Th (1970) Zum Problem der verkehrsbedingten Bleirückstände in der Vegetation. Strasse und Verkehr 56:32–34

Keller Th (1974) The use of peroxidase activity for monitoring and mapping air pollution areas. Eur J For Pathol 4:11–19

Keller Th (1981) Folgen einer winterlichen SO_2-Belastung für die Fichte. Gartenbauwissenschaft 46:170–178

Keller Th, Schwager H (1977) Air pollution and ascorbic acid. Eur J For Pathol 7:338–350

Kender WJ, Shaulis NJ (1976) Vineyard management practices influencing oxidant injury on "Concord" grapes. J Am Soc Hortic Sci 101:129–132

Kender WJ, Taschenbert EF, Shaulis NJ (1973) Benomyl protection of grapevines from air pollution injury. Hortic Sci 8:396–398

Kendrick JB, Middleton JT, Darley EF (1953) Predisposing effects of air temperature and nitrogen supply upon injury to some herbaceous plants fumigated with peroxides derived from olefins. Abstr Phytopathology 43:588

Kendrick JB, Middleton JT, Darley EF (1954) Chemical protection of plants from ozonated olefin (smog) injury. Abstr Phytopathology 44:494–495

Kendrick JB, Darley EF, Middleton JT (1962) Chemotherapy for oxidant and ozone induced plant damage. Int J Air Water Pollut 6:391–402

Khatamian H, Adedipe NO, Ormrod DP (1973) Soil-plant-water aspects of ozone phytotoxicity in tomato plants. Plant Soil 38:531–541

Kickert RN, Miller PR, Taylor OC, McBride JR, Barbieri J, Arkley R, Cobb F Jr, Dahlsten D, Wilcox WE, Wenz J, Parmeter JR Jr, Luck RF, White M (1977) Photochemical air pollutant effects on mixed conifer ecosystems. A progress report. EPA-600/3-77-058. US Environ Protect Agency, Corvallis, Oregon

Kinman RN (1975) Water and wastewater disinfection with ozone: a critical review. CRC Crit Rev Environ Control 5:141–152

Kisser J (1966) Forstliche Rauchschäden aus der Sicht des Biologen. Mitt Forstl. Bundesversuchsanst Mariabrunn 73:7–48

Klingaman GL, Link CB (1975) Reduction air pollution injury to foliage of *Chrysanthemum morifolium* ramat. using tolerant cultivars and chemical protectants. J Am Soc Hortic Sci 100:173–175

Klingensmith MJ (1969) The effect of certain benzazole compounds on plant growth and development. Am J Bot 48:40–45

Knabe W (1970) Kiefernwaldverbreitung und Schwefeldioxid-Immissionen im Ruhrgebiet. Staub Reinhalt Luft 30:32–35

Knabe W (1983) Immissionsökologische Waldzustanderfassung in Nordrhein-Westfalen. IWE (1979). Fichten und Flechten als Zeiger der Waldgefährdung durch Luftverunreinigungen. Forsch Berat Reihe C, H 37. Landwirtschaftsverlag, Münster-Hiltrup

Knabe W, Brandt CS, Haut H van, Brandt CJ (1973) Nachweis photochemischer Luftverunreinigungen durch biologische Indikatoren in der Bundesrepublik Deutschland. Proc 3rd Int Clean Air Congr. VDI-Verlag, Düsseldorf, A110–A114

Knapp CE, Fieldhouse DJ (1970) Alar and Folicote sprays for reducing ozone injury on four solanaceous genera. HortScience 5:338

Knudson LL, Tibbitts TW, Edwards GE (1977) Measurement of ozone injury by determination of leaf chlorophyll concentration. Plant Physiol 60:606–608

Kochhar M, Blum U, Reinert RA (1980) Effects of O_3 and (or) fescue on ladino clover: interactions. Can J Bot 58:241–249

Kohler A (1978) Wasserpflanzen als Bioindikatoren. Beih Veröff Naturschutz Landschaftspfl Baden-Württemberg 11:259–281

Kohut RJ (1975) The interaction of O_3 and PAN on hybrid poplar and pinto bean. Cent Environ Stud, Pa State Univ, CAES Publ 429–476

Kohut RJ, Davis DD (1978) Response of pinto bean to simultaneous exposure to ozone and PAN. Phytopathology 68:567–569

Kohut RJ, Davis DD, Merrill W (1976) Response of hybrid poplar to simultaneous exposure to ozone and PAN. Plant Dis Rep 60:777–780

Koiwai A, Kisaki I (1976) Effect of ozone on photosystem II of tobacco chloroplasts in the presence of piperonyl butoxide. Plant Cell Physiol 17:1199–1207

Koning HW De, Jegier Z (1968a) A study on the effects of ozone and sulfur dioxide on the photosynthesis and respiration of *Euglena gracilis*. Atmos Environ 2:321–326

Koning HW De, Jegier Z (1968b) Quantitative relation between ozone concentration and reduction of photosynthesis of *Euglena gracilis*. Atmos Environ 2:615–616

Koning HW De, Jegier Z (1970) Effects of sulfur dioxide and ozone on *Euglena gracilis*. Atmos Environ 4:357–361

Koritz HG, Went FW (1953) The physiological action of smog on plants. I. Initial growth and transpiration studies. Plant Physiol 28:50–62

Kormelink JR (1967) Effects of ozone on fungi. MS thesis, Univ Utah, Salt Lake City

Koukol J, Dugger WM Jr (1967) Anthocyanin formation as a response to ozone and smog treatment in *Rumex crispus* L. Plant Physiol 42:1023–1024

Koukol J, Dugger WM Jr, Palmer RL (1967) Inhibitory effect of peroxyacetyl nitrate on cyclic photophosphorylation by chloroplasts from Black Valentine bean leaves. Plant Physiol 42:1419–1422

Krämer F (1976) Erste Untersuchungen zur Erstellung eines Bundesbelastungskatasters (Pb, Zn, Cd, Cu) in Raume Duisburg, Dinslaken. Schriftenr Landesanst Immissions- und Bodennutzungsschutz Landes Nordrhein-Westfalen Essen 39:45–48

Kramer W (1982) Das Tannensterben. Forstarchiv 53:128–132

Krause CR, Jensen KF (1978) Microtopographical changes in hybrid poplar leaves associated with air pollution exposure. Scan Electr Microsc 2:755–759

Krause CR, Jensen KF (1979) Surface changes on hybrid poplar leaves exposed to ozone and sulfur dioxide. Scan Electr Microsc 3:77–80

Krause CR, Weidensaul TC (1978 a) Effects of ozone on the sporulation, germination, and pathogenicity of *Botrytis cinerea*. Phytopathology 68:195–198

Krause CR, Weidensaul TC (1978 b) Ultrastructural effects of ozone on the host parasite relation of *Botrytis cinerea* and *Pelargonium hortorum*. Phytopathology 68:301–307

Krause GHM (1974) Zur Aufnahme von Zink und Cadmium durch oberirdische Pflanzenorgane. Dissertation, Univ Bonn

Kremer D, Howell RK (1974) Phenylalanine ammonia lyase induced by ozone. Proc Am Phytopathol Soc 1, Div Meet Abstr 152

Kress LW (1972) Response of hybrid poplar to sequential exposures of ozone and PAN. Center for Air Environment Studies. Pa State Univ CAES Publ No 259–72

Kress LW (1976) Relative sensitivity of eighteen genetic lines of *Pinus taeda* L. to ozone. Proc Am Phytopathol Soc 3:328

Kress LW, Miller JE (1981) Impact of ozone on soybean yield. Argonne National Laboratory. Radiol Environ Res Div Annu Rep Ecol, January–December, ANL-80-115, Part III, pp 11–14

Krivopishin IP (1973) Effect of ozone on shell microflora and the results of chicken egg incubation (in Russian). Tr Vses Nauchno-Issled Tekhnol Inst Ptitsevod 37:39–44

Kühl U, Wagner H-M (1970) Untersuchungen über die Wirkung von Photooxidantien auf *Petunia hybrida*. Staub Reinhalt Luft 30:382–383

Kulagin YZ (1973) Gas tolerance of plants and preadaption. Sov J Ecol 4:128–131

Küppers K (1985) Wirkungen von Ethen auf Modellpflanzengemeinschaften. Verhandl Ges Ökologie (in press)

Kuss FR (1950) The effect of ozone on fungus sporulation. MS thesis, Univ New Hampshire, Durham

Landesanstalt für Umweltschutz Baden-Württemberg (1981) Zwischenbericht über SO_2-Messungen im Schwarzwald. Ber 83/81

Lang KJ (1981) Anatomische Befunde im Naßkernbereich der Wurzeln von *Abies alba*. Forstwiss Centralbl 100:180–183

Larsen RI, Heck WW (1976) An air quality data analysis system for interrelating effects, standards, and needed source reductions: Part 3. Vegetation injury. J Air Pollut Control Assoc 26:325–333

Laurence JA (1981) Effects of air pollutants on plant-pathogen interactions. Z Pflanzenkrankh Pflanzenschutz 87:156–172

Laurence JA, Wood FA (1978) Effects of ozone on infection of soybean by *Pseudomonas glycinea*. Phytopathology 68:441–445

LeBlanc F, Rao DN (1975 a) A review of the literature on bryophytes with respect to air pollution. Bull Soc Bot Fr 121: Coll Bryol 237–255

LeBlanc F, Rao DN (1975 b) Effects of air pollutants on lichens and bryophytes. In: Mudd JB, Kozlowski TT (eds) Responses of plants to air pollution. Academic Press, London New York, pp 237–272

Ledbetter MC, Zimmerman PW, Hitchcock AE (1959) The histopathological effects of ozone on plant foliage. Contrib Boyce Thompson Inst 20:275–282

Lee EH, Bennett JH (1982) Superoxide dismutase. A possible protective enzyme against ozone injury in snap beans (*Phaseolus vulgaris* L.). Plant Physiol 69:1444–1449

Lee JJ, Lewis RA (1978) Zonal air pollution system: Design and performance. In: Preston EM, Lewis RA (eds) The bioenvironment impact of a coal-fired power plant. Third Interim Report, Colstrip, Montana. US Environ Protect Agency, Corvallis Environ Res Lab, EPA-600/3-78-021, pp 322–344

Lee TT (1966) Chemical regulation of ozone susceptibility in *Nicotiana tabacum*. Can J Bot 44:487–496

Lee TT (1968) Effect of ozone on swelling of tobacco mitochondria. Plant Physiol 43:133–139

Lefohn AS, Benedict HM (1982) Development of a mathematical index that describes ozone concentration, frequency, and duration. Atmos Environ 16:2529–2532

Leone IA (1976) Response of potassium-deficient tomato plants in atmospheric ozone. Phytopathology 66:734–736

Leone IA, Brennan E (1969 a) The importance of moisture in ozone phytotoxicity. Atmos Environ 3:399–406

Leone IA, Brennan E (1969 b) Sensitivity of begonias to air pollution. Hortic Res 9:112–116

Leone IA, Brennan E (1970) Ozone toxicity in tomato as modified by phosphorus nutrition. Phytopathology 60:1521–1524

Leone IA, Brennan E (1975) Variable effects of ozone on pinto bean internodes. Phytopathology 65:666–669

Leone IA, Brennan E, Daines RH (1966) Effect of nitrogen nutrition on the response of tobacco to ozone in the atmosphere. J Air Pollut Control Assoc 16:191–196

Leopold AC (1964) Plant growth and development. McGraw-Hill Book Company, New York

Letchworth MB, Blum U (1977) Effects of acute ozone exposure on growth nodulation and nitrogen content of Ladino clover. Environ Pollut 14:303–312

Leuning R, Neumann HH, Thurtell GW (1979 a) Ozone uptake by corn (*Zea mays* L.): A general approach. Agric Meteorol 20:115–135

Leuning R, Unsworth MH, Neumann HN, King KM (1979 b) Ozone fluxes to tobacco and soil under field conditions. Atmos Environ 13:1155–1163

Levitt J (1980) Responses of plants to environmental stresses, 2nd edition. Academic Press, London New York

Lewis E, Brennan E (1977) A disparity in the ozone response of bean plants grown in a greenhouse, growth chamber, or open-top chamber. J Air Pollut Control Assoc 27:889–891

Lewis E, Brennan E (1978) Ozone and sulfur dioxide mixtures cause a PAN-type injury to petunia. Phytopathology 68:1011–1014

Linzon SN (1958) The influence of smelter fumes on growth of white pine in the Sudbury region of Canada. Can Dep Agric Publ, Ontario Dep Lands

Linzon SN (1967) Histological studies of symptoms in semimature-tissue needle blight of eastern white pine. Can J Bot 45:133–143

Linzon SN (1973) The effects of air pollution on forests. 4th Joint Chem Eng Conf, Vancouver, BC

Linzon SN, Heck WW, MacDowall FDH (1975) Effects of photochemical oxidants on vegetation. In: Photochemical air pollution: Formation, transport, effects. Nat Res Counc, Canada, pp 89–142

Lorenc-Plucińska G (1979) The effect of ozone on photosynthesis and respiration in Scots pines differing in resistance to this gas. Arbor Kornickie 24:329–338

Loucks OL, Williams WT (1980) Estimating forest growth losses due to oxidants and pollutant-insect interactions. In: Loucks OL (ed) Crop and forest losses due to current and project emissions from coal-fired power plants in the Ohio River Basin. US Environ Protect Agency, Off Res Dev, Washington Prep Ohio River Basin Energy Study (OREBS), pp 191–209

Lumis GP, Ormrod DP (1978) Effects of ozone on growth of four woody ornamental plants. Can J Plant Sci 58:769–773

Maas EV, Hoffman GH, Rawlins SL, Ogata G (1973) Salinity-ozone interactions on pinto bean: Integrated response to ozone concentration and duration. J Environ Qual 2:400–404

MacDowall FDH (1965 a) Predisposition of tobacco to ozone damage. Can J Plant Sci 45:1–12

MacDowall FDH (1965 b) Stages of ozone damage to respiration of tobacco leaves. Can J Bot 43:419–427

MacDowall FDH (1966) The relation between dew and tobacco weather fleck. Can J Plant Sci 46:349–353

MacDowall FDH, Cole AFW (1971) Threshold and synergistic damage to tobacco by ozone and sulfur dioxide. Atmos Environ 5:553–559

MacDowall FDH, Vickery LS, Runeckles VC, Patrick ZA (1963) Ozone damage to tobacco in Canada. Can Plant Dis Surv 43:131–151

MacDowall FDH, Mukammal EI, Cole AFW (1964) Direct correlation of air-polluting ozone and tobacco weather fleck. Can J Plant Sci 44:410–417

MacKnight ML (1968) The effect of ozone on stomatal aperture and transpiration. MA thesis, Univ Utah, Salt Lake City

MacLean DC, Schneider RE (1976) Photochemical oxidants in Yonkers, New York: Effects on yield of bean and tomato. J Environ Qual 5:75–78

Magdycz WP (1972) The effects of concentration and exposure time on the toxicity of ozone to the spores of *Botrytis cinerea*. MS thesis, Univ Massachusetts, Waltam

MAGS (Minister für Arbeit, Gesundheit und Soziales) and MELF (Minister für Ernährung, Landwirtschaft und Forsten) des Landes Nordrhein-Westfalen (1980) Umweltbelastung durch Thallium, Düsseldorf

Málek J (1981) Problematik der Ökologie der Tanne (*Abies alba* Mill.) und ihr Sterben in der ČSSR. Forstwiss Centralbl 100:170–174

Malhotra SS, Blauel RA (1980) Diagnosis of air pollutant and natural stress symptoms on forest vegetation in western Canada. Inf Rep NOR-X-228, North For Res Cent, Can For Serv. Environ Can, Edmonton, Alberta

Mandl RH, Weinstein LH, McCune DC, Keveny M (1973) A cylindrical, open-top chamber for the exposure of plants to air pollutants in the field. J Environ Qual 2:371–376

Manning WJ (1975) Interactions between air pollutants and fungal, bacterial, and viral plant pathogens. Environ Pollut 9:87–90

Manning WJ (1978) Chronic foliar ozone injury: Effects on plant root development and possible consequences. Calif Air Environ 7:3–4

Manning WJ, Feder WA (1976) Effects of ozone on economic plants. In: Mansfield TA (ed) Effects of air pollutants on plants. Cambridge Univ Press, Cambridge, pp 47–60

Manning WJ, Vardaro PM (1974a) Sensitivity of bean cultivars to low levels of ozone and ambient oxidants. Annu Proc Am Phytopathol Soc 1:78

Manning WJ, Vardaro PM (1974b) Ozone and *Pyrenochaeta lycopersici:* effects on growth and development of tomato plants. Abstr Phytopathology 64:583

Manning WJ, Feder WA, Perkins I, Glickman M (1969) Ozone injury and infection of potato leaves by *Botrytis cinerea*. Plant Dis Rep 53:691–693

Manning WJ, Feder WA, Papia PM, Perkins I (1971) Influence of foliar ozone injury on root development and root surface fungi of pinto bean plants. Environ Pollut 1:305–312

Manning WJ, Feder WA, Perkins I (1972) Sensitivity of spinach cultivars to ozone. Plant Dis Rep 56:832–833

Manning WJ, Feder WA, Vardaro PM (1973a) Benomyl in soil and response of pinto bean plants to repeated exposures to a low level of ozone. Phytopathology 63:1539–1540

Manning WJ, Feder WA, Vardaro PM (1973b) Reduction of chronic ozone injury on poinsettia by benomyl. Can J Plant Sci 53:833–835

Manning WJ, Feder WA, Vardaro PM (1974) Suppression of oxidant injury by benomyl: Effects on yield of bean cultivars in the field. J Environ Qual 3:1–3

Markowski A, Grzesiak S (1974) Influence of sulphur dioxide and ozone on vegetation of bean and barley plants under different soil moisture conditions. Bull Acad Pol Sci Ser Sci Biol 22:875–888

Maschke J (1981) Moose als Bioindikatoren von Schwermetall-Immissionen. Bryophytorum Bibliotheca, vol 22. Cramer, Vaduz

Materna J (1972) Einfluß niedriger Schwefeldioxydkonzentrationen auf die Fichte. Mitt Forstl Bundesversuchsanst Wien 97:219–231

Materna J, Jirgle J, Kucera J (1969) Vysledky mereni koncentraci kyslicniku siriciteho v lesich Krusnych hor. Ochr Ovzdusi 1:84–93

Matsunaka S (1973) The utilization of biological indicators in the monitoring systems for photochemical air pollution control. In: References on photochemical air pollution in Japan. Air Qual Bur, Environ Agency, pp 282–297

Matsushima J (1971) On composite harm to plants by sulphurous acid gas and oxidant. Industrial Public Damage 7:218–224

McCarthy JJ, Smith CH (1974) A review of ozone and its application to domestic wastewater treatment. J Am Water Works Assoc 66:718–725

McCarty RE, Pittman PR, Tsuchiya Y (1972) Light-dependent inhibition of photophosphorylation by N-ethylmaleimide. J Biol Chem 247:3048–3051

McClenahen JR (1978) Community changes in a deciduous forest exposed to air pollution. Can J For Res 8:432–438

McCool PM, Menge JA, Taylor OC (1977) Influence of ozone and HCl gas on citrus and the vesicular-arbuscular mycorrhizal fungus *Glomus fasciculatus*. Proc Am Phytopathol Soc 4:57

McCool PM, Menge JA, Taylor OC (1979) Effects of ozone and HCl gas on the development of the mycorrhizal fungus *Glomus fasciculatus* and growth of "Troyer" Citrange. J Am Soc Hortic Sci 104:151–154

McCord JM, Keele BB Jr, Fridovich I (1971) An enzyme-based theory of obligate anaerobiosis: the physiological function of superoxide dismutase. Proc Nat Acad USA 68:1024–1027

McCormick R (1963) Changes in herbaceous plant community during a three year period following exposures to ionizing radiation gradients. 1st Natl Symp Radioecol, Proc Radioecol. Reinhold, New York, pp 271–275

McCune DC (1969) Establishment of air quality criteria with reference to the effects on atmospheric fluorine on vegetation. Am Pet Inst, New York, Air Qual Monogr No 69-3

McLaughlin SB, Taylor GE Jr (1981) Relative humidity: Important modifier of pollutant uptake by plants. Science 211:167–169

Meidner H, Mansfield TA (1968) Physiology of stomata. McGraw-Hill, London

Meiners JP, Heggestad HE (1979) Evaluation of snap bean cultivars for resistance to ambient oxidants in field plots and to ozone in chambers. Plant Dis Rep 63:273–277

Menser HA Jr (1963) The effects of ozone and controlled-environment factors on four varieties of tobacco, *Nicotiana tabacum* L. PhD thesis, Univ Maryland, College Park

Menser HA (1964) Response of plants to air pollutants. III. A relation between ascorbic acid levels and ozone susceptibility of light preconditioned tobacco plants. Plant Physiol 39:564–567

Menser HA (1974) Response of sugar beet cultivars to ozone. J Am Soc Sugar Beet Technol 18:81–86

Menser HA, Heggestad HE (1966) Ozone and sulfur dioxide synergism: injury to tobacco plants. Science 153:424–425

Menser HA, Hodges GH (1970) Effects of air pollutants on Burley tobacco cultivars. Agron J 62:265–269

Menser HA, Hodges GH (1972) Oxidant injury to shade tobacco cultivars developed in Connecticut for weather fleck resistance. Agron J 64:189–192

Menser HA, Street OE (1962) Effects of air pollution, nitrogen levels, supplemental irrigation, and plant spacing on weather fleck and leaf losses of Maryland tobacco. Tob Sci 6:165–169

Menser HA, Heggestad HE, Street OE (1963) Response of plants to air pollutants. II. Effects of ozone concentration and leaf maturity on injury to *Nicotiana tabacum*. Phytopathology 53:1304–1308

Menser HA, Hodges GH, McKee CG (1973) Effects of air pollution on Maryland (type 32) tobacco. J Environ Qual 2:253–358

Metzler JT, Pell EJ (1980) The impact of peroxyacetyl nitrate on conductance of bean leaves and on associated cellular and foliar symptom expression. Phytopathology 70:934–938

Middleton JT (1956) Response of plants to air pollution. J Air Pollut Control Assoc 6:1–4

Middleton JT (1961) Photochemical air pollution damage to plants. Annu Rev Plant Physiol 12:431–448

Middleton JT, Paulus AO (1956) The identification and distribution of air pollutants through plant response. Arch Ind Health 14:526–532

Middleton JT, Kendrick JB Jr, Schwalm HW (1950) Injury to herbaceous plants by smog or air pollution. Plant Dis Rep 34:245–252

Middleton JT, Kendrick JB Jr, Darley EF (1953) Air pollution injury to crops. Calif Agric 7:11–12

Middleton JT, Kendrick JB Jr, Darley EF (1955) Airborne oxidants as plant damaging agents. In: Proc 3rd Natl Air Pollut Symp, Pasadena, Ca, pp 191–198

Middleton JT, Crafts AS, Brewer RF, Taylor OC (1956) Plant damage by air pollution. Calif Agric 6:9–12

Middleton JT, Darley EF, Brewer RF (1958) Damage to vegetation from polluted atmospheres. J Air Pollut Control Assoc 9:9–15

Mies E, Zöttl HW (1982) Zur Standortabhängigkeit der Tannen im Südschwarzwald. Allg Forstz 37:1296–1298

Mik G De (1973) Een gefoelige microbiologische detector voor luchtverontreiniging. Chemie Tech 28:209–212

Mik G De, Groot I De (1973) The survival of *Escherichia coli* in the open air in different parts of The Netherlands. In: Airborne transmission and airborne infection. Concepts and methods. Pres 6th Int Symp Aerobiol, Enschede, Netherlands. Halsted Press, Wiley, New York Toronto, pp 155–158

Millecan AA (1971) A survey and assessment of air pollution damage to California vegetation in 1970. Calif Dep Agric

Miller CA, Davis DD (1981) Response of pinto bean plants exposed to O_3, SO_2, or mixtures at varying temperatures. HortScience 16:548–550

Miller JR, Sprugel DG, Muller RN, Smith HJ, Xerikos PB (1980) Open-air fumigation system for investigating sulfur dioxide effects on crops. Phytopathology 70:1124–1128

Miller PM, Rich S (1968) Ozone damage on apples. Plant Dis Rep 52:730–731

Miller PM, Tomlinson H, Taylor GS (1976) Reducing severity of ozone damage to tobacco and beans by combining benomyl or carboxin with contant nematicides. Plant Dis Rep 60:433–436

Miller PR (1973a) Oxidant-induced community change in a mixed conifer forest. In: Naegele JA (ed) Air pollution damage to vegetation. Adv Chem Ser 122:101–117

Miller PR (1973 b) Susceptibility to ozone of selected western [US] conifers. Abstr 2nd Int Congr Plant Pathol No 0579

Miller PR, Kickert RN (1980) Terrestrial ecosystem impacts by gaseous air pollutants. Paper Pres Int Conf: Air pollutants and their effects on the terrestrial ecosystem, May 10–17. 1980, Banff, Alberta, Canada

Miller PR, McBride JR (1975) Effects of air pollutants on forests. In: Mudd JB, Kozlowski TT (eds) Response of plants to air pollutants. Academic Press, London New York, pp 195–235

Miller PR, White M (1977) Ecosystems. In: National Research Council (ed) Ozone and other photochemical oxidants. Nat Acad Sci, Washington DC, pp 586–642

Miller PR, Parmeter JR Jr, Taylor OC, Cardiff EA (1963) Ozone injury to the foliage of ponderosa pine. Phytopathology 53:1072–1076

Miller PR, Parmeter JR Jr, Flick BH, Martinez CW (1969) Ozone dosage response of ponderosa pine seedlings. J Air Pollut Control Assoc 19:435–438

Miller PR, Kickert RN, Taylor OC, Arkley RJ, Cobb FW Jr, Dahlsten DL, Gersper PJ, Luck RF, McBride JR, Parmeter JR Jr, Wenz JM, White M, Wilcox WW (1977) Photochemical oxidant air pollution effects on a mixed conifer forest ecosystem. A progress report. US Environ Protect Agency, EPA-600/3-77-104

Miller VL, Howell RK, Caldwell BE (1974) Relative sensitivity of soybean genotypes to ozone and sulfur dioxide. J Environ Qual 3:35–37

Mitchell A, Ormrod DP, Dietrich HF (1979) Ozone and nickel effects on pea leaf cell ultrastructure. Bull Environ Contamin Toxicol 22:379–385

Miyake Y, Uno Y, Kume H (1974) Influence of stomatal behavior on ozone injury of tobacco leaves. Tob Sci 18:28–30

Möhring K (1983) Wurzel-, Holz- und Kronenschäden sowie Vitalitätsschwund bei älterer Fichte in einem nordwestdeutschen Mittelgebirgsrevier. Ein Beitrag zur Bestandsaufnahme offensichtlicher Immissionsschäden. Forstarchiv 43:8–13

Montes RA, Blum U, Heagle AS (1982) The effect of ozone and nitrogen fertilizer on tall fescue, ladino clover, and fescue-clover mixture. I. Growth, regrowth, and forage production. Can J Bot 60:2745–2752

Mooi J (1981) Influence of ozone and sulphur dioxide on defoliation and growth of poplars. Mitt Forstl Bundesversuchsanst Wien 137:47–51

Moriondo F, Covassi F (1981) Tannensterben in Italien. Forstwiss Centralbl 100:168–170

Moskowitz PD, Coveney EA, Mediros WH, Morris SC (1982) Oxidant air pollution: A model for estimating effects on US vegetation. J Air Pollut Control Assoc 32:155–160

Mosley AR, Rowe RC, Weidensaul TC (1978) Relationship of foliar ozone injury to maturity classification and yield of potatoes. Am Potato J 55:147–153

Moyer JW, Cole H Jr, Lacasse NL (1974a) Reduction of ozone injury on *Poa annua* by benomyl and thioallophonate-ethyl. Phytopathology 64:584

Moyer JW, Cole H Jr, Lacasse NL (1974b) Suppression of naturally occurring oxidant injury on azalea plants by drench or foliar spray treatment with benzimidazole or oxathiin compounds. Plant Dis Rep 58:136–138

Mudd JB (1975) Peroxyacyl nitrates. In: Mudd JB, Kozlowski TT (eds) Responses of plants to air pollution. Academic Press, London New York, pp 97–119

Mudd JB (1980) Physiological and biochemical effects of ozone and sulfur dioxide. In: United Nations (ed) Symposium on the effects of air-borne pollution on vegetation. Warszawa, Poland, pp 80–89

Mudd JB (1982) Effects of oxidants on metabolic function. In: Unsworth MH, Ormrod DP (eds) Effects of gaseous air pollution in agriculture and horticulture. Butterworth, London, pp 189–203

Mudd JB, Dugger WM Jr (1963) The oxidation of reduced pyridine nucleotides by peroxyacyl nitrates. Arch Biochem Biophys 102:52–58

Mudd JB, Freeman BA (1977) Reaction of ozone with biological membranes. In: Lee SD, Peirano B (eds) Biochemical effects of environmental pollutants. Ann Arbor Sci Publ, Ann Arbor, Mich, pp 97–133

Mülder D (1983) Möglichkeiten der Forstbetriebe, sich Immissionsbelastungen waldbaulich anzupassen bzw. deren Schadwirkungen zu mildern. Materialien zur Umweltforschung, herausgegeben vom Rat von Sachverständigen für Umweltfragen, Kohlhammer, Stuttgart Mainz

Mumford RA, Lipke H, Laufer DA, Feder WA (1972) Ozone-induced changes in corn pollen. Environ Sci Technol 6:427–430

Murphy CE Jr, Sinclair TR, Knoerr KK (1977) An assessment of the use of forests as sinks for the removal of atmospheric sulfur dioxide. J Environ Qual 6:388–396

Musselman RC, Oshima RJ, Gallavan RE (1983) Significance of pollutant concentration distribution in the response of red kidney beans to ozone. J Am Soc Hortic Sci 108:347–351

Nagy R (1959) Application of ozone from sterilamp in control of mold, bacteria, and odors. In: Ozone chemistry and technology. Adv Chem Ser 21:57–65

Nakamura H, Ota Y (1977) Investigation on injury to rice plants from photochemical oxidants in Japan. Proc Int Clean Air Congr 4:104–105

Nakamura H, Saka H (1978) Photochemical oxidants injury in rice plants. III. Effect of ozone on physiological activities in rice plants. Jpn J Crop Sci 47:707–714

Nash TH III, Sigal LL (1979) Gross photosynthetic response of lichens to short-term ozone fumigations. Bryologist 82:280–285

Nash TH III, Sigal LL (1980) Sensitivity of lichens to air pollution with an emphasis on oxidant air pollutants. In: Symposium on effects of air pollutants on Mediterranean and temperate forest ecosystems, June 22–27, 1980, Riverside, California, USA, pp 117–124

NATO/CCMS (1974) Effects on vegetation. In: Air quality criteria for photochemical oxidants and related hydrocarbons, No 29, Bruxelles

Naveh Z, Chaim S, Steinberger EH (1978) Atmospheric oxidant concentrations in Israel as manifested by foliar injury in Bel W3 tobacco plants. Environ Pollut 16:249–262

Neely GE, Tingey DT, Wilhour RG (1977) Effects of ozone and sulfur dioxide singly and in combination on yield, quality, and N-fixation of alfalfa. In: Dimitriades B (ed) International conference on photochemical oxidant pollution and its control. EPA-600/3-77-001b, EPA, Research Triangle Park, North Carolina, pp 663–673

Nicholson CR, Skelly JM (1977) The response of 12 clones of eastern white pine (*Pinus strobus* L.) to ozone and nitrogen dioxide. Proc Am Phytopathol Soc 4:84

Nielsen DG, Terrell LE, Weidensaul TC (1977) Phytotoxicity of ozone and sulfur dioxide to laboratory fumigated scotch pine. Plant Dis Rep 8:699–703

Niklfeld H (1967) Pflanzensoziologische Beobachtungen im Rauchschadensgebiet eines Aluminiumwerkes. Centralbl Gesamte Forstwes 84:318–329

Nobel PS (1974) Ozone effects on chlorophylls a and b. Naturwissenschaften 61:80–81

Noble RD, Jensen KF (1980) Effects of sulfur dioxide and ozone on growth of hybrid poplar leaves. Am J Bot 67:1005–1009

Noble WM (1955) Air pollution effects. Pattern of damage produced on vegetation by smog. J Agric Food Chem 3:330–332

Noble WM (1965) Smog damage to plants. Lasca Leaves 15:2–4

Noland TL, Kozlowski TT (1979) Influence of potassium nutrition on susceptibility of silver maple to ozone. Can J For Res 9:501–503

Nouchi I (1979) Effects of ozone and PAN concentrations and exposure duration on plant injury. Taiki Osen Gakkaischi 14:489–496

Nouchi I, Aoki K (1979) Morning glory as a photochemical oxidant indicator. Environ Pollut 18:289–303

Nouchi I, Iljima T, Oodairi T (1975) Effects of peroxyacetyl nitrate (PAN) on vegetation. I. Herbaceous plants PAN injury symptoms. J Jpn Soc Air Pollut 9:635–643

Nouchi I, Mayumi H, Yamazoe F (1984) Foliar injury response of petunia and kidney bean to simultaneous and alternate exposures to ozone and PAN. Atmos Environ 18:453–460

Obländer W, Wörth R, König E, Braunger H, Schröter H (1983) Ergebnis und Interpretation von zwei-jährigen Schwefeldioxid-Immissionsmessungen an Tannenbeobachtungsflächen im Schwarzwald und in angrenzenden Wuchsgebieten. Allg Forst- Jagdztg 154:175–180

O'Connor JA, Parbery DG, Strauss W (1975) The effects of phytotoxic gases on native Australian plant species. Part 2. Acute injury due to ozone. Environ Pollut 9:181–192

Oertli JJ (1959) Effects of salinity on susceptibility of sunflower plants to smog. Soil Sci 87:249–251

O'Gara PJ (1922) Sulfur dioxide and fume problems and their solutions. Abstr J Ind Eng Chem 14:744

Ogata G, Maas EV (1973) Interactive effects of salinity and ozone on growth and yield of garden beet. J Environ Qual 2:518–520

Olszyk DM, Tibbitts TW (1981) Stomatal response and leaf injury of *Pisum sativum* L. with SO_2 and O_3 exposures. Plant Physiol 67:539–544

Olszyk DM, Tibbitts TW (1982) Evaluation of injury to expanded and expanding leaves of peas exposed to sulfur dioxide and ozone. J Am Soc Hortic Sci 107:266–271

Ordin L (1962) Effect of peroxyacetyl nitrate on growth and cell wall metabolism of *Avena* coleoptile sections. Plant Physiol 37:603–608

Ordin L, Hall MA (1967) Studies on cellulose synthesis by a cell-free oat coleoptile enzyme system: Inactivation by airborne oxidants. Plant Physiol 42:205–212

Ordin L, Propst B (1962) Effect of airborne oxidants on biological activity of indoleacetic acid. Bot Gaz 123:170–175

Ordin L, Hall MA, Katz M (1967) Peroxyacetyl nitrate-induced inhibition of cell wall metabolism. J Air Pollut Control Assoc 17:811–815

Ordin L, Hall MA, Kindinger JI (1969) Oxidant-induced inhibition of enzymes involved in cell wall polysaccharide synthesis. Arch Environ Health 18:623–626

Ordin L, Garber MJ, Kindinger JI, Whitmore SA, Greve LC, Taylor OC (1971) Effect of peroxyacetyl nitrate (PAN) in vivo on tobacco leaf polysaccharide synthetic pathway enzymes. Environ Sci Technol 5:621–626

Ormrod DP (1977) Cadmium and nickel effects on growth and ozone sensitivity of pea. Water Air Soil Pollut 8:263–270

Ormrod DP (1978) Pollution in Horticulture. Fundamental aspects of pollution control and environmental science, Elsevier, Amsterdam Oxford New York

Ormrod DP (1982) Air pollutant interactions in mixtures. In: Unsworth MH, Ormrod DP (eds) Effects of gaseous air pollution in agriculture and horticulture. Butterworth, London, pp 307–331

Ormrod DP, Adedipe NO, Hofstra G (1971) Responses of cucumber, onion, and potato cultivars to ozone. Can J Plant Sci 51:283–288

Oshima RJ (1973) Effect of ozone on a commercial sweet corn variety. Plant Dis Rep 57:719–723

Oshima RJ (1974a) A viable system of biological indicators for monitoring air pollutants. J Air Pollut Control Assoc 24:576–578

Oshima RJ (1974b) Development of a system for evaluating and reporting economic crop losses caused by air pollution in California. II. Yield Study. IIA. Prototype ozone dosage-crop loss conversion function. Dep Food Agric, Sacramento, California. Final Rep Calif Air Resourc Board Agreement ARB2-704

Oshima RJ (1978) The impact of sulfur dioxide on vegetation: A sulfur dioxide-ozone response model. Calif Air Resource Board, Final Rep ARB Agreement A6-162-30

Oshima RJ, Taylor OC, Braegelmann PK, Baldwin DW (1975) Effect of ozone on the yield and plant biomass of a commercial variety of tomato. J Environ Qual 4:463–464

Oshima RJ, Poe M, Braegelmann PK, Baldwin DW, Way V van (1976) Ozone dosage-drop loss function for alfalfa: a standardized method for assessing crop losses from air pollutants. J Air Pollut Control Assoc 26:861–865

Oshima RJ, Braegelmann PK, Baldwin DW, Way V van, Taylor OC (1977) Responses of five cultivars of fresh market tomato to ozone: A contrast of cultivar screening with foliar injury and yield. J Am Soc Hortic Sci 102:286–289

Oshima RJ, Bennett JP, Braegelman PK (1978) Effect of ozone on growth and assimilate partitioning in parsley. J Am Soc Hortic Sci 103:348–350

Oshima RJ, Braegelmann PK, Flagler RB, Teso RR (1979) The effects of ozone on the growth, yield, and partitioning of dry matter in cotton. J Environ Qual 8:474–479

Otto HW, Daines RH (1969) Plant injury by air pollutants: Influence of humidity on stomatal aperture and plant response to ozone. Science 163:1209–1210

Pan CH, Gast JH, Estes FL (1961) A comparative procedure for evaluating antimicrobial activity of gaseous agents. Appl Microbiol 9:45–51

Papple DJ, Ormrod DP (1977) Comparative efficacy of ozone-injury suppression by benomyl and carboxin on turf grasses. J Am Soc Hortic Sci 102:792–796

Parmeter JR Jr, Bega RV, Neff T (1962) A chlorotic decline of ponderosa pine in southern California. Plant Dis Rep 46:269–273

Patton RL (1981) Effects of ozone and sulfur dioxide on height and stem specific gravity of *Populus* hybrids. For Serv Res Pap NE-471. For Serv, US Dep Agric, Northeastern For Exp Stn, Broomal, Pa

Peak MJ, Belser WI (1969) Some effects of the air pollutant peroxyacetyl-nitrate upon deoxyribonucleic acid and upon nucleic acid bases. Atmos Environ 3:385–397

Pearson RG, Drummond DB, McIlveen WD, Linzon SN (1974) PAN-type injury to tomato crops in southwestern Ontario. Plant Dis Rep 58:1105–1108

Pearson LC, Henriksson E (1981) Air pollution damage to cell membranes in lichens. 2. Laboratory experiments. Bryologist 84:515–521

Pell EJ (1974) Effectiveness of benomyl in reducing peroxyacetyl nitrate damage of pinto bean foliage. Pa State Univ Cent Air Environ Stud Dep Plant Pathol, 211 Buckhout Lab, Univ Park, Pa

Pell EJ (1976) Influence of benomyl soil treatment on pinto bean plants exposed to peroxyacetyl nitrate an ozone. Phytopathology 66:731–733

Pell EJ, Brennan E (1973) Changes in respiration, photosynthesis, adenosine 5-triphosphate, and total adenylate content of ozonated pinto bean foliage as they relate to symptom expression. Plant Physiol 51:378–381

Pell EJ, Gardner W (1979) Enhancement of peroxyacetyl nitrate injury to petunia foliage by benomyl. HortScience 14:61–62

Pell EJ, Pearson NS (1983) Ozone induced reduction in quantity of ribulose-1,5-biphosphate carboxylase in alfalfa foliage. Plant Physiol 73:185–187

Pell EJ, Weissberger WC (1976) Histopathological characterization of ozone injury to soybean foliage. Phytopathology 66:856–861

Pell EJ, Lukezic FL, Levine RG, Weissberger WC (1977) Response of soybean foliage to reciprocal challenges by ozone and a hypersensitive-response-inducing pseudomonad. Phytopathology 67:1342–1345

Pell EJ, Weissberger WC, Speroni JJ (1979) Altered tuber quality of potato plants exposed to ozone. Phytopathology 69:1041–1042

Pell EJ, Weissberger WC, Speroni JJ (1980) Impact of ozone on quantity and quality of greenhouse-grown potato plants. Environ Sci Technol 14:568–571

Pellissier M, Lacasse NL, Cole H Jr (1972 a) Effectiveness of benomyl and benomyl-folicote treatments in reducing ozone injury to pinto beans. J Air Pollut Control Assoc 22:722–725

Pellissier M, Lacasse NL, Cole H Jr (1972 b) Effectiveness of benzimidazole, benomyl, and thiabendazole in reducing ozone injury to pinto beans. Phytopathology 62:580–582

Pellissier M, Lacasse NL, Ercegovich CD, Cole H Jr (1972 c) Effects of hydrocarbon as emulsion sprays in reducing visible injury to *Phaseolus vulgaris* "Pinto III". Plant Dis Rep 56:6–9

Pelz E (1963) Untersuchungen über die Fruktifikation rauchgeschädigter Fichtenbestände. Arch Forstwes 12:1066–1077

Pelz E, Materna J (1964) Beiträge zum Problem der individuellen Rauchhärte von Fichte. Arch Forstwes 13:177–210

Perchorowicz JT, Ting IP (1974) Ozone effects in plant cell permeability. Am J Bot 61:787–793

Perner E (1962) Elektronenmikroskopische Befunde über Kristallgitterstrukturen im Stroma isolierter Spinatchloroplasten. Port Acta Biol Ser A 6:359–372

Perner E (1963) Kristallisationserscheinungen im Stroma isolierter Spinatchloroplasten guter Erhaltung. Naturwissenschaften 50:134–135

Pollanschütz J (1975) Zuwachsuntersuchungen als Hilfsmittel der Diagnose und Beweissicherung bei Forstschäden durch Luftverunreinigungen. Allg Forstztg 86:187–192

Posthumus AC (1976) The use of higher plants as indicators of air pollution in The Netherlands. In: Kärenlampi L (ed) Proceedings of the Kuopio meeting on plant damages caused by air pollution, Kuopio, pp 115–120

Posthumus AC (1977) Experimentelle Untersuchungen der Wirkung von Ozon und PAN auf Pflanzen. VDI-Ber 270:153–161

Posthumus AC (1980) Elaboration of a community methodology for the biological surveillance of the air quality by the evaluation of the effects on plants. Commiss Eur Commun Publ Eur 6642 EV

Posthumus AC, Tonneijck AEG (1982) Monitoring of effects of photooxidants on plants. In: Steubing L, Jäger HJ (eds) Monitoring of air pollutants by plants. Junk, The Hague, pp 115–119

Pratt GC, Krupa SV (1981) Soybean cultivar Hodgson response to ozone. Phytopathology 71:1129–1132

Preston EM (1979) The ecological implications of chronic sulfur dioxide exposure for native grasslands. 72nd Annu Meet Air Pollut Control Assoc, Cincinnati, Ohio, June

Preston EM, Gullett TL (eds) (1978) The bioenvironmental impact of a coalfired power plant. 4th Interim Rep, Colstrip, Montana, USEPA, Corvallis Environ Res Lab, Corvallis, Oregon

Price H, Treshow M (1972) Effects of ozone on the growth and reproduction of grasses. In: Proc Int Clean Air Conf. Clean Air Soc Aust N Z, pp 275–280

Prinz B, Brandt CJ (1978) Study on the impact of the principal atmospheric pollutants on the vegetation. A review prepared on behalf of the European Communities. Publ Eur 6644. Landesanst Immissionsschutz Landes Nordrhein-Westfalen, Essen

Prinz B, Scholl G (1978) Erhebungen über die Aufnahme und Wirkung gas- und partikelförmiger Luftverunreinigungen im Rahmen eines Wirkungskatasters. Schriftenr Landesanstalt Immissionsschutz Landes Nordrhein-Westfalen, Essen, 46:25–29

Prinz B, Krause GHM, Stratmann H (1979) Thalliumschäden in der Umgebung der Dyckerhoff Zementwerke AG in Lengerich/Westfalen. Staub Reinh Luft 39:457–462

Prinz B, Krause GHM, Stratmann H (1982) Waldschäden in der Bundesrepublik Deutschland. LIS-Ber Nr 28 Landesanstalt Immissionsschutz Landes Nordrhein-Westfalen, Essen

Proctor JTA, Ormrod DP (1977) Response of celery to ozone. HortScience 12:321–322

Provilaitis B (1967) Gene effects for tolerance to weather fleck in tobacco. Can J Genet Cytol 9:327–334

Provilaitis B, White FH (1966) Delcrest 66, a new variety of bright tobacco. Can J Plant Sci 46:457–458

Puckett KJ, Nieboer E, Flora WP, Richardson DHS (1973) Sulphur dioxide: Its effect of photosynthetic ^{14}C fixation in lichens and suggested mechanisms of phytotoxicity. New Phytol 72:141–154

Rabe R (1978) Bioindikation von Luftverunreinigungen aufgrund der Änderung von Enzymaktivität und Chlorophyllgehalt von Testpflanzen. Diss Bot 45. Cramer, Vaduz

Rao DN, Robitaille G, LeBlanc F (1977) Influence of heavy metal pollution on lichens and bryophytes. J Hattori Bot Lab 42:213–239

Reckendorfer P (1952) Ein Beitrag zur Mikrochemie des Rauchschadens durch Fluor. Die Wanderung des Fluors in pflanzlichem Gewebe, I. Teil: Die unsichtbaren Schäden. Pflanzenschutzberichte 9:33–55

Rehbock N (1982) Der Stand der Luftschadstoff-Problematik in der Forstwirtschaft. Allg Forstztg 37:1179, 1182–1183, 1186–1187

Rehfuess KE (1981 a) Über die Wirkungen der sauren Niederschläge in Waldökosystemen. Forstwiss Centralbl 100:363–381

Rehfuess KE (1981 b) Waldböden-Entwicklung, Eigenschaften und Nutzung. Parey, Hamburg Berlin

Rehfuess KE (1983) Eine Arbeitshypothese über die Fichtenerkrankung in den Hochlagen des Bayerischen Waldes. Tagungsber Symp Saurer Regen – Waldschäden, Jülich 27.–28. Januar 1983, pp 34–35

Rehfuess KE, Bosch C, Pfannkuch E (1982) Nutrient imbalances in coniferous stands in southern Germany. Int workshop on growth disturbances of forest trees, Jyväskyla/Finland, 10–13 October 1982

Reich PB, Amundson RG, Lassoie JP (1982) Reduction in soybean yield after exposure to ozone and sulfur dioxide using a linear gradient exposure technique. Water Air Soil Pollut 17:29–36

Reichmann H, Streitz H (1983) Fortschreitende Bodenversauerung und Waldschäden im industrienahen Stadtwald Wiesbaden. Forst- Holzwirt 13:322–328

Reilly JJ, Moore LD (1982) Influence of selected herbicides on ozone injury in tobacco (*Nicotiana tabacum*). Weed Sci 30:260–263

Reinert RA (1975) Monitoring, detecting, and effects of air pollutants on horticultural crops, sensitivity of genera and species. HortScience 10:495–500

Reinert RA (1980) Assessment of crop productivity after chronic exposure to ozone or pollutant combinations. In: 73rd Annu Meet Air Pollut Control Assoc, Montreal, Canada. Pap 80-26.4. Air Pollut Control Assoc, Pittsburgh

Reinert RA, Gooding GV Jr (1978) Effect of ozone and tobacco streak virus alone and in combination on *Nicotiana tabacum*. Phytopathology 68:15–17

Reinert RA, Gray TN (1981) The response of radish to nitrogen dioxide, sulfur dioxide, and ozone, alone and in combination. J Environ Qual 10:240–243

Reinert RA, Henderson WR (1980) Foliar injury and growth of tomato cultivars as influenced by ozone dose and plant age. J Am Soc Hortic Sci 105:322–324

Reinert RA, Nelson PV (1979) Sensitivity and growth of twelve elatior begonia cultivars to ozone. HortScience 14:747–748

Reinert RA, Nelson PV (1980) Sensitivity and growth of five elatior begonia cultivars to SO_2 and O_3, alone and in combination. J Am Soc Hortic Sci 105:721–723

Reinert RA, Spurr HW Jr (1972) Differential effect of fungicides on ozone injury and brown spot disease of tobacco. J Environ Qual 1:450–452

Reinert RA, Weber DW (1980) Ozone and sulfur dioxide-induced changes in soybean growth. Phytopathology 70:914–916

Reinert RA, Tingey DT, Heck WW, Wickliff C (1969) Tobacco growth influenced by low concentration of sulfur dioxide and ozone. Abstr Agronomy 61:34

Reinert RA, Heagle AS, Miller JR, Geckeler WR (1970) Field studies of air pollution injury to vegetation in Cincinnati, Ohio. Plant Dis Rep 54:8–11

Reinert RA, Tingey DT, Koons CE (1971) The early growth of soybean as influenced by ozone stress. Abstr Agronomy 63:148

Reinert RA, Tingey DT, Carter HB (1972) Ozone induced foliar injury in lettuce and radish cultivars. J Am Soc Hortic Sci 97:711–714

Reinert RA, Heagle AS, Heck WW (1975) Plant responses to pollutant combinations. In: Mudd JB, Kozlowski TT (eds) Responses of plants to air pollution. Academic Press, London New York, pp 159–177

Reinert RA, Shriner DS, Rawlings JO (1982) Responses of radish to all combinations of three concentrations of nitrogen dioxide, sulfur dioxide, and ozone. J Environ Qual 11:52–57

Reiter H, Alcubilla M, Rehfuess KE (1983) Standortkundliche Studien zum Tannensterben: Ausbildung und Mineralstoffgehalte der Wurzeln von Weißtannen (*Abies alba* Mill.) in Abhängigkeit von Gesundheitszustand und Boden. Allg Forst- Jagdztg 154:82–92

Rentschler I (1973) Die Bedeutung der Wachsstruktur auf den Blättern für die Empfindlichkeit der Pflanzen gegenüber Luftverunreinigungen. In: Proc 3rd Int Clean Air Congr, Düsseldorf 1973. VDI, Düsseldorf, pp A139–A142

Rice RG, Netzer A (1982) Handbook of ozone technology and applications, vol I. Ann Arbor Sci Publ, Ann Arbor, Mich

Rich S, Hawkins A (1970) The susceptibility of potato varieties to ozone in the field. Abstr Phytopathology 60:1309

Rich S, Tomlinson H (1966) Ozone injury to conidiophores of *Alternaria solani*. Phytopathology 56:896–897

Rich S, Tomlinson H (1968) Effects of ozone on conidiophores and conidia of *Alternaria solani*. Phytopathology 58:444–446

Rich S, Turner NC (1972) Importance of moisture on stomatal behavior of plants subjected to ozone. J Air Pollut Control Assoc 22:718–721

Rich S, Ames R, Zuckel JW (1974) 1,4-oxanthiin derivatives protect plants against ozone. Plant Dis Rep 58:162–164

Richards BL, Taylor OC (1961) Status and redirection of research on the atmospheric pollutants toxic to field grown crops in southern California. J Air Pollut Control Assoc 11:125–128

Richards BL, Middleton JT, Hewitt WB (1958) Air pollution with relation to agronomic crops: V. Oxidant stipple of grape. Agron J 50:559–561

Richards BL, Taylor OC, Edmunds GF Jr (1968) Ozone needle mottle of pine in southern California. J Air Pollut Control Assoc 18:73–77

Richards GA, Mulchi CL, Hall JR (1980) Influence of plant maturity on the sensitivity of turfgrass species to ozone. J Environ Qual 9:49–53

Richards MC (1949) Ozone as a stimulant for fungus sporulation. Abstr Phytopathology 39:20

Richkind KE, Hacker AD (1980) Responses of natural wildlife populations to air pollution. J Toxicol and Environ Health 6:1–10

Ridley JD, Sims ED Jr (1966) Preliminary investigations on the use of ozone to extent the shelf-life and maintain the market quality of peaches and strawberries. S C Agric Exp Stn Res Ser No 70. Clemson Univ, Clemson

Rogers HH, Jeffries HE, Stahel EP, Heck WW, Ripperton LA, Witherspoon AM (1977) Measuring air pollutant uptake by plants: A direct kinetic technique. J Air Pollut Control Assoc 27:1192–1197

Roll-Hansen F, Roll-Hansen H (1981) Root wound infection of *Picea abies* at three localities in southern Norway. Rep Norw For Res Inst. Medd Nor Skogforsksves 36.4:1–8

Roose ML, Bradshaw AD, Roberts TM (1982) Evolution of resistance to gaseous air pollutants. In: Unsworth MH, Ormrod DP (eds) Effects of gaseous air pollution in agriculture and horticulture. Butterworth, London, pp 379–409

Ro-Paulsen H, Andersen B, Mortensen L, Moseholm L (1981) Elevated ozone levels in ambient air in and around Copenhagen indicated by means of tobacco indicator plants. Oikos 36:171–176

Rosen PM, Musselman RC, Kender WJ (1978) Relationship of stomatal resistance to sulfur dioxide and ozone injury in grapevines. Sci Hortic 8:137–142

Rosentreter R, Ahmadjian V (1977) Effect of ozone on the lichen *Cladonia arbuscula* and the *Trebouxia* phycobiont of *Cladonia stellaris*. Bryologist 80:600–605

Ross DN, Reinert RA, Halfacre RG (1976) The response of *Chrysanthemum morifolium* to chronic exposure to ozone and suppression of injury by ancymidol. HortScience 11:232

Ross LJ, Nash TH III (1983) Effect of ozone on gross photosynthesis of lichens. Environ Exp Bot 23:71–77

Rubin B, Leavitt JRC, Penner D, Saettler AW (1980) Interaction of antioxidants with ozone and herbicide stress. Bull Environ Contamin Toxicol 25:623–629

Rudolph E (1977a) Überwachung der Luftqualität in München mit ausgewählten Testpflanzen. VDI-Ber 270:175–178

Rudolph E (1977b) Exposition von Indikatorpflanzen zur Erfassung komplexer Immissionswirkungen in München. Staub Reinhalt Luft 37:467–472

Rufner, Witham FH, Cole J Jr (1975) Ultrastructure of chloroplasts of *Phaseolus vulgaris*. Phythopathology 65:345–349

Runeckles VC, Resh HM (1975a) The assessment of chronic ozone injury to leaves by reflectance spectrophotometry. Atmos Environ 9:447–452

Runeckles VC, Resh HM (1975b) Effects of cytokinins in responses of bean leaves to chronic ozone treatment. Atmos Environ 9:749–753

Runeckles VC, Rosen PM (1977) Effects of ambient ozone pretreatment on transpiration and susceptibility to ozone injury. Can J Bot 55:193–197

Rusnow P (1919) Über die Entkalkung des Bodens durch den Einfluß SO_2-haltiger Rauchgase. Centralbl Forstwiss 283–290

Saettler AW (1975) Studies on air pollution damage to beans in Michigan: Observation and breeding nurseries. In: Rep Bean Improvement Cooperative and National Dry Bean Council. Michigan State Univ, East Lansing, Mich, pp 7–11

Santamour FS Jr (1969) Air pollution studies on Platanus and American elm seedlings. Plant Dis Rep 53:482–484

Saunders PJ (1973) Effects of atmospheric pollution on leaf surface microflora. Pestic Sci 4:589–595

Schalekamp M (1982) Ozone as a sterilizing agent, its advantages and disadvantages in the treatment of water. Pres 11th Biennial Conf Int Assoc Water Pollut Res Control, Cape Town, South Africa, 29 March – 2 April 1982. Water Sci Technol 14:291–301

Scheffer F, Schachtschabel P (1982) Lehrbuch der Bodenkunde, 11th edn. Enke, Stuttgart

Scheffer TC, Hedgecock GG (1955) Injury to northwestern forest trees by sulfur dioxide from smelters. US Dep Agric Tech Bull 1117

Scholl G (1971) The rate of absorption of fluorine in plants as a criteria for emission limitation. VDI-Ber 164:39–45

Scholl G, Haut H van (1977) Erhebungen mit standardisierten Pflanzenkulturen über Belastungen durch Photooxidantien im Rhein-Ruhr-Gebiet. VDI-Ber 270:179–182

Scholz F (1980) Genökologische Wirkungen von Luftverunreinigungen aufgrund von Expositionsunterschieden im Bestand. Forstarchiv 52:58–61

Scholz F (1981) Genecological aspects of air pollution effects on northern forests. Silva Fenn 15:384–391

Schomer HA, McColloch LP (1948) Ozone in relation to storage of apples. US Dep Agric Circ No 765

Schönbeck H, Haut H van (1971) Messung von Luftverunreinigungen mit Hilfe pflanzlicher Organismen. In: Bioindicators of landscape deterioration, Praha.

Schönbeck H (1963) Beispiel für Untersuchungen von Raucheinwirkungen im Freiland durch Anwendung einer Modifikation des Fangpflanzenverfahrens. Forsch Berat R 5:47–56

Schönbeck H, Buck M, Haut H van, Scholl G (1970) Biologische Meßverfahren für Luftverunreinigungen. VDI-Ber 149:225–236

Schönborn A von, Weber E (1981) Untersuchungen über die Immissionsbelastung von Tannen- und Fichtennadeln im Bereich des Bayerischen Waldes. Forstwiss Centralbl 100:265–270

Schuck HJ (1981) Untersuchungen über die Wasserleitung in am Tannensterben erkrankten Weißtannen (*Abies alba* Mill.) Forstwiss Centralbl 100:184–190

Schuette LR (1971) Response of the primary infection process of *Erysiphe graminis* f. sp. *hordei* to ozone. PhD thesis, Univ Utah, Salt Lake City

Schütt P (1977) Das Tannensterben. Der Stand unseres Wissens über eine aktuelle und gefährliche Komplexkrankheit der Weißtanne (*Abies alba* Mill.). Forstwiss Centralbl 96:177–186

Schütt P (1981a) Die Verteilung des Tannennaßkerns in Stamm und Wurzel. Forstwiss Centralbl 100:174–179

Schütt P (1981b) Ursache und Verlauf des Tannensterbens – Versuch einer Zwischenbilanz. Forstwiss Centralbl 100:286–287

Schütt P (1982a) Aktuelle Schäden am Wald – Versuch einer Bestandsaufnahme. Holz-Zentralbl 26:369–374

Schütt P (1982b) Stand der Luftschadstoffproblematik in der Forstwirtschaft: zum Erkennen von Immissionsschäden. Allg Forstz 39:1180–1181, 1184–1185, 1188–1189

Schütt P (1982c) Das Krankheitsbild – verschiedene Baumarten, gleiche Symptome. Bild der Wissenschaft 12:86–101

Schütt P (1983) Botanische Aspekte der Forschung zum Waldsterben – Ergebnisse und Perspektiven. Tagungsber Symp Saurer Regen – Waldschäden, Jülich, 27–28 January 1983, pp 36–37

Schwenke W (1982) Mögliche Beziehungen tierischer Schädlinge zu derzeitigen Krankheitserscheinungen bei Tanne und Fichte. Allg Forstz 15:446–447

Scott DBM, Lesher EC (1963) Effects of ozone and survival and permeability of *Escherichia coli*. J Bacteriol 85:567–576

Seem RC, Cole H Jr, Lacasse NL (1972) Suppression of ozone injury to *Phaseolus vulgaris* "Pinto III" with triarimol and its monochlorophenyl cyclohexyl analogue. Plant Dis Rep 56:386–390

Seem RC, Cole H Jr, Lacasse NL (1973) Suppression of ozone to *Phaseolus vulgaris* L. with thiophanate ethyl and its methyl analogue. J Environ Qual 2:266–268

Seitschek O (1981) Verbreitung und Bedeutung der Tannenerkrankung in Bayern. Forstwiss Centralbl 100:138–148

Seitschek O (1982) Die gegenwärtige Waldschutz-Situation in Bayern. Allg Forstz 37:423–428

Serat WF, Budinger FE Jr, Mueller PK (1967) Toxicity evaluation of air pollutants by use of luminescent bacteria. Atmos Environ 1:21–32

Serat WF, Kyono J, Mueller PK (1969) Measuring the effects of air pollutants on bacterial luminescence: a simplified procedure. Atmos Environ 3:303–309

Shannon JG, Mulchi CL (1974) Ozone damage to wheat varieties at anthesis. Crop Sci 14:335–337

Shertz RD, Kender WJ, Musselman RC (1980a) Effects of ozone and sulfur dioxide on grapevines. Sci Hortic 13:37–45

Shertz RD, Kender WJ, Musselman RC (1980b) Foliar response and growth of apple trees following exposure to ozone and sulfur dioxide. J Am Soc Hortic Sci 105:594–598

Shinn JH, Clegg BR, Stuart ML, Thompson SE (1976) Exposure of field-grown lettuce to geothermal air pollution-photosynthetic and stomatal responses. J Environ Sci Health All (10 and 11):603–612

Shinn JH, Clegg BR, Stuart ML (1977) A linear-gradient chamber for exposing field plants to controlled levels of air pollutants. Reprint UCRL-80 411

Shinn JH, Clegg BR, Stuart ML (1979) A minimum field fumigation method for exposing plants to controlled gradients of air pollution levels. Contr Environ Sci Div, Lawrence Livermore Lab, Livermore, California

Shinohara T, Yamamoto Y, Kitano H, Fukada M (1973) Effects of temperature on ozone injury to tobacco. Proc Crop Sci Soc Jpn 42:418–421

Shiratori K (1973) Field survey method on damage to plants centering on crops caused by oxidants. In: References on photochemical air pollution in Japan. Air Qual Bur, Environ Agency, pp 246–258

Shriner DS (1978) Effects of simulated acidic rain on host-parasite interactions in plant diseases. Phytopathology 68:213–218

Shriner DS, Cure WW, Heagle AS, Heck WW, Johnson DW, Olson RJ, Skelly JM (1982) An analysis of potential agriculture and forest impacts of long-range transport of air pollutants. Final Rep Off Technol Assess, US Congr

Shumway LK, Weier TE, Stocking RC (1967) Crystalline structures in *Vicia faba* chloroplasts. Planta 76:182–189

Siccama TG, Smith WH (1978) Lead accumulation in a northern hardwood forest. Environ Sci Technol 12:593–594

Siegel SM (1962) Protection of plants against airborne oxidants: cucumber seedlings at extreme ozone levels. Plant Physiol 37:261–266

Sierpinski Z (1981) Rückgang der Tanne (*Abies alba* Mill.) in Polen. Eur J For Pathol 11:153–162

Sigal LL (1982) Photosynthesis of lichen species in relation to gaseous pollutants and acid rain. Bot Soc Am, Abstr Annu Meet, 8–12 August 1982, Pennsylvania State Univ, Univ Park, Pa. Bloomington, Indiana. Misc Ser Publ No 162:1

Sigal LL, Nash TH III (1983) Lichen communities on conifers in southern California mountains: an ecological survey relative to oxidant air pollution. Ecology 64:1343–1354

Sigal LL, Taylor OC (1979) Preliminary studies of the gross photosynthetic response of lichens to PAN fumigations. Bryologist 82:564–575

Sinclair WA, Costonis AC (1967) Ozone injury to eastern white pine. Arbor News 32:49–52

Skärby L (1979) Elevated ozone levels at the Swedish west coast and in southern Sweden (Skåne) using tobacco as an indicator plant. In: Eastmond M, Skärby L (eds) Report from the workshop "Ozone effects on vegetation in Europe". Swed Water Air Pollut Res Inst, Göteborg, pp 9–11

Skelly JM, Johnson JW (1979) Oxidant air pollution impact to the forest of eastern United States – A literature review. US Environ Protect Agency, Corvallis, Oregon, EPA-600/3-79-045

Skelly JM, Duchelle SF, Kress LW (1979) Impact of photochemical oxidant air pollution on eastern white pine in the Shenandoah, Blue Ridge, and Great Smoky National Parks. Proc 2nd Conf Sci Res Natl Parks, San Francisco, Calif

Skelly JM, Corghan CF, Hayes EM (1977) Oxidant levels in remote mountainous areas of southwestern Virginia and their effects on native white pine (*Pinus strobus* L.). In: Dimitriades B (ed) Proceedings of the international conference of photochemical oxidant pollution and its control, September 12–17, 1976, Raleigh, North Carolina. US Environ Protect Agency, Res Triangle Park, North Carolina, EPA-600/3-77-001b, pp 611–620

Smith WH (1970) Tree pathology. Academic Press, London New York

Smith WH (1974) Air pollution – effects on the structure and function of the temperate forest ecosystem. Environ Pollut 6:111–129

Smith WH (1980) Air pollution – A 20th century allogenic influence on forest ecosystems. Proceedings of the symposium on effects of air pollutants on Mediterranean and temperate forest ecosystems, June 22–27, 1980. Pac Southwest For Range Exp Stn, Berkeley, Gen Tech Rep PSW-43, Calif, pp 79–87

Smock RM, Watson RD (1941) Ozone in apple storage. Refrig Eng 42:97–101

Sorauer P, Ramann E (1899) Sogenannte unsichtbare Rauchbeschädigungen. Bot Centralbl 80:50–56, 106–116, 156–168, 205–216, 251–262

Spaulding P (1909) The present status of the white pine blights. US Dep Agric Bur Plant Ind Circ 35:1–12

Spierings FHFG (1967) Chronic discoloration of leaf tips of gladiolus and its relation to the hydrogen fluoride content of the air and the fluorine content of the leaves. Neth J Plant Pathol 73:25–28

SR-U, Der Rat von Sachverständigen für Umweltfragen (1978) Umweltgutachten 1978. Kohlhammer, Stuttgart Mainz

SR-U, Der Rat von Sachverständigen für Umweltfragen (1983) Waldschäden und Luftverunreinigungen. Kohlhammer, Stuttgart Mainz

Stan H-J, Schicker S, Kassner H (1981) Stress ethylene evolution of bean plants – a parameter indicating ozone pollution. Atmos Environ 15:391–395

Stanford Research Institute (1954) The smog problem in Los Angeles county. A report by Stanford Research Institute on the nature and causes of smog. Western Oil Gas Assoc, Los Angeles, Calif

Stark RW, Cobb FW Jr (1969) Smog injury, root diseases, and bark beetle damage in ponderosa pine. Calif Agric 23:13–15

Stark RW, Miller PR, Cobb FW Jr, Wood DL, Parmeter JR Jr (1968) Photochemical oxidant injury and bark beetle (Coleoptera: Scolytidae) infestation of ponderosa pine. I. Incidence of bark beetle infestation in injured trees. Hilgardia 39:121–126

Starkey TE (1975) The influence of peroxyacetyl nitrate on bean (*Phaseolus vulgaris* L.) subjected to post-exposure water stress. Center for Air Environment Studies. Pennsylvania State Univ, CAES Publ No 400–75

Starkey TE, Davis DD, Merill W (1976) Symptomatology and susceptibility of ten bean varieties exposed to peroxyacetyl nitrate (PAN). Plant Dis Rep 60:480–483

Starkey TE, Davis DD, Pell EJ, Merrill W (1981) Influence of peroxyacetyl nitrate (PAN) on water stress in bean plants. HortScience 16:547–548

Steer RP, Darnall KR, Pitts JN Jr (1969) The base-induced decomposition of peroxyacetylnitrate. Tetrahedron Lett 43:3765–3767

Steiner KG, Davis DD (1979) Variation among *Fraxinus* families in foliar response to ozone. Can J For Res 9:106–109

Stephens ER, Darley EF, Taylor OC, Scott WE (1961) Photochemical reactions products in air pollution. Int J Air Water Pollut 4:79–100

Steubing L (1977) Pflanzenökologische Untersuchungen in der Region Untermain auf photochemischen Smog. In: Regionale Planungsgemeinschaft Untermain Frankfurt/Main (ed) Lufthygienisch-meteorologische Modelluntersuchung in der Region Untermain, 6. Arbeitsbericht, pp 1–6

Steubing L (1980) Bioindicators as proof for the different immission loads in the lower Main region. MAB-Mitt H 5:58–81

Stöckhardt JA (1871) Untersuchungen über die schädlichen Einwirkungen des Hütten- und Steinkohlenrauches auf das Wachstum der Pflanzen, insbesondere der Fichte und Tanne. Tharandter Forstl Jahrb 21:218

Stoklasa J (1923) Die Beschädigung der Vegetation durch Rauchgase und Fabriksexhalationen. Urban & Schwarzenberg, München

Sung SS, Moore LD (1979) The influence of three herbicides on the sensitivity of greenhouse-grown flue-cured tobacco (*Nicotiana tabacum*) plants to ozone. Weed Sci 27:167–173

Sutton R, Ting IP (1977a) Evidence for repair of ozone induced membrane injury: Alterations in sugar uptake. Atmos Environ 11:273–275

Sutton R, Ting IP (1977b) Evidence for the repair of ozone-induced membrane injury. Am J Bot 64:404–411

Swanson ES, Thomson WW, Mudd JB (1973) The effect of ozone on leaf cell membranes. Can J Bot 51:1213–1219

Tanaka K (1976) Tree diseases in air polluted area and their usability as indicators. Annu Rep 1975, The range of tolerance of biological indicators to pollutants, 5–9

Tanaka K, Sugahara K (1980) Role of superoxide dismutase in the defense against SO_2 toxicity and induction of superoxide dismutase with SO_2 fumigation. In: Studies on the effects of air pollutants on plants and mechanisms of phytotoxicity. Res Rep Natl Inst Environ Stud No 11, Ibaraki, Japan

Taylor GE Jr (1978a) Plant and leaf resistance to gaseous air pollution stress. New Phytol 80:523–534

Taylor GE Jr (1978b) Genetic analysis of ecotypic differentiation within an annual plant species, *Geranium carolinianum* L., in response to sulfur dioxide. Bot Gaz 139:362–368

Taylor GE Jr, Murdy WH (1975) Population differentiation of an annual plant species, *Geranium carolinianum* L., in response to sulfur dioxide. Bot Gaz 136:212–215

Taylor GE Jr, Tingey DT, Ratsch HC (1982) Ozone flux in *Glycine max* (L.) Merr: Sites of regulation and relationship to leaf injury. Oecologia 53:179–186

Taylor GS (1968) Ozone injury on Bel W-3 tobacco controlled by at least two genes. Phytopathology 58:1069

Taylor GS (1970) Tobacco protected against fleck by benomyl and other fungicides. Phytopathology 60:578

Taylor GS, DeRoo HG, Waggoner PE (1960) Moisture and fleck of tobacco. Tob Sci 4:62–68

Taylor OC (1969) Importance of peroxyacetyl nitrate (PAN) as a phytotoxic air pollutant. J Air Pollut Control Assoc 19:347–351

Taylor OC (1973a) Oxidant air pollution effects on a western coniferous forest ecosystem. Task B. Historical background and proposed systems study of the San Bernardino Mountain area. Statewide Air Pollut Res Cent. US Environ Protect Agency, Washington DC, EPA-R-3-73-043 NTIS: PB 228 332/3GA. Calif Univ, Riverside

Taylor OC (1973b) Oxidant air pollution effects on a western coniferous forest ecosystem. Task C. Study site selection and on-site collection of background information. Statewide Air Pollut Res Cent, Riverside, NTIS: PB 228 333/1GA. Calif Univ, Riverside

Taylor OC (1973c) Acute response of plants to aerial pollutants. Air pollution damage to vegetation. Adv Chem Ser 122:9–20

Taylor OC (1974a) Oxidant air pollution effects on a western coniferous forest ecosystem. Task D. Annu Prog Rep Task D, 15 June 1973 – 14 June 1974. 1978. Statewide Air Pollut Res Cent, Riverside, NTIS: PB 281 858/1GA. EPA-600/3-78-052D. Calif Univ, Riverside

Taylor OC (1974b) Air pollution effects influenced by plant-environmental interaction. In: Dugger M (ed) Air pollution effects on plant growth. Am Chem Soc Symp Ser 3, Washington DC, pp 1–7

Taylor OC, MacLean DC (1970) Nitrogen oxides and the peroxyacetyl nitrates. In: Jacobson JS, Hill AC (eds) Recognition of air pollution injury to vegetation: A pictorial atlas. Air Pollut Control Assoc, Pittsburgh, pp E1–E14

Taylor OC, Stephens ER, Darley EF, Cardiff EA (1960) Effect of airborne oxidants on leaves of pinto bean and petunia. Proc Am Soc Hortic Sci 75:435–444

Taylor OC, Dugger WM Jr, Cardiff EA, Darley EF (1961) Interaction of light and atmospheric photochemical products ("smog") within plants. Nature (London) 192:814–816

Teige B, McManus TT, Mudd JB (1974) Reaction of ozone with phosphatidylcholine liposomes and the lytic effect of products on red blood cells. Chem Phys Lipids 12:153–171

Temple PJ (1982) Effects of peroxyacetyl nitrate (PAN) on growth of plants. Dissertation, Univ California, Riverside

Temple PJ, Taylor OC (1983) Peroxyacetyl nitrate (PAN) and implications for plant injury. Atmos Environ 17:1583–1587

Templin E (1962) Zur Populationsdynamik einiger Kiefernschadinsekten in rauchgeschädigten Beständen. Wiss Z Tech Hochsch Dresden 11:631–637

Teso RR, Oshima RJ, Carmean MJ (1979) Ozone-pesticide interactions. Calif Agric April:13–15

Thomas MD, Hendricks RH (1956) Effect of air pollution on plants. In: Magill PL, Holden FR, Ackley C (eds) Air pollution handbook. McGraw-Hill, New York, pp 1–44

Thomas MD, Hendricks RH, Hill GR (1952) Some impurities in the air and their effects on plants. In: Air pollution. Proc US Tech Conf Air Pollut. McGraw-Hill, New York, pp 41–47

Thompson CR (1968) Effects of air pollutants on lemons and navel oranges. Calif Agric 22:2–3

Thompson CR (1970) The total effect of air pollution on bearing citrus. NAPCA AP-70:12

Thompson CR, Kats G (1970) Antioxidants reduce grape yield reductions from photochemical smog. Calif Agric 24:12–13

Thompson CR, Taylor OC (1969) Effects of air pollutants on growth, leaf drop, fruit drop, and yield of citrus trees. Environ Sci Technol 3:934–940

Thompson CR, Taylor OC, Thomas MD, Ivie JO (1967) Effects of air pollutants on apparent photosynthesis and water use by citrus trees. Environ Sci Technol 1:644–650

Thompson CR, Hensel E, Kats G (1969) Effects of photochemical air pollutants on zinfandel grapes. HortScience 4:222–224

Thompson CR, Taylor OC, Richards BL (1970) Effects of photochemical smog on lemons and navel oranges. Calif Agric 24:10–11

Thompson CR, Kats G, Hensel E (1972) Effects of ambient levels of ozone on navel oranges. Environ Sci Technol 6:1014–1016

Thompson CR, Kats G, Camerson JW (1976a) Effects of ambient photochemical air pollutants on growth, yield, and ear characters of two sweet corn hybrids. J Environ Qual 5:410–412

Thompson CR, Kats G, Pippen EL, Isom WH (1976b) Effect of photochemical air pollution on two varieties of alfalfa. Environ Sci Technol 10:1237–1241

Thomson WW (1975) Effects of air pollutants on plant ultrastructure. In: Mudd JB, Kozlowski TT (eds) Responses of plants to air pollution. Academic Press, London New York, pp 179–194

Thomson WW, Swanson ES (1972) Some effects of oxidant air pollutants (ozone and peroxyacetyl nitrate) on the ultrastructure of leaf tissues. Annu Proc Electron Microsc Soc Am 30:360–361

Thomson WW, Dugger WM, Palmer RL (1965) Effects of peroxyacetyl nitrate on the ultrastructure of chloroplasts. Bot Gaz 126:66–72

Thomson WW, Dugger WM Jr, Palmer RL (1966) Effects of ozone on the fine structure of the palisade parenchyma cells of bean leaves. Can J Bot 44:1677–1682

Thomson WW, Nagahashi J, Platt K (1974) Further observation on the effects of ozone on the ultrastructure of leaf tissue. In: Dugger M (ed) Air pollution effects on plant growth. ACS Ser 3, Am Chem Soc, Washington DC, pp 83–93

Thorne L, Hanson GP (1972) Species differences in rates of vegetal ozone absorption. Environ Pollut 3:303–312

Thorne L, Hanson GP (1976) Relationship between genetically controlled ozone sensitivity and gas exchange rate in *Petunia hybrida* Vilm. J Am Soc Hortic Sci 101:60–63

Ting IP, Dugger WM Jr (1968) Factors affecting ozone sensitivity and susceptibility of cotton plants. J Air Pollut Control Assoc 18:810–813

Ting IP, Dugger WM (1971) Ozone resistance in tobacco plants: A possible relationship to water balance. Atmos Environ 5:147–150

Ting IP, Mukerji SK (1971) Leaf ontogeny as a factor in susceptibility to ozone: Amino acid and carbohydrate changes during expansion. Am J Bot 58:497–504

Tingey DT (1974) Ozone induced alterations in the metabolite pools and enzyme activities of plants. In: Dugger M (ed) Air pollution effects on plant growth. ACS Symp Ser 3. Am Chem Soc, Washington DC, pp 40–57

Tingey DT (1977) Ozone induced alterations in plant growth and metabolism. In: Dimitriades B (ed) Proceedings of international conference on photochemical oxidant pollution and its control, vol II. US Environ Protect Agency, Res Triangle Park, North Carolina, EPA-600/3-77-001b, pp 601–609

Tingey DT, Blum U (1973) Effects of ozone on soybean nodules. J Environ Qual 2:341–342

Tingey DT, Reinert RA (1975) The effect of ozone and sulfur dioxide singly and in combination on plant growth. Environ Pollut 9:117–125

Tingey DT, Taylor GE Jr (1982) Variation in plant response to ozone: A conceptual model of physiological events. In: Unsworth MH, Ormrod DP (eds) Effects of gaseous air pollutants in agriculture and horticulture. Butterworth, London, pp 111–138

Tingey DT, Heck WW, Reinert RA (1971) Effect of low concentrations of ozone and sulfur dioxide on foliage, growth, and yield of radish. J Am Soc Hortic Sci 96:369–371

Tingey DT, Reinert RA, Carter HB (1972) Soybean cultivars: Acute foliar response to ozone. Crop Sci 12:268–270

Tingey DT, Dunning JA, Jividen GM (1973a) Radish root growth reduced by acute ozone exposures. In: Proc 3rd Int Clean Air Congr, Düsseldorf, 1973. VDI, Düsseldorf, pp A154–A156

Tingey DT, Fites RC, Wickliff C (1973b) Foliar sensitivity of soybeans to ozone as related to several leaf parameters. Environ Pollut 4:183–192

Tingey DT, Fites RC, Wickliff C (1973c) Ozone alteration of nitrate reduction in soybean. Physiol Plant 29:33–38

Tingey DT, Reinert RA, Dunning JA, Heck WW (1973d) Foliar injury responses of eleven plant species to ozone/sulfur dioxide mixtures. Atmos Environ 7:201–208

Tingey DT, Reinert RA, Wickliff C, Heck WW (1973e) Chronic ozone or sulfur dioxide exposures, or both, affect the early vegetative growth of soybean. Can J Plant Sci 53:875–879

Tingey DT, Fites RC, Wickliff C (1975) Activity changes in selected enzymes from soybean leaves following ozone exposure. Physiol Plant 33:316–320

Tingey DT, Fites RC, Wickliff C (1976a) Differential foliar sensitivity of soybean cultivars to ozone associated with differential enzyme activities. Physiol Plant 37:69–72

Tingey DT, Standley C, Field RW (1976b) Stress ethylene evolution: A measure of ozone effects on plants. Atmos Environ 10:969–974

Tingey DT, Wilhour RG, Standley C (1976c) The effect of chronic ozone exposures on the metabolite content of ponderosa pine seedlings. For Sci 22:234–241

Tingey DT, Wilhour RG, Taylor OC (1979) The measurement of plant responses. In: Handbook of methodology for the assessment of air pollution effects on vegetation. Air Pollut Control Assoc, Pittsburgh, pp 7-1–7-35

Tingey DT, Thutt GL, Gumpertz ML, Hogsett WE (1982) Plant water stress influences ozone sensitivity of bean plants. Agric Environ 7:243–254

Todd GW (1958) Effect of ozone and ozonated 1-hexene on respiration and photosynthesis of leaves. Plant Physiol 33:416–420

Todd GW, Probst B (1963) Changes in transpiration and photosynthetic rates of various leaves during treatment with ozonated hexene or ozone gas. Physiol Plant 16:57–65

Tomlinson H, Rich S (1967) Metabolic changes in free amino acids of bean leaves exposed to ozone. Phytopathology 57:972–974

Tomlinson H, Rich S (1968) The ozone resistance of leaves as related to their sulfhydryl and adenosine triphosphate content. Phytopathology 58:808–810

Tomlinson H, Rich S (1969) Relating lipid content and fatty acid synthesis to ozone injury of tobacco leaves. Phytopathology 59:1284–1286

Tomlinson H, Rich S (1970) Disulfides in bean leaves exposed to ozone. Phytopathology 60:1842–1843

Tomlinson H, Rich S (1973) Relating ozone resistance antisenescence in beans treated with benzimidazole. Abstr Phytopathology 63:208

Tonneijck AEG (1983a) Urtica urens L. as indicator plant for peroxyacetyl nitrate (PAN). 6th Clean Air Congr, 16–20 May 1983, Paris

Tonneijck AEG (1983b) Foliar injury responses of 24 bean cultivars (Phaseolus vulgaris) to various concentrations of ozone. Neth J Plant Pathol 89:99–104

Torricelli A (1959) Drinking water purification. In: Ozone chemistry and technology. Adv Chem Ser 21. Am Chem Soc, Washington D C, pp 453–465

Townsend AM (1974) Sorption of ozone by nine shade tree species. J Am Soc Hortic Sci 99:206–208

Townsend AM, Dochinger LS (1974) Relationship of seed source and developmental stage to the ozone tolerance of Acer rubrum seedlings. Atmos Environ 8:957–964

Toyama S (1976) Effects of photochemical oxidants on plant organelles. Sholubutsu Boeki 30:223–229

Trautmann W, Krause A, Wolff-Straub R (1971) Veränderung der Bodenvegetation in Kiefernforsten als Folge industrieller Luftverunreinigungen im Raum Mannheim-Ludwigshafen. Schriftenr Vegetationskd 5:193–207

Treshow M (1975) Interactions of Air Pollutants and Plant Disease. In: Mudd JB, Kozlowski TT (eds) Responses of plants to air pollution. Academic Press, London New York, pp 307–334

Treshow M (1980) Interactions of air pollutants and plant disease. Proceedings of symposium on effects of air pollutants on Mediterranean and temperate forest ecosystems, June 22–27. Pac Southwest For Range Exp Stn, Berkeley, Gen Tech Rep PSW 43, pp 103–109

Treshow M, Stewart D (1973) Ozone sensitivity of plants in natural communities. Biol Conserv 5:209–214

Treshow M, Harner FM, Price HE, Kormelink JR (1969) Effects of ozone on growth, lipid, metabolism, and sporulation of fungi. Phytopathology 59:1223–1225

Trimble JL, Skelly JM, Tolin SA, Orcutt DM (1982) Chemical and structural characterization of the needle epicuticular wax of two clones of Pinus strobus differing in sensitivity to ozone. Phytopathology 72:652–656

Tukey HB (1980) Some effects of rain and mist on plants, with implications of acid precipitation. In: Hutchinson TC, Havas M (eds) Effects of acid precipitation on terrestrial ecosystems. NATO Conf Ser I. Ecology, vol IV. Plenum Press, New York, pp 141–150

Turner NC, Rich S, Tomlinson H (1972) Stomatal conductance, fleck injury, and growth of tobacco cultivars varying in ozone tolerance. Phytopathology 62:63–67

Turrell FM (1942) A quantitative morphological analysis of large and small leaves of alfalfa with species reference to internal surface. Am J Bot 29:400–415

UBA – Umweltbundesamt (1977) Luftqualitätskriterien für Cadmium. Ber 4/77. Schmidt, Berlin

UBA – Umweltbundesamt (1981) Luftreinhaltung '81. Entwicklung- Stand- Tendenzen. Schmidt, Berlin

UBA – Umweltbundesamt (1982) Monatsber Meßnetz des Umweltbundesamtes 7:1–13

Uhring J (1978) Leaf anatomy of petunia in relation to pollution damage. J Am Soc Hortic Sci 103:23–27

Ulrich B (1980) Die Wälder in Mitteleuropa. Meßergebnisse ihrer Umweltbelastung, Theorie ihrer Gefährdung, Prognose ihrer Entwicklung. Allg Forstz 36:1198–1202

Ulrich B (1981a) Bodenchemische und Umweltaspekte der Stabilität von Waldbeständen. Schriftenr Forstl Fak, Univ Göttingen 69:19–29

Ulrich B (1981b) Eine ökosystemare Hypothese über die Ursache des Tannensterbens *Abies alba* Mill. Forstwiss Centralbl 100:228–236

Ulrich B (1981c) Destabilisierung von Waldökosystemen durch Akkumulation von Luftverunreinigungen. Forst- Holzwirt 36:525–532

Ulrich B (1982) Gefahren für das Waldökosystem durch saure Niederschläge. In: Immissionsbelastungen von Waldökosystemen. Landesanstalt für Ökologie, Landschaftsentwicklung und Forstplanung Nordrhein-Westfalen, Recklinghausen, pp 9–25

Ulrich B, Pankrath J (eds) (1983) Accumulating air pollutants in forest ecosystems. Reidel, Dordrecht

Ulrich B, Mayer R, Khanna PK (1979) Deposition von Luftverunreinigungen und ihre Auswirkungen in Waldökosystemen im Solling. Schriftenr Forstwiss Fak Univ Göttingen Niedersächs Forstl Versuchsanst 58

US Department of Agriculture (1973) Silvicultural systems for the forest types of the United States. US Dep Agric, Agric Handb 566:1–114

US Department of Health, Education and Welfare (1970) Air quality criteria for photochemical oxidants. Nat Air Pollut Control Administr Publ No AP-63, Washington

Vargo RH, Pell EJ, Smith SH (1978) Induced resistance to ozone injury of soybean by tobacco ringspot virus. Phytopathology 68:715–719

VDI (Verein Deutscher Ingenieure) (1978a) Messen der Wirkdosis. Verfahren der standardisierten Graskultur. Richtlinie 3792, Bl 1. VDI, Düsseldorf

VDI (Verein Deutscher Ingenieure) (1978b) Maximale Immissionswerte für Schwefeldioxid. Richtlinie 2310. VDI, Düsseldorf

VDI (Verein Deutscher Ingenieure) (1982) Messen der Immissionswirkdosis. Messen der Immissionswirkdosis von gas- und staubförmigem Fluorid in Pflanzen nach dem Verfahren der standardisierten Graskultur. Richtlinie 3792, Bl 2. VDI, Düsseldorf

Verkroost M (1974) The effect of ozone on photosynthesis and respiration of *Scenedesmus obtusisculus* Chod., with a general discussion of effects of air pollutants in plants. Meded Landbouwhogesch Wageningen, 74–19

Vogl M, Börtitz S, Polster H (1965) Physiologische und biochemische Beiträge zur Rauchschadenforschung. 6. Mitt. Definitionen von Schädigungsstufen und Resistenzformen gegenüber der Schadgaskomponente SO_2. Biol Zentralbl 84:763–777

Vokk R (1977) Changes in the fractional phospholipid composition of *Escherichia coli* under the effects of ozone. Eesti NSV Tead Akad Toim Biol 26:203–207

Vos NE De, Hill RR, Hepler RW, Pell EJ, Craig R (1980) Inheritance of peroxyacetyl nitrate resistance in petunia. J Am Soc Hortic Sci 105:157–160

Wagner F (1981) Ausmaß und Verlauf des Tannensterbens in Ostbayern von 1975 bis 1980. Forstwiss Centralbl 100:148–160

Walker EK (1967) Evaluation of foliar sprays for control of weather fleck on flue-cured tobacco. Can J Plant Sci 47:99–108

Walker EK, Vickery LS (1961) Influence of sprinkler irrigation on the incidence of weather fleck on flue-cured tobacco in Ontario. Can J Plant Sci 41:281–287

Wammsley L, Ashmore MR, Bell JNB (1980) Adaptation of radish *Raphanus sativus* L. in response to continuous exposure to ozone. Environ Pollut Ser A 23:165–177

Watson RD (1942) Ozone as a fungicide. PhD thesis, Cornell Univ, Ithaca, N Y

Weaver GM, Jackson HO (1968) Relationship between bronzing in white beans and phytotoxic levels of atmospheric ozone in Ontario. Can J Plant Sci 48:561–568

Wedeck H (1980) Über Änderungen in der Zusammensetzung der Ackerunkrautvegetation durch Flugasche-Immissionen im Raum Weisweiler bei Düren. Decheniana 133:180–196

Weinstein LH, McCune DC (1971) Effects of fluoride on agriculture. J Air Pollut Control Assoc 21:410–413

Weinstein LH, McCune DC, Aluisio AL, Leuken P van (1975) The effects of sulphur dioxide on the incidence and severity of bean rust and early blight of tomato. Environ Pollut 9:145–155

Wentzel KF (1962) Konkrete Schadwirkungen der Luftverunreinigung in der Ruhrgebietslandschaft. Nat Landsch 37:118–124

Wentzel KF (1963) Waldbauliche Maßnahmen gegen Immissionen. Allg Forstz 18:101–106

Wentzel KF (1965) Die Winterfrost-Schäden 1962/63 in Koniferen-Kulturen des Ruhrgebietes und ihre vermutlichen Ursachen. Forstarchiv 36:49–59

Wentzel KF (1968) Empfindlichkeit und Resistenzunterschiede der Pflanzen gegenüber Luftverunreinigungen. Forstarchiv 39:189–194

Wentzel KF (1981) Maximale Immissionswerte zum Schutze der Wälder. Mitt Forstl Bundesversuchsanst Wien 137:175–180

Wentzel KF (1982) Ursachen des Waldsterbens in Mitteleuropa. Allg Forstz 37:1365–1368

Wentzel KF (1983) Waldbauliche Erfahrungen zur Erkennung der Immissionswirkungsmechanismen. Allg Forst- Jagdz 154:181–185

Wert SL, Miller PR, Larsh RN (1970) Color photos detect smog injury to forest trees. J For 68:536–539

Wesely ML, Eastman JA, Cook DR, Hicks BB (1978) Daytime variations of ozone eddy fluxes to maize. Boundary-Layer Meteorol 15:361–373

West DC, McLaughlin SB, Shugart HH (1980) Simulated forest response to chronic air pollution stress. J Environ Qual 9:43–49

Westman WE (1977) How much are nature's services worth? Science 197:960–964

Westman WE (1979) Oxidant effects on Californian coastal sage scrub. Science 205:1001–1003

Wieler A (1905) Untersuchungen über die Einwirkung schwefeliger Säure auf die Pflanzen. Bornträger, Berlin

Wilhour RG (1970a) The influence of temperature and relative humidity on the response of white ash to ozone. Abstr Phytopathology 60:579

Wilhour RG (1970b) The influence of ozone on white ash (*Fraxinus americana* L.). Cent Air Environ Stud, Pa State Univ CAES Publ No 188–71

Wilhour RG, Neely GE (1977) Growth response of conifer seedlings to low ozone concentrations. In: Dimitriades B (ed) International conference on photochemical oxidants and its control. US Environ Protect Agency, Res Triangle Park, NC EPA-600/3-77-001b, pp 635–645

Wilkinson TG, Barnes RL (1973) Effects of ozone on $^{14}C_2$ fixation patterns in pine. Can J Bot 9:1573–1578

Williams WT (1979) Long distance transport of air pollution into Sequoia, Kings Canyon, Yosemite, and Lassen National Parks. 2nd Conf Sci Res Natl Parks, November 26–30, San Francisco, Calif

Williams WT (1980) Air pollution disease in the Californian forests. A baseline for smog disease on ponderosa and jeffrey pines in the Sequoia and Los Padres National Forests, California. Environ Sci Technol 14:179–182

Williams WT, Brady M, Willison SC (1977) Air pollution damage to the forests of the Sierra Nevada Mountains of California. J Air Pollut Control Assoc 27:230–234

Wislicenus H (1901) Vorträge, gehalten auf der Hauptversammlung des Vereins Deutscher Chemiker. Z Angew Chem 28:689–712

Witheridge WN, Yaglou CP (1939) Ozone in ventilation – its possibilities and limitations. Trans Am Soc Heat Ventil Eng 45:509–520

Wittman O, Fetzer DK (1983) Aktuelle Bodenversauerung in Bayern. Bayer Staatsminist Landesentwickl Umweltfragen. Stockert, München

Wolak J (1971) Studies on the industrioclimax in Poland: Methods for the identification and evaluation of air pollutants injurious to forests. Proc XV. IUFRO Congress, Wien

Wolak J (ed) (1977) Relationship between increase in air pollution toxicity and elevation above ground. Wyd Inst Badasczy Lesnictwa, Warsaw

Wood FA, Coppolino JB (1972) The influence of ozone on deciduous forest tree species. Mitt Forstl Bundesversuchsanst Mariabrunn 97:233–253

Wood FA, Drummond DB (1974) Response of eight cultivars of chrysanthemum to peroxyacetyl nitrate. Phytopathology 64:897–898

Wood FA, Coppolino JB, Davis D, Drummond D, Wilhour R (1969) Interim progress report on air pollution effects. Cent Air Environ Stud, Pa State Univ

Wood T, Bormann FH (1975) Increases in foliar leaching caused by acidification of an artificial mist. Ambio 4:169–171

Wood T, Bormann FH (1976) Short-term effects of a simulated acid rain upon the growth and nutrient relations of *Pinus strobus* L. In: Dochinger LS, Seliga TA (eds) Proceedings of the 1st international symposium on acid precipitation and the forest ecosystem, May 1975, Columbus, Ohio. USDA For Serv Gen Tech Rep NE-23, pp 815–825

Woodwell GM (1970) Effects of pollution on the structure and physiology of ecosystems. Science 168:429–433

Wrischer M (1973) Protein crystalloids in the stroma of bean plastids. Protoplasma 77:141–150

Yamazoe F, Mayumi H (1977) Vegetation injury from interaction of mixed air pollutants. In: Kasuga S, Suzuki N, Yamada T, Kimura G, Inagaka K, Onoe K (eds) 4th Int Clean Air Congr, Tokyo, Japan, May 16–20, 1977. Jpn Un Air Pollut Prevent Assoc, pp 106–109

Youngner VB, Nudge FJ (1980) Air pollution oxidant effects on cool-season and warm-season turf grasses. Agron J 72:169–170

Zahn R (1963) Untersuchungen über die Bedeutung kontinuierlicher und intermittierender Schwefeldioxideinwirkung für die Pflanzenreaktion. Staub Reinhalt Luft 23:343–352

Zech W (1968) Kalkhaltige Böden als Nährsubstrat für Koniferen. Dissertation, Univ München

Zech W, Popp E (1983) Magnesiummangel, einer der Gründe für das Fichten- und Tannensterben in NO-Bayern. Forstwiss Centralbl 102:50–55

Zelac RE, Cromroy HL, Bolch WE, Dunavant BG, Bevis HA (1971) Inhaled ozone as a mutagen. I. Chromosome aberrations induced by Chinese hamster lymphocytes. Environ Res 4:262–282

Zelitch I (1969) Stomatal control. Annu Rev Plant Physiol 20:329–350

Ziegler I (1973) The effect of air polluting gases on plant metabolism. In: Coulston F, Korte F (eds) Environmental quality and safety. Thieme, Stuttgart; Academic Press, London New York, pp 182–208

Zöttl HW (1983) Zur Frage der toxischen Wirkung von Aluminium auf Pflanzen. Allg Forstz 38:206–208

Zöttl HW, Mies E (1983) Die Fichtenerkrankung in Hochlagen des Südschwarzwaldes. Allg Forstz 154:110–114

Zöttl HW, Stahr K, Keilen K (1977) Bodenentwicklung und Standorteigenschaften im Gebiet des Bärhaldegranits (südlicher Hochschwarzwald). Allg Forst- Jagdztg 148:185

Subject Index

Ecological Studies

Analysis and Synthesis

Editors: W.D.Billings, F.Golley, O.L.Lange, J.S.Olson, H.Remmert

Springer-Verlag
Berlin
Heidelberg
New York
Tokyo

Ecological Studies

Analysis and Synthesis

Editors: W.D.Billings, F.Golley, O.L.Lange, J.S.Olson, H.Remmert

Springer-Verlag
Berlin
Heidelberg
New York
Tokyo